Cockroaches as Models for Neurobiology: Applications in Biomedical Research

Volume I

Editors

Ivan Huber, Ph.D.
Professor
Department of Biological and Allied Health Sciences
Fairleigh Dickinson University
Florham Park - Madison Campus
Madison, New Jersey

Edward P. Masler, Ph.D.
Research Physiologist
Insect Reproduction Laboratory
Agricultural Research Service
U.S. Department of Agriculture
Beltsville, Maryland

B. R. Rao, Ph.D.
Professor
Department of Biological Sciences
East Stroudsburg University
East Stroudsburg, Pennsylvania

CRC Press
Taylor & Francis Group
Boca Raton London New York

CRC Press is an imprint of the
Taylor & Francis Group, an **informa** business

First published 1990 by CRC Press
Taylor & Francis Group
6000 Broken Sound Parkway NW, Suite 300
Boca Raton, FL 33487-2742

Reissued 2018 by CRC Press

Library of Congress Cataloging-in-Publication Data

Cockroaches as models for neurobiology: applications in biomedical
  research/editors, Ivan Huber, Edward P. Masler, B. R. Rao.
      p.   cm.
  Includes bibliographical references.
  ISBN 0-8493-4838-2 (v. 1) — ISBN 0-8493-4839-0 (v. 2)
  1. Neurobiology—Research—Methodology.   2. Plectoptera
(Cockroach)—Physiology.   I. Huber, Ivan, 1931-   .   II. Masler,
Edward P,, 1948-   .   III. Rao, B. R., 1936-   .
  QP357.C63   1990
595.7'22'0724—dc20   89-22149

A Library of Congress record exists under LC control number: 89022149

Publisher's Note
The publisher has gone to great lengths to ensure the quality of this reprint but points out that some imperfections in the original copies may be apparent.

Disclaimer
The publisher has made every effort to trace copyright holders and welcomes correspondence from those they have been unable to contact.

ISBN 13: 978-1-315-89165-1 (hbk)
ISBN 13: 978-1-351-07075-1 (ebk)

Visit the Taylor & Francis Web site at http://www.taylorandfrancis.com and the
CRC Press Web site at http://www.crcpress.com

THE EDITORS

Ivan Huber, Ph.D., is Professor of Biology at Fairleigh Dickinson University, Madison, New Jersey.

Dr. Huber received his A.B. degree in Zoology from Cornell University, Ithaca, New York in 1954 and his Doctorate in Entomology from the University of Kansas, Lawrence in 1968.

Dr. Huber is a member of the Entomological Society of America, The New York Entomological Society, the Society for the Study of Evolution, the Society of Systematic Zoology, and the honorary society Sigma Xi. He has been the recipient of research grants from the Merck Institute for Therapeutic Research.

Dr. Huber has written on allomones in chigger mites and the ecological genetics of flour beetles, but most of his work has been on cockroaches. He has published articles on the taxonomy, population biology, and oviposition behavior of this group. He currently is studying learning and memory in cockroaches.

Edward P. Masler, Ph.D., is a Research Physiologist in the Insect Reproduction Laboratory, Agricultural Research Service, U.S. Department of Agriculture, Beltsville, Maryland. He received his A.B. in Biology from St. Anselm College, Manchester, New Hampshire (1970) and an M.S. in Genetics from the University of New Hampshire, Durham (1973). He was awarded a Ph.D. in Biology from the University of Notre Dame, South Bend, Indiana (1978), where he became interested in the physiology of insect development and reproduction. He was awarded a postdoctoral fellowship to the Roche Institute of Molecular Biology, where he spent 2 years in the Department of Biochemistry studying vitellogenic protein structure and modification in the developing oocyte. A postdoctoral appointment to the Department of Entomology, Cornell University, Ithaca, New York followed, where he investigated neuropeptides responsible for ecdysteroid and vitellogenin production. Dr. Masler has been a Research Physiologist with the U.S. Department of Agriculture since 1982 and currently is interested in the biochemistry and molecular biology of development and reproduction in selected dipteran and lepidopteran species. Special emphasis is placed on the identities and roles of neuropeptides in metamorphosis and ovarian maturation. Memberships include the International Brain Research Organization, the International Society of Invertebrate Reproduction and Development, The Society for Developmental Biology, the Society for Neuroscience, and the Society of Sigma Xi.

Balakrishna R. Rao, Ph.D., is Professor of Biology at East Stroudsburg University, East Stroudsburg, Pennsylvania. He received his B.S. in Agriculture from Banaras Hindu University, Varanasi, India in 1957, his M.S. in Agriculture from Karnatak University, Dharwad, Karnataka, India in 1959, and a Ph.D. in Entomology from The Ohio State University, Columbus in 1964. His doctoral work in insect physiology culminated in his dissertation, entitled "Trypsin Activity Associated with the Reproductive Development in the Female Tampa Cockroach, *Nauphoeta cinerea* Oliv.". Dr. Rao had been a Postdoctoral Fellow at The Johns Hopkins University, the University of Connecticut, and The Marine Biological Laboratories, Woods Hole, Massachusetts prior to becoming Associate Professor of Biology at East Stroudsburg University. He is a member of the Entomological Society of Pennsylvania, the Entomological Society of America, Sigma Xi, the the Commonwealth of Pennsylvania University Biologists. Dr. Rao's current interests include reproductive physiology and digestive enzymes in cockroaches as well as mosquito reproductive physiology.

CONTRIBUTORS

Volume I

David W. Alsop, Ph.D.
Department of Biology
Queens College, CUNY
Flushing, New York

Moray Anderson, Ph.D.
Department of Zoology and
  Comparative Physiology
University of Birmingham
Birmingham, England

David J. Beadle, Ph.D.
School of Biological and Molecular
  Sciences
Oxford Polytechnic
Headington, Oxford
England

William J. Bell, Ph.D.
Department of Entomology
University of Kansas
Lawrence, Kansas

Isabel Bermudez, Ph.D.
School of Biological and
  Molecular Sciences
Oxford Polytechnic
Headington, Oxford
England

Jonathan M. Blagburn, Ph.D.
Institute of Neurobiology
University of Puerto Rico
San Juan, Puerto Rico

Jean-Jacques Callec, Ph.D.
Laboratory of Animal Physiology
University of Rennes I
Rennes, France

Yesu T. Das, Ph.D.
Innovative Scientific Services, Inc.
Piscataway, New Jersey

Charles R. Fourtner, Ph.D.
Department of Biological Sciences
State University of New York
Buffalo, New York

Ayodhya P. Gupta, Ph.D.
Department of Entomology
Rutgers University
New Brunswick, New Jersey

Ivan Huber, Ph.D.
Department of Biological and
  Allied Health Sciences
Fairleigh Dickinson University
Florham Park — Madison Campus
Madison, New Jersey

Bernard Hue, Ph.D.
Laboratory of Physiology
URA CNRS 611
University of Angers
Angers, France

Charles J. Kaars, Ph.D.
Department of Biological Sciences
State University of New York
Buffalo, New York

E. P. Masler, Ph.D.
Agricultural Research Service
U.S. Department of Agriculture
Beltsville, Maryland

Thomas A. Miller, Ph.D.
Department of Entomology
University of California
Riverside, California

Marcel Pelhate, Ph.D.
Laboratory of Physiology
URA CNRS 611
University of Angers
Angers, France

Yves Pichon, Ph.D.
Laboratory of Cellular Neurobiology
Department of Biophysics
C.N.R.S.
Gif-sur-Yvette, France

Robert M. Pitman, Ph.D.
Department of Biology and
  Preclinical Medicine
Gatty Marine Laboratory
St. Andrews, Fife
Scotland

B. R. Rao, Ph.D.
Department of Biological Sciences
East Stroudsburg University
East Stroudsburg, Pennsylvania

David B. Sattelle, Ph.D.
Unit of Insect Neurophysiology
  and Pharmacology
Department of Zoology
University of Cambridge
Cambridge, England

Karel Sláma, Ph.D.
Insect Chemical Ecology Unit
Institute of Organic Chemistry
Czechoslovak Academy of Sciences
Prague, Czechoslovakia

C. S. Thompson, Ph.D.
Department of Zoology
University of Toronto
Toronto, Ontario
Canada

S. S. Tobe, Ph.D.
Department of Zoology
University of Toronto
Toronto, Ontario
Canada

Hiroshi Washio, Ph.D.
Laboratory of Neurophysiology
Mitsubishi-Kasei Institute
  of Life Sciences
Machida, Tokyo
Japan

CONTRIBUTORS

Volume II

Michael E. Adams, Ph.D.
Division of Toxicology and Physiology
Department of Entomology
University of California
Riverside, California

Cynthia A. Bishop, Ph.D.
Department of Psychology
Stanford University
Stanford, California

Benjamin J. Cook, Ph.D.
Veterinary Toxicology and Entomology
 Research Laboratory
Agricultural Research Service
U.S. Department of Agriculture
College Station, Texas

Roger G. H. Downer, Ph.D.
Department of Biology
University of Waterloo
Waterloo, Ontario
Canada

Franz Engelmann, Ph.D.
Department of Biology
University of California
Los Angeles, California

Derek W. Gammon, Ph.D.
Agricultural Research Division
American Cyanamid Co.
Princeton, New Jersey

Ivan Huber, Ph.D.
Department of Biological and Allied
 Health Sciences
Fairleigh Dickinson University
Florham Park — Madison Campus
Madison, New Jersey

Larry L. Keeley, Ph.D.
Laboratories for Invertebrate
 Neuroendocrine Research
Department of Entomology
Texas Agricultural Experiment Station
Texas A&M University
College Station, Texas

Manfred J. Kern, Ph.D.
Pflanzenschutzforschung-Biologie
Hoechst Aktiengesellschaft
Frankfurt am Main, Federal Republic
 of Germany

Michael K. Leung, Ph.D.
Multidisciplinary Center for the Study of
 Aging and Chemistry/Physics Program
SUNY/College at Old Westbury
Old Westbury, New York

Michael I. Mote, Ph.D.
Department of Biology
Temple University
Philadelphia, Pennsylvania

Michael O'Shea, Ph.D.
Laboratory of Neurobiology
University of Geneva
Geneva, Switzerland

Terry L. Page, Ph.D.
Department of Biology
Vanderbilt University
Nashville, Tennessee

Susan M. Rankin, Ph.D.
Department of Entomology
Texas A&M University
College Station, Texas

Coby Schal, Ph.D.
Department of Entomology
Cook College
Rutgers University
New Brunswick, New Jersey

Berta Scharrer, Ph.D.
Department of Anatomy and
 Structural Biology
Albert Einstein College of Medicine
Bronx, New York

Günter Seelinger, Ph.D.
Institute for Zoology
University of Regensburg
Regensburg, Federal Republic
 of Germany

Alan F. Smith, Ph.D.
Department of Entomology
Cook College
Rutgers University
New Brunswick, New Jersey

Rajindar S. Sohal, Ph.D.
Department of Biological Sciences
Southern Methodist University
Dallas, Texas

George B. Stefano, Ph.D.
Multidisciplinary Center for the Study of
 Aging and Biological Sciences Program
SUNY/College at Old Westbury
Old Westbury, New York

Renée M. Wagner, Ph.D.
Veterinary Toxicology and Entomology
 Research Laboratory
Agricultural Research Service
U.S. Department of Agriculture
College Station, Texas

Jane L. Witten, Ph.D.
Laboratory of Neurobiology
University of Geneva
Geneva, Switzerland

Stephen Zawistowski, Ph.D.
Division of Social Sciences
St. John's University
Staten Island, New York

Sasha N. Zill, Ph.D.
Department of Anatomy
Marshall University School of Medicine
Huntington, West Virginia

TABLE OF CONTENTS

Volume I

TABLE OF CONTENTS

Volume II

VI. Sense Organs, Plasticity, and Behavior

PROLOGUE

Ivan Huber

Some years ago, I toured the animal rooms at the College of Pharmacy, St. John's University, Staten Island, NY. In casual conversation, the curator commented that in a few years he expected his rats and mice to be replaced by insects or other invertebrates. The prime reasons for this were the ever rising cost of maintaining facilities for mammals and the increasingly stringent government regulations because of agitation by animal rights activists. It occurred to me then that my favorite research subject, the cockroach, would make an admirable substitute for mammalian models in many biomedical research programs.[1] There are a number of very good reasons to consider using cockroaches for these purposes, and it is appropriate at this point to mention them. Cockroaches are large enough to handle, yet do not require much room. Large numbers of them can be raised very inexpensively. Furthermore, many species with a variety of reproductive strategies are already available in culture. Cockroaches are particularly resistant to disease and trauma. A circumstance which may pacify antivivisectionists is that insects are said to be insensitive to pain.[2,3] Prospective users should be encouraged by the fact that considerable literature already exists dealing with cockroach physiology and anatomy (for references, see Introduction), so it should not be necessary to establish parameters for a completely new organism. Finally, but most significantly, in neurobiology, at least, many functional analogies can be made between cockroach and mammalian systems.[4]

In the fall of 1984, a small workshop entitled "Cockroaches as Models in Biomedical Research" was organized for an Eastern Branch Meeting of the Entomological Society of America. My intention was to explore, in a public forum, the potential of cockroaches in biomedicine. Response was positive and gratifying. Karl Maramorosch of Rutgers University later suggested that the presentations could form the nucleus of a book, and for this I thank him. The other organizers of the workshop (E. P. Masler, Agricultural Research Service, U.S. Department of Agriculture, Beltsville, MD and B. R. Rao, East Stroudsburg University, East Stroudsburg, PA) agreed with me that such a volume would be of benefit to illustrate the use of insects as alternative experimental animal systems. We realized that such an undertaking rapidly could become unmanageable if some limitations were not set. As we examined the subjects addressed by the workshop, it became clear that all the papers dealt in some way with neurobiology. We therefore chose this as the topic of our book, inviting the speakers and other contributors to participate.

The chapters in this book are concerned partly, and in some cases, solely, with methods. We believe this will be especially useful to those for whom this book is primarily intended, workers in the biomedical sciences. We hope, through this volume, to persuade more researchers to adopt the cockroach as their experimental animal. Since many of these investigators may be more familiar with mammals than with insects, it is our intention not only to review neurobiological areas in which cockroaches could serve as models, but also to provide a source of information on many aspects of work with cockroaches. For those progressive readers who already are using insects in their research, we trust that they, too, will find something of value in these pages.

During the 16th century, animals, including insects (a source of remedies), were studied as an essential part of the pharmacopeia. As a result, a close connection between entomology and medicine developed and persisted for the next 200 years. Marcello Malpighi (1628—1694), the Italian physician and professor of medicine, was the first to separate the two disciplines because he considered the aims of animal anatomists to be different from those of the medical profession.[5] It is my hope that after reading this book researchers will reassociate the two fields to their mutual benefit.

REFERENCES

1. Committee on Models for Biomedical Research, *Models for Biomedical Research: A New Perspective*, Board on Basic Biology, Commission on Life Sciences, National Research Council, National Academy Press, Washington, D.C., 1985.
2. **Wigglesworth, V. B.,** Do insects feel pain?, *Antenna,* No. 1, 8, 1980.
3. **Eisemann, C. H., Jorgensen, W. K., Merritt, D. J., Rice, M. J., and Cribb, B. W.,** Do insects feel pain? A biological review, *Experientia,* 40, 164, 1984.
4. **Scharrer, B.,** Insects as models in neuroendocrine research, *Annu. Rev. Entomol.,* 32, 1, 1987.
5. **Beier, M.,** The early naturalists and anatomists during the Renaissance and seventeenth century, in *History of Entomology,* Smith, R. F., Mittler, T. E., and Smith, C. N., Eds., Annual Reviews, Palo Alto, CA, 1973, 81.

INTRODUCTION

E. P. MASLER

Insects present an extremely diverse and rich source of material for numerous aspects of biomedical research. It was precisely this diversity that forced us to limit our focus to a single suborder, the cockroaches, and to a single theme, neurobiology. As indicated in the Prologue, cockroaches possess a number of attractive features which make them suitable for laboratory experimentation. To recapitulate somewhat, they are a convenient size, inexpensive to rear and maintain, produce a number of generations per year, and are easy to sex, even at hatching. There is also a substantial basic literature[1-6] which emphasizes their physiology.

Our concentration on neurobiology results from a number of observations, not least of which is the explosive growth of this field as a research discipline over the last two decades. Registration figures for the annual meetings of the Society for Neuroscience show an increase from 1,400 in 1971 to 8,610 in 1983 and to over 12,000 in 1987. This increase cannot avoid placing demands upon available resources. It is prudent, therefore, to consider alternative biological systems for some research efforts.

Cockroaches have been used for numerous studies in insect physiology. Their size facilitates electrophysiological recording through the anatomical localization of single neurons and the location, excision, and manipulation of other organs such as the brain, hemolymph, gonads, Malpighian tubules, and fat body (analogous to the vertebrate liver). In addition, the amenability of insects, including cockroaches, to experimental manipulation makes them suitable for a number of approaches, ranging from neuroendocrine and metabolic studies to perception and behavior.

It was our intention with these volumes to introduce the reader to the use of cockroaches in biomedical research, to provide an overview of the anatomy and biology of the cockroach, and to present current research results and applications. Authors were free to construct their chapters according to individual expertise and judgment. Each author was aware that the volume was intended to provide the impetus for the adoption of insect systems by researchers in other disciplines. To this end, each section includes discussion and review of individual areas of research and the treatment of relevant practical methods and techniques. When appropriate, authors cite specific instances where biomedical research may be served by studies with cockroaches. In other cases, the author may draw on research reports with other insects in an effort to complete discussion of a topic, illustrate with examples an experimental approach which may be possible with cockroaches but has not yet been reported, or identify limitations of the cockroach as an experimental tool.

Two features of these volumes should be noted. First, the six sections address three distinct but interrelated considerations of the cockroach as an experimental tool. Basic biology, anatomy, and rearing are featured in Section I. Sections II through IV deal with specific aspects of individual systems as they relate to neurobiology. Neural structure and neurohemal organs are described here, along with signal transmission and cellular communication. Chemical and electrical signals, including neurotransmitters and peptide hormones, are also discussed. In Sections V and VI, the integration of neural and neuroendocrine signals in controlling higher level processes, the regulation of internal processes of metabolism and reproduction, and the transduction of external signals are considered.

The second feature to note is the integration between individual chapters. Authors were encouraged to cross-reference other chapters as much as possible, allowing the reader who selects a chapter title of particular interest to be referred to other chapters where related subjects can be found. This should be of value to those already somewhat familiar with the subject as well as those approaching insect neurobiology for the first time. Although the volume is arranged in logical sequence from general to specific and integrative aspects of the cockroach neurobiological system, it nevertheless can be entered from almost any section or chapter. Above all, it should be regarded as a resource manual as well as a review of the literature. Although various analogies between insect and vertebrate systems are discussed (e.g., the corpus allatum-corpus cadiacum complex vs. the vertebrate hypothalamo-hypophyseal system, fat body vs. liver, neuroendocrine systems, sensory and motor innervation and proprioception, and so forth) and authors speculate freely on potential uses for cockroach systems, it is hoped that the reader will recognize new opportunities for adaptation of insect systems to the unique problems of his or her research interests.

REFERENCES

1. **Beier, M.**, Blattariae (Schaben), in *Handbuch der Zoologie*, Vol. 4, Walter de Gruyter, Berlin, 1974, 1.
2. **Bell, W. J.**, *The Laboratory Cockroach*, Methuen, New York, 1982.
3. **Bell, W. J. and Adiyodi, K. G.**, Eds., *The American Cockroach*, Chapman and Hall, New York, 1981.
4. **Cornwell, P. B.**, *The Cockroach*, Vol. 1, Hutchinson and Co., London, 1968.
5. **Cornwell, P. B.**, *The Cockroach*, Vol. 2, Associated Business Programmes, London, 1976.
6. **Guthrie, D. M. and Tindall, A. R.**, *The Biology of the Cockroach*, St. Martin's Press, New York, 1968.

Section I. Biology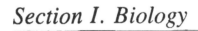

Chapter 1

BIOLOGY OF THE COCKROACH

William J. Bell

There is at present an emphasis on finding and deploying invertebrates to serve as models for vertebrate systems. In her recent paper on this topic, Scharrer[1] states that "the current focus is on lower forms for several reasons, including relative simplicity, greater accessibility, and applicability for certain experimental procedures. But first and foremost, the relevance of information gained from the use of such models has been established by the demonstration of remarkable structural and functional parallels" between insects and vertebrates. The research presented in this volume certainly attests to the parallelisms between cockroach and vertebrate biology and suggests that the cockroach will continue to reign high among potential candidates as an invertebrate model system. What is it about the cockroach that has led to its prevailing acceptability?

First, we should narrow the subject to include only those cockroach species that are used in biomedical research. Unfortunately, cockroaches are often referred to as though they were a totally homogeneous grouping, when actually there are more than 4000 species of cockroaches comprising the suborder Dictyoptera, and they are quite diverse. Cockroaches are placed in five families which belong to two major phyletic lines (Blattoidea and Blaberoidea) separated on the basis of reproductive strategies and morphology[2] (Figure 1). Females of the Blattoidea (families Cryptocercidae and Blattidae) are oviparous, producing hard, rigid egg cases which are dropped shortly after their formation; embryogenesis proceeds in the deposited ootheca. Common examples of the Blattidae are in the genera *Periplaneta* (the American cockroach, *P. americana*) and *Blatta* (the Oriental cockroach, *B. orientalis*). Several different oviposition mechanisms characterize the Blaberoidea (families Blaberidae, Polyphagidae, and Blattellidae). Females in the Polyphagidae rotate the ootheca and then carry it attached externally at the genital vestibulum. Species in the family Blaberidae, typified by *Blaberus craniifer, Blaberus discoidalis, Nauphoeta cinerea*, and *Leucophaea maderae*, are ovoviviparous. They rotate the egg case and retract it internally into the uterus, where embryogenesis occurs. *Diploptera punctata* represents the most highly evolved reproductive mode within the Blaberidae because a highly nutritious "milk", rich in protein, is transferred from the mother to the embryo. This transfer occurs from the wall of the uterus via pleuropodia (embryonic first abdominal segment appendages).[2a] Thus, this species is a truly viviparous one. Family Blattellidae includes only oviparous species, but some (e.g., the German cockroach, *Blattella germanica*) carry their oothecae externally throughout embryogenesis in an apparent intermediate condition between oviparity and ovoviviparity.

Cockroaches can be found in almost all habitat types where insects occur, but they are most diverse in warm, humid tropical regions.[3] Many of the Polyphagidae (*Arenivaga, Polyphaga*) inhabit desert regions of North Africa, Asia, and the southwestern U.S., and many have adopted unique behavioral and physiological strategies to maintain water balance, such as vertical movements through sand.[4-6] Aquatic species are the least studied of all cockroaches, although they have been observed inhabiting streams, pools, and ephemeral water-filled bromeliads.[7-8]

Of more than 50 species recorded in man-made habitats, four *Periplaneta*, two *Blatta*, *Blattella*, and several blaberid genera (*Blaberus, Eublaberus, Leucophaea*, and *Nauphoeta*) are most commonly given a domiciliary status.[9-10] Most domiciliary cockroaches are of African or Indo-Malayan origin,[11] and they have spread through trade routes (especially by air and sea) as commerce with tropical regions has proliferated.

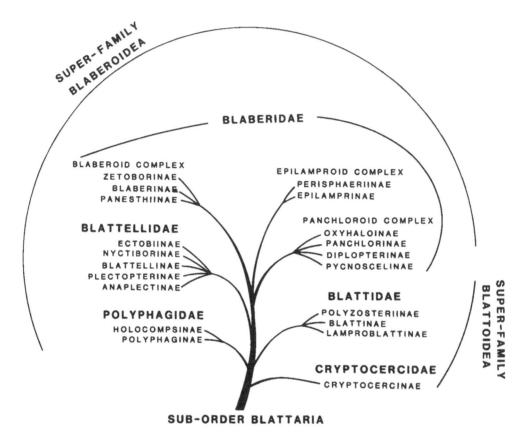

FIGURE 1.   Phyletic relationships of the families and subfamilies of the cockroaches according to McKittrick.[2] The Blattaria are regarded by the author as a suborder of the Dictyoptera, the other two suborders being the termites and mantises.

Cavernicolous cockroach species have been studied with respect to social interactions, mating systems, and natural history; the degree of specialization to cave habitats varies in cockroaches from accidental habitation of caves by outdoor or domiciliary species to restriction to such habitats accompanied by adaptations that may be morphological (loss of vision and wings), behavioral (no circadian rhythmicity), or physiological (narrow range of temperature or humidity tolerance).[12]

The majority of studies on outdoor cockroaches are the result of systematic collection, thus providing excellent distribution data, but little biological information. Our knowledge of the behavioral ecology of cockroaches in forests and grasslands has increased vastly over the past decade, however. Studies of tropical habitats where cockroaches are most diverse and abundant indicate a large variety of habitats where cockroaches are found, including leaf litter, dead leaves trapped above ground, caves, hollow trees, rotting logs, pools and streams, nests of other animals, flowers, at various altitudes, and, of course, in homes.

It is not necessary to visit the tropics to observe cockroaches in their natural habitat. In temperate regions there are cockroaches living in woods and grasslands, *Parcoblatta* in North America and *Ectobius* in northern Europe and Asia. Although these species are successful in winter diapause and other mechanisms for coping with temperate climates, they cannot live in our houses. They may fly in occasionally in the late spring, but they do not colonize the bathroom or the kitchen. In fact, most of the species used in research are very good fliers, although they do not usually fly unless the temperature is quite high. *Blaberus,* for example, causes considerable commotion when it flies into poolside lights at tropical resorts.

With the exception of a few species that inhabit caves, nests of ants or birds, or burrows in the desert, most adult cockroaches spend the day hidden under bark or in subterranean or aerial leaf litter, and then at dusk they begin to move onto the vegetation and, in many species, to migrate vertically to a preferred height.[13] They are found more often on leaves or stems than on large trunks, probably because vibrations of predators can be perceived more easily on a leaf or stem than on a surface that does not transmit vibrations as well. The sensory mechanisms for detecting vibrations from the substratum and the air have been studied extensively. When disturbed, cockroaches jump off the leaf or run quickly to a hiding place.

Cockroaches play major roles in their niche. In the tropics they are preyed upon by army ants which periodically raid hollow trees inhabited by *Blaberus* species;[12] they comprise 92% of the diet of Australian net-casting spiders,[14] and they are captured by scorpions in deserts.[5] There is also evidence of cockroaches being preyed upon by toads, birds, geckos, bats, and opossums.[9,12] In turn, cockroaches turn over considerable quantities of nitrogen-rich foods and also may be important in transporting fungus spores that act upon leaf litter. They consume huge quantities of leaf litter, contributing to the efficient return of the nutrients locked up in leaves back into the soil.

Anyone who has watched a cockroach closely would have to admit that they are intriguing beasts. We are impressed by their stealthy walk and their boldness in scurrying across the kitchen table. However, few scientists select the cockroach as their experimental animal because of its charms. The cockroach most often is selected because it is relatively large and hardy and is easy to obtain and culture — and because of its prolific breeding qualities. Hardiness is certainly a major advantage for the cockroach as compared to many other invertebrates; this hardiness can be manifested in various ways, all of which contribute to the usefulness of the beast in experimental work. One way that cockroaches show their resiliency is in their response to experimental manipulations, such that after injections or transplantation therapy the wound heals quickly and there is no immune response to intraspecific tissues. Many commonly used species also show broad tolerance to environmental conditions. Most species used in research are of tropical origin, and they must be kept warm and humid. However, they can tolerate a relatively wide range of temperature and humidity conditions, especially compared to many tropical insects that have rather narrow tolerances for temperature and humidity. For example, *Periplaneta americana* adults are active between 15 and 37°C, although they prefer the middle of this range. Perhaps the most astounding example of cockroach hardiness is their ability to survive without food and water. Female *P. americana* at 27°C and 30 to 40% relative humidity survive for up to 42 d with no food or water and up to 90 d with water, but with no food.[15] It is not surprising, then, that Alsop writes in Chapter 2 of this work that cockroaches are easier to rear than many other invertebrates.

Cockroaches are primitive on the family tree of insects, and as such most of the functional systems of cockroaches are fairly unspecialized compared to representatives of higher insect orders. Lack of specialization may be a good trait for a laboratory animal in that general rules can be tested rather than highly specific ones. Thus, the nervous system of the cockroach is less centralized and cephalized than those of Hymenoptera, Diptera, Coleoptera, and Lepidoptera, and the digestive system is designed to handle a variety of food types rather than specific types such as fluids in mosquitoes and aphids or leaves in grasshoppers and lepidopterous larvae. In addition to being unspecialized, some species of cockroaches, such as *Blaberus* and *Leucophaea,* are quite large. This attribute is a distinct advantage in research because collecting tissue or cells for extraction can be accomplished efficiently and quickly. Since individual organs such as the gut, brain, and antenna are large enough to penetrate with electrodes, there are excellent possibilities for electrophysiological recording and for sampling of fluids. Size often correlates with accessibility of physiological systems. The cockroach has been used extensively to study endocrine and ovarian physiology because the corpus allatum and the ovary are exposed and operated upon so easily.

A major advantage of cockroaches is the diversity within this closely related taxon. It is no accident, for example, that among the five most common laboratory species we find representatives characteristic of oviparous, ovoviviparous, and viviparous reproduction. Early workers studying various aspects of cockroach reproduction chose representatives of these three reproductive strategies in order to better understand how control mechanisms with the same components could bring about very different types of reproduction.[16] In addition, because the reproductive cycle of cockroaches is controlled by the neuroendocrine system, which includes feedback components, it is analogous in many ways to the mammalian reproductive cycle.

Many of the characteristics of the four or five species commonly maintained in the laboratory relate to preadaptations of certain species for surviving in domestic habitats and to natural selection which has occurred since cockroaches joined human households when cavemen set up housekeeping. Unfortunately, we know nothing about the extent of natural selection that has operated on cockroaches over the past several thousand years. This is because we have not been able to locate natural populations of *Blattella germanica, Blatta orientalis,* or *Supella longipalpa,* and, except for cave habitats, most outdoor populations of *P. americana* are still associated with man. Some of the nondomestic species, such as *Blaberus craniifer* and *Blaberus discoidalis,* have been studied extensively in natural habitats, while others such as *L. maderae* have not. Nevertheless, we can speculate about the pre-adaptations that may have allowed some species of cockroaches to survive in our houses and/or in the laboratory. Like rats and mice, most cockroach species are nocturnal, a characteristic that offers considerable protection since natural cockroach predators are largely absent in the domestic environment and because humans are active during the day. To be sure, fewer cockroaches would survive if they had diurnal habits. Again, we must be careful not to generalize because there are some diurnal species in the tropics; in fact, *Euphyllodromia angustata,* a strikingly beautiful bronze, red, and blue insect, is only active during the day.

Because we now find some species of cockroaches in caves and because the conditions under which they live probably have been the same for millions of years, we assume that caves have provided habitats for cockroaches for a long time. Cockroaches such as *Periplaneta* and *Blaberus* feed on bat guano and on pieces of fruit dropped by bats onto the floor of a cave or hollow tree.[17] Caves, which in some ways are similar to man-made structures, are ecologically stable habitats. In fact, it has been hypothesized that the association between cockroaches and man originated when cockroaches followed man's food into caves.[18] Because cavernicolous cockroaches (and domiciliary species) associate with man's shelter and food even when man is not present, it seems more likely that caves were already inhabited by cockroaches when man abandoned his arboreal habitats.[12] Associations with other cavernicolous animals (e.g., bats, porcupines, opossums) may have preadapted cockroaches to later associations with man. Thus, it seems likely that cockroaches were already living in caves when humans took up similar residences and that the two species were sufficiently compatible that the travels of man allowed cockroaches to exploit new areas by depending on humans rather than on bats for food and safe housing.

Another reason why a few select species are able to survive in homes and in the laboratory is related to their general nutritional requirements. Most cockroach species have special nutritional requirements that would be impossible to supply on a large scale in the laboratory. For example, *Cryptocercus punctulatus* lives in social family groups within rotting logs, and adults ingest wood and digest cellulose using enzymes produced by protozoans within their guts.[19] Nymphs are fed by trophallaxis.[20] Many of the species used for experimentation, however, are largely omnivorous; at least they can survive on laboratory chow, which is supremely nutritious, but certainly quite different from the diets of most cockroach species.

Anecdotal observations of cockroaches in our houses add to the notion that they are omnivorous. *P. americana,* for example, has been observed (in India) to ingest paper, boots, hair, bread, fruit, clothing, and book bindings, as well as (in Taipei) banana, fish, peanuts, bean cake, old rice, putrid sake, oil paper, animal hides, and crepe de Chine. Cockroaches are really not omnivores, however, in the sense that they "eat everything". As with most so-called omnivores, cockroaches actually balance their diet by selecting among available types of foods[21] rather than trying to obtain all required nutrients from one type of food. Honeybees, for example, obtain everything they need from nectar and pollen, and they never eat leaves or seeds. On the other hand, female *Xestoblatta hamata* select among various types of food according to vitellogenic requirements.[22] Attempts to rear species such as *X. hamata* in the laboratory generally have been unsuccessful.

An important characteristic of laboratory cockroaches is that they can withstand and in many cases thrive in crowded conditions. Often cockroaches in containers are literally on top of one another, and they seem quite able to carry on feeding, mating, and production of offspring under these circumstances. Interestingly, under very crowded conditions the mean size of individuals decreases, allowing more individuals to be packed into a small space. If and when National Institutes of Health guidelines are finally issued for cockroach husbandry, they surely will allow a measure of crowding because some species are unhealthy when living in small groups and/or when they are continually transferred to clean cages. As with many other characteristics of laboratory-reared cockroaches, the ability to withstand crowding and/or the natural tendency to aggregate are not typical of cockroaches in general. Although species that inhabit caves often live in somewhat crowded conditions, individuals of most species are not at all crowded in nature. Researchers who study agonistic or mating behaviors of cockroaches in the laboratory are invariably amazed when they are unable to observe these activities in nature. In one 3-year study of cockroach behavior only four incidences of agonistic behavior were recorded, while in laboratory cages agonistic behavior occurs nearly continuously among males. The low frequency of agonistic behavior in nature seems to result from the low frequency of interactions between adults. Most of the time we find individual cockroaches when we search for them at night in the rain forest. Small groups are sometimes observed feeding and two individuals may be copulating, but they are never crowded.

There are a few minor disadvantages to using cockroaches in research. Some species secrete foul-smelling odors (although most do not), but this does not suggest that cockroaches are dirty. In fact, cockroaches often spend more than 50% of their time cleaning their antennae, legs, feet, and wings. Another disadvantage is the ease with which susceptible people become allergic to cockroach (usually cuticular) antigens. It is quite possible, however, that the incidence of allergies to cockroaches is no greater than that to rats, mice, or moths.

In addition to learning more about vertebrates through the use of cockroaches, studying model systems leads to a better understanding of the evolutionary history of biological phenomena and elucidates the general applicability of basic principles. These objectives are important ones, since research goals should not always be so anthropocentric. At least as interesting is knowledge about how man is related to other organisms on earth, and, whereas the comparison is quite complete at a morphological level, it is severely lacking at the physiological and biochemical levels.

One of the most remarkable conclusions we can derive from this book is that studies of cockroaches as biomedical models, driven to a large extent by grant funding possibilities, have inadvertently produced a large body of knowledge about insect physiology. This information has opened new vistas in entomology and will in the future provide ideas for the development of new pest management schemes. Whenever we gain information about how insects work, we expose vulnerable points in their biochemistry or physiology which can be exploited by imaginative entomologists. The results have gone far in dismissing the myth that insects are "simple"; indeed, if we continue to find all of the biochemical complexity in the cockroach that is typical of the vertebrate, we will soon have to say that insects are "complex". If that happens we may have to seek a simple vertebrate as a model system to better understand complex animals such as cockroaches!

REFERENCES

1. **Scharrer, B.,** Insects as models in neuroendocrine research, *Annu. Rev. Entomol.*, 32, 1, 1987.
2. **McKittrick, F. A.,** Evolutionary study of cockroaches, *Mem. Cornell Agric. Exp. Stat.*, 389, 1, 1964.
2a. **Stay, B. and Coop, A.,** Developmental stages and chemical composition in embryos of the cockroach, *Diploptera punctata*, with observations on the effect of diet, *J. Insect Physiol.*, 19, 147, 1973.
3. **Guthrie, D. M. and Tindall, A. R.,** *The Biology of the Cockroach*, Edward Arnold, London, 1968.
4. **Edney, E. B., Haynes, S., and Gibo, D.,** Distribution and activity of the desert cockroach *Arenivaga investigata* (Polyphagidae) in relation to microclimate, *Ecology*, 55, 420, 1974.
5. **Hawke, S. D. and Farley, R. D.,** Ecology and behavior of the desert burrowing cockroach, *Arenivaga* sp., *Oecologia*, 11, 263, 1973.
6. **Cohen, A. C. and Cohen, J. L.,** Microclimate, temperature and water relations of two species of desert cockroaches, *Comp. Biochem. Physiol.*, 69, 165, 1981.
7. **Takahashi, R.,** Observations on the aquatic cockroach, *Opistoplatia maculata*, *Dobutsugaku Zasshi*, 38, 89, 1926.
8. **Albuquerque, R. S. I., Tibana, R., Jerberg, J., and Rebordões, A. M. P.,** Contribuição para o conhecimento ecològico de *Poeciloderrhis cribosa* (Burmeister) e *Poeciloderrhis verticalis* (Burmeister) com um estudo sobre a genitalia externa (Dictyoptera: Blattariae), *Rev. Bras. Biol.*, 36, 239, 1976.
9. **Roth, L. M. and Willis, E. R.,** The biotic associations of cockroaches, *Smithson. Misc. Collect.*, 141, 1, 1960.
10. **Vargas, M. and Fisk, F. W.,** Two new records of roaches invading houses in Costa Rica, *J. Med. Entomol.*, 10, 411, 1973.
11. **Rehn, J. A. G.,** Man's uninvited fellow traveller — the cockroach, *Sci. Mon.*, 61, 265, 1945.
12. **Schal, C., Gauthier, J.-Y., and Bell, W. J.,** Behavioural ecology of cockroaches, *Biol. Rev. Cambridge Philos. Soc.*, 59, 209, 1984.
13. **Schal, C. and Bell, W. J.,** Vertical community structure and resource utilization in neotropical forest cockroaches, *Ecol. Entomol.*, 11, 411, 1986.
14. **Austin, A. D. and Blest, A. D.,** The biology of two Australian species of dinopid spider, *J. Zool. (London)*, 189, 145, 1979.
15. **Willis, E. R. and Lewis, N.,** The longevity of starved cockroaches, *J. Econ. Entomol.*, 50, 432, 1957.
16. **Roth, L. M. and Willis, E. R.,** An analysis of oviparity and viviparity in the Blattaria, *Trans. Am. Entomol. Soc. (Philadelphia)*, 83, 221, 1958.
17. **Gautier, J.-Y.,** Etude comparée de la distribution spatiale et temporelle des adultes de *Blaberus atropos* et *B. colosseus* (Dictyoptères) dans cinq grottes de l'ile de Trinidad, *Biol. Behav.*, 9, 237, 1974.
18. **Chopard, L.,** La biologie des Orthoptères, *Encycl. Entomol.*, A20, 1, 1938.
19. **Cleveland, L. R., Hall, S. R., Sanders, E. P., and Collier, J.,** The wood-feeding roach, *Cryptocercus*, its protozoa, and the symbiosis between protozoa and roach, *Mem. Am. Acad. Arts Sci.*, 17, 185, 1934.
20. **Seelinger, G. and Seelinger, U.,** On the social organisation, alarm and fighting in the primitive cockroach *Cryptocercus punctulatus* Scudder, *Z. Tierpsychol.*, 61, 315, 1983.
21. **Cohen, R. W., Heydon, S. L., Waldbauer, G. P., and Friedman, S.,** Nutrient self-selection by the omnivorous cockroach *Supella longipalpa*, *J. Insect Physiol.*, 33, 77, 1987.
22. **Schal, C.,** Behavioral and Physiological Ecology and Community Structure of Tropical Cockroaches (Dictyoptera: Blattaria), Ph.D. thesis, University of Kansas, Lawrence, 1983.

Chapter 2

COCKROACH CULTURING

David W. Alsop

TABLE OF CONTENTS

I. INTRODUCTION

The primary aim in culturing cockroaches is to achieve a steady, high output of healthy specimens with a minimal expenditure of time, space, and money. Such conditions have rarely been met, primarily because of a lack of information about most species' requirements and behavior, and about optimal culturing methods and conditions. As a result, in most places the culturing of these insects has been on a purely ad hoc basis, with little attempt being made to accurately assess the cost and health of the specimens, or of the actual output of the cultures. The reader interested in insect rearing is referred to the volume edited by King and Leppla.[1] The contributions deal with such topics as the genetics of reared insects, especially the decline of variability and performance and ways to guard against both; actual rearing systems for a diverse selection of insects intended for a variety of uses; and the management of insect-rearing programs, including data-processing techniques, systems management, and quality control.

Most of the literature on the culturing of cockroaches concerns only the three main pest species, *Periplaneta americana, Blatta orientalis,* and *Blattella germanica* (see, for example, References 2 and 3). There are, however, nearly 4000 other species, some of which may well prove to be even better subjects for a variety of investigations than these pests. Thus, it is fortunate that over the years a few investigators have devoted some effort to devising methods for raising many different species under standardized conditions at low cost in limited spaces and with relatively little expenditure of time. The chief purpose of this chapter, then, is to present the culturing methods and materials which these researchers have found to be most useful, with the hope that investigators will not only be able to better their culturing of the three domiciliary pest species, but will also be drawn to investigate others.

II. OBTAINING SPECIES FOR CULTURING

A. GOVERNMENT REQUIREMENTS

The importation and interstate movement of cockroaches is regulated by the United States Department of Agriculture, Animal and Plant Health Inspection Service, Plant Pest and Quarantine Section (USDA-APHIS-PPQ). Persons wishing to obtain permits for interstate movement and foreign importation must apply to USDA-APHIS, Federal Center Building, Hyattsville, MD 20782. An application must be filled out and acted on affirmatively by APHIS-PPQ and a pertinent state regulatory agency before permits are issued.

B. SOURCES AND THEIR REQUIREMENTS

Periplaneta americana can be purchased from many of the larger biological suppliers (e.g., Carolina Biological Supply Company, Connecticut Valley Biological Company, Nebraska Scientific, Wards Natural Science). Of the above companies, only Carolina Biological also sells *Blattella germanica* and one of the *Blaberus* spp. (probably *B. craniifer*). *P.*

americana, Blattella germanica, and *Blatta orientalis* can also be obtained from the USDA-ARS Live Insects Laboratory, Building 476, Beltsville Agricultural Center, Beltsville, MD 20705. Delays of 4 weeks to 6 months may be expected when obtaining species from this source.[3] Other species are available from a number of university sources (those of the author included), and here, also, considerable delays are often to be expected.

All sources require permits and as a general rule will not ship during the cold winter months or very hot summer months. Only commercial suppliers can be expected to provide immediate shipment, and even they have limitations. Most noncommercial culturers have tightly controlled cultures that provide only the specimens they need. Thus, considerable lead times are often required to build up the excess numbers needed to provide additional specimens to other parties. Finally, unless cultures are being exchanged between individuals, the requestor should, at a minimum, expect to pay for the postage and handling of the shipment, and as a matter of courtesy large numbers of specimens should neither be requested nor expected.

III. PHYSICAL REQUIREMENTS FOR CULTURING

A. INCUBATORS AND ASSOCIATED FACILITIES

Ideally, cockroaches should be cultured in a sealed constant-temperature incubator with light and humidity controls and containing both a sink and a work table. USDA-APHIS-PPQ requires that incubators have an insect-proof door and that they be sealed, negative-pressure containments. Thus, air must be actively, rather than passively, exhausted from the room. The racks for holding cages should have shelf spacing allowing 2 to 3 in. above and around the cage for air circulation. Open-back metal racks are best for this purpose, with those made of stainless steel being preferred because they do not rust and are easily cleaned. Traps and/or poison baits should be placed under the racks so that cockroaches escaping from cages are prevented from infesting other cages and the surrounding area. Besides the incubator, other requirements for culturing are (1) a large outside sink for washing waterers and cages, (2) a freezer for disposing of cultures and waterer plugs, and (3) storage space for extra cages, waterers, and animal food.

B. LIGHT, TEMPERATURE, AND HUMIDITY REGIMES

The great majority of cockroaches are subtropical or tropical in origin and therefore are best raised at temperatures of 26 ± 2°C, 40 to 60% relative humidity, and day lengths approximating 12 h light:12 h darkness.

1. Temperature

Temperatures over 29°C are near the upper thermal limits of many species, particularly cavernicolous ones, and many species do not do well below 21°C. In all cases, growth is considerably slowed at lower temperatures and at temperatures >30°C.

2. Humidity Control

In general, cockroaches require large amounts of water to prevent death by dessication. Maintaining 40 to 60% relative humidity requires that humidification be provided in the winter and dehumidification in the summer. At <40% relative humidity, waterers dry out rapidly and the nymphs of many species die of dessication. On the other hand, while most species do very well and actually prefer >60% relative humidity, two problems become dominant at higher humidities. First, the acaroid mites associated with virtually all species rarely become numerous enough to cause discernable damage at <60% relative humidity, but can rapidly build to numbers great enough to annihilate a culture at higher humidities.[4] Mold growth on food, waterers, harborage materials, and carcasses also rapidly becomes rampant at high humidities. Here it is important to note that, in addition to controlling the humidity in the incubator, it is also necessary to provide adequate ventilation within the cages so that their internal humidity does not reach higher values.

3. Lighting

All species do well on a 12 h light/12 h dark cycle. The vast majority of species are most active at night, and a light-dark cycle is necessary to entrain their many circadian rhythms. Considerable variation in light intensity is permissible, and it can be held at a comfortable level for the investigator if the cages are provided with enough harborage material so that the cockroaches can hide when they wish.

IV. NUTRITIONAL REQUIREMENTS

The best food for raising cockroaches is dried (pelleted) dog food. For most species this is all that need be provided. It is one of the most completely fortified of all animal feeds, and many species eat it entirely, while others (particularly many of the blattellids) honeycomb it in a variety of patterns, an indication that they apparently do not require or are avoiding certain of its components. On the other hand, most species do not grow well on any of the many laboratory chows (rabbit, mouse, rat, etc.), presumably because these are apparently lacking one or more substances required for their optimal growth. Also, while most species are detritivores that will eat almost anything (including each other) when hungry, it does not necessarily mean that they will grow well on whatever is presented to them (i.e., bread, lettuce, etc.) or that it is necessary to provide them with a variety of different foods. Indeed, "minimal" diets such as bread and/or lettuce are merely a means of slowly starving them to death.

Two precautions are necessary when using dried dog food: (1) preventing it from becoming a route for the introduction of pests into the cultures and (2) keeping it from getting wet. Pest introduction (mainly beetles) can be prevented by freezing the food or storing it in a low-humidity cold room. Keeping it dry is necessary because (1) it molds rapidly when wet, and certain species will not eat it in this condition; and (2) when wet it attracts and becomes a favored breeding ground for fungus flies (Phoridae) which often become a major pest problem in cultures.

A few species, all live-bearing blaberids, seemingly do better when additionally provided with certain fruit supplements (e.g., banana peels and melon rinds) and/or brewer's yeast. The matter of additional supplementation is one of the greatest areas of uncertainty in cockroach culturing, since it is not at all clear which species require various types of dietary additives (beyond those present in dog food) for optimal growth. This lack of knowledge is one of the reasons that certain species have never been cultured successfully.

V. WATERERS

All cockroaches require water provided *ad lib*. While nearly all of them can survive for relatively long periods without food, virtually none can do so without being given fairly constant access to water. Since the majority of time spent in caring for cultures is devoted to watering the insects, a major aim has been to develop low-cost, high-capacity waterers that are easy to prepare, are of long-term usage, and which prevent water from spilling or siphoning into the cage.

Because of the great variety of cage sizes (see Section VI below), two general types of long-term waterers have been developed for culturing: *horizontal* ones for smaller cages and *vertical* ones for the larger cages (Figures 1 and 2). There are a number of variations on these, including a special type of vertical waterer developed for culturing species that produce

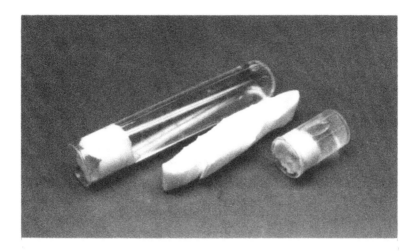

FIGURE 1. Horizontal waterers: 200-mm culture tube, left; 15-dram vial, right; folded strip of absorbent paper wadding, center.

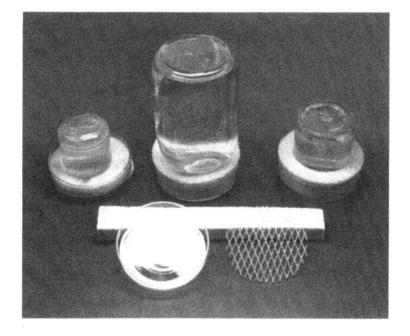

FIGURE 2. Vertical waterers. Background: assembled waterers — special waterer for oothecal moisture uptake, left; quart jar, center; pint jar, right. Foreground: additional materials — deep petri dish cover and disk of plastic mesh leaning against an expanded compressed sponge strip.

oothecae which must acquire water from a substrate in order for development to occur (Figure 3).

A. HORIZONTAL WATERERS
Materials include

• Plastic vials, 15 dram, and/or 35 × 200 mm culture tubes

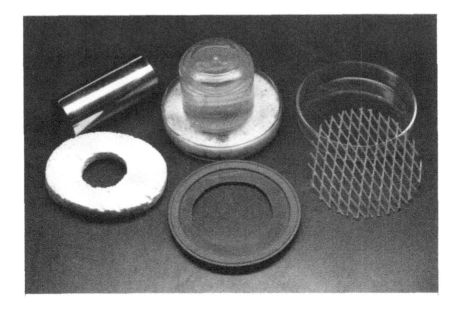

FIGURE 3.   Special waterer for oothecal moisture uptake. Clockwise from center top: assembled waterer; petri dish and disk of plastic mesh; cutout coffee can lid that fits over waterer so that it can be inverted for refilling without losing oothecae; oval sponge with hole cut in it; piece of sink drain tailstock used to cut holes in dried sponges.

- Nonsterile absorbant paper wadding
- Racks for the tubes
- Paper cutter

Waterers of this type are used mainly in small or low cages, but also should be employed in addition to large vertical waterers (Section V.B below) in large cages if the nymphs are very small or have difficulty climbing up the edge of the vertical waterers. Two basically different types of horizontal waterers have been employed: (1) pieces of sponge or wads of cotton in a plastic or glass petri dish and (2) large-diameter tubular containers fitted with absorbent plugs. While the first type offers a large access surface from which the cockroaches may drink, it requires fairly constant attention because the amount of water held is only that which can be taken up by the sponge or cotton. The second type is greatly preferred because the tubes can hold substantial amounts of water and, depending upon their volume and the number of individuals in a cage, will last for as long as 3 weeks before having to be replaced.

Tubular waterers can be made from any of a variety of plastic vials or culture tubes (Figure 1). The best ones are those having mouth diameters of 30 to 35 mm because they hold much greater amounts of water than waterers with smaller diameters and they offer a plug face accessible to all but the largest of the cockroaches. Plastic snap-cap vials (15 dram short, 33-mm opening and 15 dram long, 29-mm opening) make useful waterers for very small cages. These can be bought in gross quantities from most medical supply houses and a variety of other sources, and they will last for a long time if cared for properly. In larger cages it is best to use large glass culture tubes (200 × 35 mm) because of their far greater capacities. These tubes, together with the stainless steel racks necessary to hold them, can be obtained from Bellco Glass, Inc., P. O. Box B, 340 Erudo Road, Vineland, NJ 08360.

In terms of cost, preparation time, and quality control (i.e., not getting "leakers"), nonsterile absorbent paper wadding is by far the best material to use to make plugs for large-diameter horizontal waterers. It is much less expensive than using wads of adsorbent cotton for plugs, and it yields far fewer leakers. Also, while sponge plugs can be used safely in vials with small mouth openings (e.g., 10 to 15 mm), they become expensive and time-consuming to make, and they almost invariably leak when placed in large-diameter tubes.

Nonsterile absorbent paper wadding (formerly called Cellu-Cotton) is a multilayered product that can be obtained from certain paper companies and medical supply houses, usually in the form of 24-in.-high (5-lb) rolls. The rolls are first cut in half with a cross-cut saw or band saw to yield two 12-in.-high rolls. These half rolls are then unwound, and the wadding is cut into either 14-in. (for 29- to 30-mm-opening tubes) or 16-in. (for 35-mm-opening tubes) lengths with a paper cutter. The resulting lengths are folded in thirds and cut crosswise with the paper cutter into ca. 1-in.-wide strips (Figure 1). The yield from each folded length should be about 12 to 13 strips for tubes with 29- to 30-mm openings and 10 to 11 strips for those with 33- to 35-mm openings.

Plugs are formed by rolling the strips to approximately the same diameter as the waterer opening. These are then pushed firmly with a slight twisting motion into already filled waterers, and any exuded water is shaken off. They can be inserted dry into large tubes, but because of their great absorbency should be wetted thoroughly before being inserted into the shorter vials. If the insertion is done properly, the surface of the plug will remain uniformly moist as long as there is any water in the tube, but will not permit it to leak into the cage. Leakage is most often caused by (1) having insufficiently long lengths of wadding in the roll, (2) winding the wadding rolls so tightly that they cannot fully expand against the walls of the waterers, and (3) culturing species that actively burrow into the surface of the plugs (e.g., *Panchlora* and *Pycnoscelus* spp.). When culturing burrowing species, the best way to avoid leakage is to cut the wadding strips into ca. $1^1/_2$-in. widths so that the plugs are sufficiently deep not to be broached before the water in the tube is exhausted.

Horizontal waterers of this type can be prepared very quickly. A total of 20 to 30 lengths of wadding are cut and folded at a time, then stored for use as needed. In a few minutes, 40 strips can be cut with a paper cutter and then rolled and inserted into filled waterers at the rate of 10 to 15 per minute. As well as being much less expensive than using cotton to make the plugs, this is far faster, neater, and yields virtually no leakers.

Plugs are removed from spent waterers most easily with large (ca. 12-in.) forceps which are forced down into the tubes with their points open on either side of the plugs, then squeezed and withdrawn. Dried wadding is very hard and should be moistened before being removed. Spent plugs should always be frozen to kill any roaches, oothecae, and/or pests present in them before discarding them.

B. VERTICAL WATERERS
Materials include

- Half pint, pint, or quart mason jars, or the equivalent
- Lids of deep (200 × 20 mm) glass petri dishes
- Compressed sponge
- Open-mesh plastic laboratory matting
- Unglazed tiles (6-in. square)
- Paper cutter

Large waterers (Figure 2) are best made from any of a variety of jars or bottles having approximately 65-mm outside neck diameters (e.g., mason jars) because strips of compressed sponge can then be used as the water distributors and the lids of deep petri dishes (glass, 100 × 20 mm) as the waterer bases. These lids are readily available and can be obtained separately from the bottoms through most laboratory suppliers. Bottles having mouth openings with other diameters require bases of different (usually more difficult to obtain) sizes.

The best water distributor is a strip of sponge wrapped around the bottle neck and held in place by the petri dish. Using pads of cotton or paper wadding alone in the dish directly leads to an unstable situation in which the slightest movement of the cage can tip the bottle over. Sheets of compressed sponge ($^{1}/_{8}$ in. thick, expanding to ca. 1 in.) are obtainable from a number of suppliers. When dry they are easily cut with a paper cutter. Strips ($10^{1}/_{2} \times 1^{3}/_{16}$ in.) are cut for mason jars (other jars or dishes require different sizes), then wetted and squeezed dry several times before being used. It is also necessary to insert a disk of open-mesh plastic padding (of the type widely sold for use on laboratory sinks and drainboards) between the mouth of the bottle and the petri dish to ensure a free flow of water between the bottle and the sponge.

Waterers of this type usually do not have to be refilled for several weeks. They take little time and effort to set up: the jar is filled to the top with water, a strip of thoroughly wetted sponge is wrapped around its neck, a plastic mesh disk is laid across the mouth of the jar atop the sponge, and a petri dish is forced down over the mesh and sponge. Almost no water is lost when inverting the jar if the petri dish has been pushed all the way down to the mouth of the jar and the sponge packed up against it.

Large waterers (1 qt or more) used in large cages should be placed on 6-in. squares of unglazed tile rather than directly on the cage floor or harborage material. The tiles provide a stable base for the waterers, serve to elevate them above the debris that inevitably builds up in the cages, and help to keep the cage and harborage materials from getting wet.

C. SPECIAL WATERER FOR OOTHECAL WATER UPTAKE

Many species produce oothecae having such an initially low water content that an exogenous source of moisture must be provided if development is to occur. This is a widespread occurrence in the cockroaches, occurring in some or all members of the Cryptocercidae, Lamproblattinae (Blattidae), Polyphagidae, and Blattellidae.[5] Species with this requirement are among the most difficult of all cockroaches to culture.

Many different moistened materials have been used in attempts to meet this need, including tubes and dishes of sand, sphagnum moss, filter paper, cotton, absorbant paper wadding, and sponges. The major problems they have presented have been (1) the almost constant attention required to keep the substrates moist and (2) the growth of molds on the substrate and oothecae. While the waterer described below is not a final answer to culturing all species with this requirement, it has been used to successfully culture a fair number of blattellids that actively select a wet substrate for oothecal deposition. Its main advantages are that it requires only weekly attention and only rarely supports mold growth.

This waterer has a wide, moist surface. It consists of (1) a ca. 5-oz baby food (or similar) bottle ($2^{1}/_{4}$ in. diameter, $2^{3}/_{4}$ in. high) as the reservoir; (2) a deep petri dish cover (see Section V.B above) as the base plate; (3) a $^{1}/_{2}$-in.-thick, $3^{1}/_{4} \times 4^{1}/_{4}$ in. oval cellulose sponge as the water distributor; and (4) a disk of woven plastic window screening to be placed beneath the sponge (Figure 3). In addition, it is also necessary to have a plastic cover with an overall diameter slightly greater than that of the petri dish, having a central hole slightly larger than that of the bottle, for use when recharging the waterer.

The oval sponges are widely available. They have the distinct advantage of having many large surface pores into which many species will insert their oothecae. Before use they must be washed thoroughly to remove the humifactants in them, and then dried. Holes are bored through the middle of the dried sponges with a $1^{3}/_{4}$-in.-diameter metal cutter with notched edges. (Note: it is almost impossible to cut holes in sponges which are wet!) Assembling the waterers is straightforward: the plastic mesh is placed in the petri dish and a wet sponge pushed down on top of it; this is then inverted and pushed down over an already filled bottle, and the waterer is then turned upright. There is virtually no water loss while doing this, particularly if the sponge has been saturated thoroughly.

If mold growth becomes a problem, as happens with only certain species, the sponge should be soaked initially in a solution of 0.1% methyl *para*-hydroxybenzoate (sold under the trade names Nipagin M, Methylparaben, and Tegosept M). To make up the 0.1% solution, a 20% stock solution of methyl *para*-hydroxybenzoate in 70% ethyl alcohol is prepared first. The stock solution is then added to distilled water to produce the final concentration. Furthermore, if the oothecae show signs of supporting mold growth, they should be sprayed lightly with the solution at 1- to 3-d intervals.

These waterers last for 1 week or longer before having to be refilled. They are never taken completely apart until the sponge degenerates. Instead, to refill them, the plastic cover is first slipped over the bottle down onto the petri dish, and the entire assembly is inverted. The water bottle is then pulled out of the hole in the sponge, filled, and reinserted, and the waterer is then turned upright. This sequence prevents the loss of oothecae loosely deposited atop and along the sides of the sponge, and it permits moisture to be supplied continuously to oothecae which require more than 1 week to develop.

D. WATERER CLEANING

Other than the paper wadding, all of the waterer materials discussed above can be used repeatedly. Vials, culture tubes, jars, petri dishes, and plastic disks are soaked in a strong detergent solution for a few days and then washed. Sponge strips are soaked in hot water, rinsed and squeezed dry a few times, and then either immediately reinserted into waterers or dried out for reuse at a later time. Because the sponge deteriorates slowly, the strips usually have to be replaced after four to six uses. The plastic mesh disks deteriorate more slowly and should be disposed of only when they become brittle. Thus, the ongoing expense of making waterers is quite low since, if properly taken care of, all of the other materials last for many years.

VI. CAGES FOR SMALL- AND LARGE-SCALE CULTURING

The type and size of the cage chosen to house a particular species are dependent on the size of the individuals, their initial numbers, their spacing requirements, whether they prefer to rest on horizontal or vertical surfaces, and the final colony size desired. While many different types of cages have been used, ranging in size from small and large plastic boxes, through various types and sizes of glass jars and aquaria, to large metal containers and plastic or metal garbage cans, the types which have been found to be of overall greatest use are

1.   Round acrylic containers (6 in. diameter, $2^1/_2$ in. high) with tightly fitting lids (Figure 4)
2.   Rigid rectangular acrylic "refrigerator chests" ($12^3/_8 \times 10^1/_{16} \times 3^7/_8$ in.) (Figure 4)
3.   Wide-mouth gallon bottles, battery jars of a number of sizes (including ones that can be made by removing the necks from ca. 1-gal glass chemical bottles), and tall cylindrical plastic containers (Figure 4)
4.   Aluminum or galvanized steel cages, usually of a large size (e.g., ca. $10 \times 14 \times 24$ in.)
5.   Plastic garbage cans (20 to 30 gal) having semirigid lids that can be clamped tightly against the rim of the can by heavy steel handles

These containers have the following features in common: (1) long-term durability, (2) ease of cleaning, (3) resistance to corrosion or being eaten by the cockroaches, (4) tightly fitting lids or the ability to have cheese cloth stretched tightly across their openings, and (5) all can be provided with a means of permitting airflow into the cage so that elevated humidity does not cause the food to become moldy or mites to become a serious problem in the colony.

FIGURE 4.   Plastic cages: type 1 (right, front), type 2 (left), type 3 (right, rear). See Section VI of text for details.

Type 1 cages are mainly used for starting colonies of small- to medium-sized species from a few individuals, particularly when they are of the sort which does best under crowded or semicrowded conditions. Type 2 cages are ideal for long-term culturing of small- to medium-sized species that favor horizontal spacing and do not otherwise require special care. Type 3 containers are the best for most blattellids (this also includes large metal canisters), while type 4 are best for culturing medium- to large-sized blaberids and very large blattellids (e.g., *Nyctibora* spp.). The large garbage cans (type 5) are primarily used for mass rearing of small- to medium-sized species that tolerate or do best on vertical surfaces.

Durable acrylic containers of a variety of sizes and shapes that are ideal for making type 1, 2, and 3 cages are available from Tri-State Plastics (Crown Hill Industries, Inc., 4779 Upper Valley Pike, Urbana, OH 43078). Battery jars of various sizes can be obtained from most glass suppliers or can be made from large ($\geq$2 qt) bottles if a proper glass cutoff wheel is available. Various sizes of rectangular aluminum or galvanized metal cages must be fabricated, while plastic garbage cans are relatively inexpensive and widely available. Very light plastic containers (such as the widely available shoe boxes, etc.) should not be used because they are not durable, they crack easily, and they usually do not have tightly fitting lids. Similarly, while glass aquaria can be used, they are not recommended for large-scale culturing because of their expense, weight, and high cost of repair.

Provision for air circulation within the cage is a necessity no matter what type of cage is used. Battery jars and other type 3 containers that have large openings are best covered with several layers of cheese cloth held in place with bands made from cloth-covered elastic tape. Although the elastic tape is more expensive than rubber bands, its use is highly recommended because it lasts far longer than rubber bands, which age rapidly and have a disconcerting tendency to break when least expected, allowing the cockroaches to escape. If plastic cage tops are used, merely punching holes in them with a heated needle or rod is not sufficient. Instead, sizable openings equivalent to one half or more of the cage top must be made and then covered with fairly fine-mesh metal screening (Figure 4). Holes can be easily cut in rigid plastics with a high-wattage (300-W) soldering iron whose tip has been cut and ground to a knife-like blade, and in soft plastics with tin shears. The screening, which should be stainless steel to withstand repeated washing, is then sealed around the edge of the openings by being melted into the plastic with a lower-wattage (150-W) soldering iron. When metal containers are used, the screening, usually with a heavier $^3/_8$-in.-mesh hardware cloth supporting layer, must be attached to the lids by welding or riveting. Breaks in plastic cages are best repaired by fusing screen across the splits with a hot soldering iron.

All cages must have some provision made for the placement of food and waterers, and they must also have some form of bedding material placed in them (see Section VII below). In addition, a band of petroleum jelly should always be applied around the inside of the upper part of the cage to keep the cockroaches from crawling onto the lid or cheese cloth. This can be obtained cheaply in 1- to 5-lb containers and is applied most easily and neatly to the cages with a 1-in.-wide paint brush.

VII. HARBORAGE MATERIALS

Nearly all cockroaches thrive when provided with hiding spaces that approximate their body height. Additionally, most have a decided preference for resting on either horizontal or vertical surfaces. The harborage materials used must take these necessities into account and in addition (1) should be low in cost or extremely durable; (2) should facilitate, rather than hinder, the transfer of cultures; (3) should not interfere with or be harmed by the waterers; and (4) for many blattids, must be of a material that they can excavate and to which they can then cement and cover their oothecae.

The materials which have proved to be the best for meeting these requirements are

1. Large-sized (8 × 11 in.) and medium-sized (5 × 11 in.) papier-mâché separators from fluorescent light tube cartons
2. Tempered sheet Masonite® in several configurations
3. Hardward cloth ($^1/_4$-in. mesh)
4. Paper toweling and wadding, both of which are always used as adjuncts to other materials
5. Pelleted laboratory bedding ($^1/_8$-in. size; a corncob product)

The materials *not* recommended for use are wood shavings, shredded paper, and rolls of corrugated cardboard. Many species find the oils from the wood shavings objectionable. Furthermore, shavings and shredded paper tend to get knocked into the waterers and act as wicks, turning the entire bedding into a wet mass that molds rapidly. Additionally, it is very difficult to remove nymphs from them when the cultures have to be transferred. Rolls of corrugated cardboard tend to be eaten by the cockroaches and then collapse; also, because of the air circulation problems they present, they tend to mold rapidly.

Large papier-mâché separators are ideal harborage materials for raising most blattids, blaberids, and many blattellids, particularly larger ones such as *Nyctibora* and *Xestoblatta* species. These separators cost nothing and can be obtained in large quantities wherever there is extensive use of fluorescent lighting. They provide a multitude of spaces and of horizontal and vertical surfaces, and they also admirably meet requirement 4 above. Most species do well when provided with nothing more than horizontal stacks formed either by having alternating layers running crosswise to each other or by having them all run in the same direction, but offset so that the grooves of one layer sit atop the flat crests of the one beneath it.

Some of the smaller burrowing blaberids such as the *Panchlora* and *Pycnoscelus* spp. do best when provided with much closer spacing, a requirement that is met most easily by placing sheets of multilayered paper wadding (see Section V above) between the separator layers. On the other hand, some of the very large blaberids that inhabit caves and hollow trees (e.g., the two *Archimandrita* spp. and *Blaberus* spp. such as *B. colosseus*, *B. giganteus*, and *B. parabolicus*) do well only when provided with extensive vertical resting surfaces. This requirement is met most easily by stapling paired separators to the sides of short lengths of 1 × 2 in. furring strips and then standing them upright in large type 4 cages. A 2- to 3-in. layer of the pelleted laboratory bedding should be placed in these cages (into which the small nymphs can burrow), and a block of wood or a brick must be used to elevate the waterer above this bedding because it molds rapidly when damp.

Because of their smaller size, most blattellids require much closer spacing than can be provided by the papier-mâché separators. The ideal harborage material for raising them is a Masonite® spacer with staple legs. The Masonite® spacers are cut into squares or octagons that will fit loosely into battery jars or similar type 3 containers and which have four or more $^3/_8$-in. staples driven into them as spacing legs. These are then stacked in the containers, with a piece of paper toweling between each layer to provide the very tight spaces favored by the tiny nymphs.

Cornwell[6] recommended the use of rolls of corrugated paper in battery jars or large can containers for blattelids and others. Their use presents many problems. The rolls must be prepared by inserting in them some form of spacer material such as a strip of the corrugated paper (this requires unwinding and rewinding the roll) so that the resulting spiral has its layers held about $^3/_8$ in. apart. They also must be supported above the floor of the cage by wood or metal spacers so that the cockroaches can move easily from one layer to the next. Additionally, care must be taken with waterer placement to ensure that no water reaches the rolls, since they collapse and mold rapidly when wet. For these reasons, even though the initial cost and time expended in making the Masonite® separators is greater than that involved in using paper rolls, the former are best for long-term culturing in battery jars because of their great durability and ease of cleaning and because they also permit a much easier transfer of cultures from one cage to another.

When mass rearing blattids such as *Blatta orientalis* and the various *Periplaneta* spp., the required spacing must be between $^3/_4$ and 1 in., something that is difficult to maintain if corrugated paper is used. It also does not last long because both of these genera excavate the paper for oothecal attachment. A much better material for culturing such species is standing lengths of $^1/_4$- or $^3/_8$-in.-mesh hardware cloth that have been folded back and forth, leaving about 1 in. between the layers (see Section VIII.E). The screens must be supported above the floor of the cage (usually a plastic garbage can), like the paper rolls, but when using them there is no danger of collapse due to a waterer spill or of the paper being eaten by the cockroaches.

VIII. SETTING UP CAGES

The primary requirements for setting up cages are to provide adequate harborage and a means for keeping both harborage and food away from the waterers. Each type of cage (see Section VI) has its own requirements. Since most can be used for both continuing and age-class culturing, the methods for doing the latter are discussed separately in Section VIII.F below.

A. TYPE 1 CAGES

A type 1 cage makes use of a 15-dram vial waterer and a papier-mâché separator cut two "U"s wide and then in half crosswise. The separator will hold the vial firmly against the side of the cage. Dog food can be placed alongside or on top of the separator, but always must be kept away from the waterer face.

B. TYPE 2 CAGES

These cages use 35 × 200 mm culture tubes as waterers. While a large papier-mâché separator can be employed to hold the culture tube against the long wall of the cage, this is satisfactory only for culturing small blattellids such as *Supella* and *Shawella* spp. For the larger blattid genera (*Blatta, Neostylopyga,* and *Periplaneta*) and, in general, all but the largest blattellid (*Nyctibora*) and blaberid (*Archimandrita, Blaberus,* etc.) genera, a much better solution is to use a Masonite® base provided with wood block legs, which firmly pins the culture tube against the side of the cage, keeps it from sliding back and forth, and also permits the cockroaches to pass freely beneath it (Figures 5 to 7).

These bases are very durable, with some in the author's laboratory having been used for more than 10 years. (They are simply cleaned by first being frozen and then having the oothecae and other detritus scraped from them with a spatula and scrub brush.) The dimensions of the base and the positioning and lengths of the leg blocks are given in Figure 7. Wood glue and short flathead nails should be used to attach the blocks to the base. If many separators are to be made, it is best to first make up a jig that holds the base and permits accurate positioning of the legs. This allows completed bases to be prepared in 1 min or less and thus, in the long run, amply makes up for whatever time was spent making the jig.

The cage is prepared by first putting a base into it, then a filled waterer, and then some food down the back so that it slides under the base. A large papier-mâché separator is then placed on top of the base and shoved as far forward as possible. More food is added at the back, and another separator is added on top of it with its "V" bases sitting on the flattened top of the first separator. One great advantage of this type of arrangement is that waterers are easy to remove and replace; another is that colonies can be enlarged or transferred merely by taking the top separator off and transferring it to another freshly prepared cage.

C. TYPE 3 CAGES

Cages of this type are most useful for raising the smaller blattellids (e.g., *Blattella*). Stacks of Masonite® disks with staple legs are layered in the containers. If vials are used as waterers, they should be supported in "M"-shaped saddles made from hardware cloth to keep them from shifting. The food should be placed in a plastic or aluminum cup or in a small open-top box of hardware cloth to keep it from coming into contact with the waterer.

The special vertical waterer (Section V.C) should be employed if a longer time between watering is desired or if blattellids, which require a moist substrate for oothecal water uptake, are being cultured. Because these waterers are heavy and tend to slide around on the Masonite® surface, they must be confined in position. One way is to use three short wood blocks attached to the uppermost Masonite® spacer. An alternative is to attach to a spacer (with wood glue) a ¹/₂-in.-thick layer of polyurethane foam with a hole in it large enough to hold the waterer. The hole is made most neatly with the rim of a 1-lb coffee can, which easily cuts through the plastic when heated, leaving an opening only slightly larger than the petri dish base of the waterer. Because of the space taken up by such waterers, it is best to use rectangular hardware cloth holders for the food (Figure 8).

D. TYPE 4 CAGES

Larger cages such as these are ideal for raising the large blaberids, and overall they are easier to handle and require much less space than the type 5 cages (see Section E below). They are sufficiently large to allow the use of the 1-qt mason jar waterers and provide sufficient space to raise even the largest cockroaches. The waterer is placed on a ceramic tile and surrounded with 7- to 8-in.-high, square, large-mesh wire fence which keeps the papier-mâché harborage material from falling onto the waterer surface. Fences are best created from the vinyl-coated fencing rolls made for garden and lawn edges, widely available in hardware stores. This material has approximately a 2¹/₂-in. spacing between horizontals

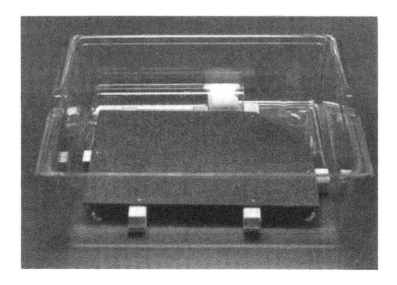

FIGURE 5.   Type 2 cage with Masonite® base and 200-mm culture tube in place.

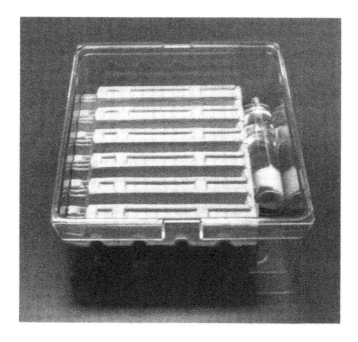

FIGURE 6.   Same as Figure 5, with large papier-mâché separator placed on top of the Masonite® base.

and verticals so that it can be bent easily to form a $7^1/_2$-in. square and can be cut to the required height. As with the type 2 cages, the best harborage material is the large or small papier-mâché separator. In these larger cages, however, the layers of separators should be placed at a 90° angle to each other because they tend to be shifted greatly by the large-bodied cockroaches. An alternative arrangement (still with a 1-qt waterer within a fence) used for the very largest species (*Archimandrita* spp., *Blaberus colosseus,* and *B. giganteus*) has been given in Section VII.

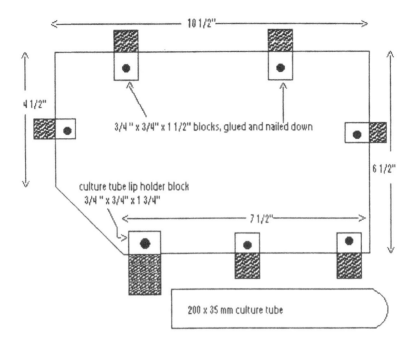

FIGURE 7. Masonite® base for use with $12^3/_8 \times 10^1/_{16} \times 3^7/_8$ in. acrylic refrigerator chests.

FIGURE 8. Waterer holder for type 3 cages. Hole was cut in polyurethane with a heated 1-lb coffee can. Food holder is made from $^1/_2$-in.-mesh hardware cloth.

E. TYPE 5 CAGES

Large garbage can containers are best for the mass rearing of blattids (*Blatta*, *Eurycotis*, *Periplaneta*) which do well on vertical surfaces. Other than waterers, the only materials employed in these cages are $^1/_4$- or $^3/_8$-in.-mesh hardware cloth and either two bricks or two 8-in. lengths of 2 × 4 in. wood. First, a circular fence of $^1/_4$- or $^3/_8$-in.-mesh hardware cloth is made that just fits within the container and extends to within about 6 in. of the top. This should have a number of holes cut into it so that roaches can pass from one side to the other. Next, a length of hardware cloth whose width is 1 in. less than the height of the container is folded back and forth into a zig-zag that fills the space within the outer circular fence, with the folds being made around a $1^1/_2$-in. wood form to give proper spacing between the layers. This inner ziz-zag portion is supported on the bricks or pieces of wood to allow the passage of roaches between the layers. A 5-in.-wide piece of hardware cloth is then placed on top of the zig-zag portion and firmly attached to both it and the outer circle. This piece will support the waterer, which should be placed on a tile. No special feeders are necessary as the roaches are fed by pouring dog food down between the vertical leaves of the zig-zag.

F. AGE-CLASS CULTURING

Setting up cultures to produce large numbers of adults that are all approximately the same age requires either very large cultures or considerable lead time. The reason, very simply, is that one has only two alternatives: either the cultures must be of such a size that a sufficient number of freshly molted adults can be collected within a very short period of time, or it must begin with obtaining fresh oothecae (blattids), ootheca-bearing females (*Blattella* spp.), pregnant females (blaberids), or recently eclosed nymphs (all groups). In any event, it always requires building the cultures to a large size in holding cages (types 2 to 4) so that sufficient numbers become available.

If the age classes are to be generated in terms of birthdate, the best methods are to have a good ootheca collector or a means of collecting freshly eclosed nymphs. For blattids which actively attach their oothecae to surfaces, the best way to do this is to have a number of standing population cages with papier-mâché separator harborage (this must be added to the type 4 cages; see Section VIII.E above). New separators ·.re added to the cages. When a sufficient number of oothecae have been collected, adults and nymphs are shaken off the separator, which is then moved to a newly prepared cage (usually type 2) with a waterer. The ootheca-bearing separators are moved to new cages at 3- to 5-d intervals, with all nymphs appearing during that interval being left together within a cage. If this is done with five to ten standing-population cages as the sources, then outputs of several hundred nymphs per interval can be sustained for some time.

Blattella spp. must be handled differently because the females bear their oothecae until just prior to eclosion of the nymphs. A very successful method for collecting age-classed nymphs is to place either only ootheca-bearing females or a mixture of males and females in an aluminum-screened cage that is within a larger cage. The nymphs eclosing from the oothecae are very small and easily slip through the screening into the larger cage, where they can be collected at whatever interval desired. With such a system, depending on the number of females and holder cages employed, up to several hundred nymphs born within a short time of each other can be collected daily.

As all blaberids are live bearers, the best method for obtaining nymphs is to confine adults within a cage with a raised hardware mesh floor through which the adults cannot squeeze. In this circumstance, it is best to have nothing more than food and water in the upper cage so that the nymphs, instead of finding hiding places, will tend to slip through the mesh to the floor below. Unless nymphs are collected daily, a means of providing them with water must be employed, as they tend to dessicate rapidly.

No matter which of the nymph-collecting methods is used, a central problem remains: not all nymphs will grow at the same rate. Thus, there should be no expectation that they will all become adults at the same time. For this reason, if large numbers of adults of known age are required, it is best to start with a number of large age-classed nymphal cultures in order to be able to remove on a daily basis a sufficient number of adults that have emerged within a short time of each other.

IX. TRANSFERRING CULTURES

All cultures must be transferred at intervals into fresh cages, both as a sanitation measure and because the population density within them becomes too high. For most species, the best way to judge when this should be done is to note the rate of water usage: in general, it is when a 300-mm horizontal waterer or a vertical waterer becomes depleted in 1 week or less.

For all but the large blaberids and species in garbage cans, the best way to set up a new culture is to take the top papier-mâché separator or several Masonite® separators from the old cage and place it or them in a new cage. If the population densities have become extremely high, or if the species has started to eat the papier-mâché or it has become moldy, then the uppermost layer should be discarded and a new separator placed in the old cage, left for a period of time, and then transferred to the new cage. For species in garbage cans, placing separators in the can and allowing the roaches to crawl onto them is the best method for easily obtaining a sufficient number to transfer to a new cage.

Large blaberids (*Archimandrita* spp., *Blaberus* spp., etc.) are handled differently. Here, 10 to 15 pregnant females, an equal number of males, and ca. 30 nymphs of assorted sizes are picked by hand from the old cage and transferred to a new, already set up cage. If pelleted bedding has been used in the cage, then it is almost always necessary to use a $^1/_4$-in.-mesh sieve to obtain a sufficient number of small nymphs from the bedding.

The timing for the transfer of large *Gromphadorhina* spp. and large blaberids given vertical surfaces on which to rest (*Archimandrita* spp., *Blaberus giganteus,* etc.) is slightly different from the above. For species resting on vertical surfaces, the colony should be split or transferred when the adults occupy approximately half of the vertical surfaces because larger nymphs require a vertical surface on which to molt successfully and are forced to attempt to do so on horizontal surfaces when the adult density is too high. The large *Gromphadorhina* spp. seemingly represent a unique case, in that a colony can become senescent if the adult density becomes too high. A sure sign that this is occurring is to have a culture containing a large number of adults and small nymphs, but virtually no mid-sized ones. There are two reasons for this:

1. The adults are large and heavy bodied and have long, thick spines on their legs. Under crowded conditions, while walking around they inadvertently puncture the cuticle of young nymphs, particularly freshly molted ones. When the adult density is very high the nymphal death rate is so great as to almost eliminate any new recruitment to adulthood.
2. At high densities the females either self-abort their oothecae or have them partially or entirely eaten by other adults while they are being extruded prior to rotation and withdrawal into the internal incubating uterus. As a result, with this genus the colony must be divided or a new one started as soon as the relative number of mid-sized nymphs starts to fall.

X. CULTURE AND SPECIMEN DISPOSAL

Deep-freezing is the best method for disposal of specimens and cultures. It avoids the use of noxious chemicals and is far better than autoclaving, which always results in a wet, malodorous sludge that is difficult to dispose of properly. Freezing at $-30°C$ for 24 to 48 h kills everything and yields a relatively dry mass that easily can be scraped into a plastic bag and disposed of with the regular garbage. Before washing, the petroleum jelly should be scraped from the rim of the cage with a spatula, and any that remains is removed with a paper laboratory wipe (tissue). This eliminates leaving jelly films on the cages during washing and subsequent fouling (and blockage) of sinks and drain lines. The cages are then either filled with a strong, hot detergent solution or submerged in a tank of detergent solution, allowed to soak, and then washed. If metal cages are being used, they should then be wiped dry to prevent rusting or other corrosion.

Masonite® separators should not be wetted. Instead, after being frozen they should be scraped clean with a spatula and then placed in a 70°C oven and baked for 1 d or more to kill any mold spores. All waterers, bricks, unglazed tiles, etc. are soaked in detergent solutions and then washed and rinsed. Paper waterer plugs are disposed of after being frozen. If sponges are used, they should be washed thoroughly and then baked dry in a 45°C oven.

XI. PESTS AND PEST CONTROL

The major "pest" problems that can occur in cockroach cultures are mold growth, phorid flies, acaroid mites, spiders, beetles, and, in multispecies cultures, other cockroaches.

A. MOLDS

The occurrence of molds in cultures is primarily a manifestation of high-humidity conditions, either in the incubator or locally within a cage. Dog food that becomes moist or wetted (e.g., by a waterer spill) rapidly becomes moldy. While many species will eat moldy food, others will not. If mold growth becomes prolific, small nymphs may become entrapped in it and die. Thus, any cage in which there has been a waterer spill must be cleaned out immediately.

The only cure for mold growth is to lower the overall humidity and/or to thoroughly clean out the cages as outlined above. This is not merely a matter of proper sanitation because molds are direct attractants of phorid flies.

B. PHORID FLIES

These are small, hump-backed flies with an absolute avidity for wet food and fresh cockroach carcasses. While primarily attracted by molding food, their larvae require a semiliquid medium for growth, so that in addition to carcasses they will start to grow on waterers and any other source of moisture in an incubator when they become numerous. Although they seemingly present no problem to the cockroaches, these flies can rapidly become so numerous that it becomes difficult to work in an incubator without inhaling them.

Preventing invasion by phorids is a matter of eternal vigilance. Fly paper and *Drosophila* traps charged with molding dog food can be used as early-warning systems, but with traps of these types the absence of flies does not mean that the cultures have not been invaded. All cultures should be examined weekly, particularly during the summer months. If flies are seen in a cage, it should be removed from the incubator *immediately*; then either the cage should be disposed of directly by freezing, or some individuals should be removed from it in a hood (or outdoors) prior to freezing of the cage.

It takes prodigious efforts to exterminate these flies once they become established in an incubator. For all intents and purposes, most or all cultures must be restarted. Individuals and/or oothecae must be put in clean, fly-proof cages and then moved to a new location where they can be observed to ensure that no flies have been carried over. This must be done for every cage in the incubator, with all infested cages and their waterers being frozen. Finally, when everything has been removed from the incubator, its temperature should be raised to 40°C or more for 1 week to ensure that all adults, larvae, and pupae of the flies have been killed. This general sanitation procedure also kills any other pests in the incubator.

C. ACAROID MITES

Most cockroaches have or can become infested with acaroid mites. Small as they are, if they become numerous enough, they can seriously debilitate or annihilate a culture. Mite populations increase prodigiously at temperatures >28°C and at >60% relative humidity. While it is probable that they can never be totally eliminated, they usually can be controlled to a low level merely by keeping temperature and humidity levels in the incubator and within the cages below these values and by practicing proper cage sanitation. This is yet another reason why the insides of the cages must be kept dry and waterer spills avoided. It also is why cages must be cleaned at intervals to prevent a buildup of carcasses, since these are humidity traps and can greatly elevate the relative humidity, and thus mite production, within a cage.

D. SPIDERS

Spiders are adventives that become established in incubators only if there are many escaping cockroach nymphs or phorid flies. They only become a real problem if they become established within a cage, since most will eat very small (nymphal) cockroaches. Fortunately, they are easily controlled by direct killing or by sucking them up with a vacuum cleaner. If a vacuum cleaner is used, its bag should be frozen after each use to kill whatever living material it contains.

E. BEETLES

Most of the beetles occurring in cultures have been brought in with and are living on dog food. Prevention of infestation is therefore a matter of ensuring that they are not furnished with this means of entry. When they become exceedingly numerous, the only way to get rid of them is to follow the procedure for the phorid flies, outlined above in Section XI.C.

Dermestids are the only other type of beetle likely to infest an incubator. In general, this only occurs when someone else in the same building or insectory is raising them. Their larvae are voracious, avidly devouring most dry organic materials in a cage, paper and carcasses included. If very numerous, they will even begin to eat freshly molted roaches. As is the case with other beetles, they cannot be controlled to unobjectionable levels; it is necessary to eliminate them from the incubator as outlined above.

F. COCKROACHES

Several cockroach species are escape artists, particularly as nymphs, and have a propensity for establishing themselves in the cages of other species. Included in this category are *Blattella germanica*, *Blaberus discoidalis*, *Blaberus craniifer*, *Gromphadorhina portentosa (?)*, the *Periplaneta* spp., *Symploce capitata*, and, among the exotics, *Phoetalia pallida*. If left unattended, they actually can outcompete certain other species.

To a large extent, preventing escapees is primarily a matter of ensuring that the Vaseline® bands around the cage rims remain intact, that all cage tops are closed and their openings covered, and that poison traps are set around the incubator. Other measures are required when working with blattellids such as *Blattella germanica* and *S. capitata*, since the adults of both of these will fly short distances under incubator conditions. When working with these cockroaches, it is necessary to set their cages on blocks in the middle of fairly large moats filled with water to which a small amount of detergent has been added. This assures that any individual jumping from the cage will be wetted and drowned.

REFERENCES

1. **King, E. G. and Leppla, N. C., Eds.,** Advances and Challenges in Insect Rearing, Agricultural Research Service (Southern Region), U.S. Department of Agriculture, New Orleans, LA, 1984.
2. **Cornwell, P. B.,** *The Cockroach*, Vol. 1, Hutchison and Co., London, 1968.

3. **Morgan, N. O.,** Blattidae, in *Handbook of Insect Rearing,* Vol. 1, Singh, P. and Moore, R. R., Eds., Elsevier, Amsterdam, 1985, 321.

4. **Piquett, P. G. and Fales, J. H.,** Rearing Cockroaches for Experimental Purposes, ET-301, U.S. Department of Agriculture, 1952.

5. **Roth, L. M.,** Water changes in cockroach oöthecae in relation to the evolution of ovoviviparity and viviparity, *Ann. Entomol. Soc. Am.,* 60, 928, 1967.

6. **Cornwell, P. B.,** *The Cockroach,* Vol. 2, Associated Business Programmes, London, 1976, chap. 12.

Chapter 3

ANATOMY OF THE COCKROACH

A. P. Gupta, Y. T. Das, and B. R. Rao

TABLE OF CONTENTS

I. INTRODUCTION

This chapter is intended to provide basic information on the external and internal anatomy of the cockroach to a scientist in the biomedical and veterinary sciences who may not be familiar with insect anatomy and who may wish to venture into dissecting one, either to satisfy his/her scientific curiosity or with a view to using it as a model for his/her research. An entomologist or neurobiologist already familiar with insect anatomy, therefore, may find most of the contents of the chapter elementary, but then its purpose is to inform the novice rather than one with expertise in entomology.

The presentation on each organ system includes a brief description of its basic features, presented in a jargon-free style, although the important technical terms have been included. Should further details be necessary, the reader may consult the many reviews and original papers referred to in the section on each organ system. Note that sections on the nervous system and sense organs are not included because they are described in detail in other chapters in this work.

The chapter is organized into three major sections: external anatomy, including integument; dissection instructions; and internal anatomy.

II. EXTERNAL ANATOMY

The body of a cockroach, or for that matter any other insect, is composed of some 20 segments that are grouped into three easily recognizable regions: *head, thorax,* and *abdomen*. It also has a neck (cervix) which is not often visible.

A. HEAD

The head (Figure 1) is composed of six or seven segments and bears a pair of segmented *antennae*, two *compound eyes*, two simple eyes (*ocelli*), and the chewing type of *mouthparts*. Because the latter point downward, the head in cockroaches is termed *hypognathous*. In a frontal view, the head shows several grooves (*sulci*) which define various parts that form the head capsule.

The *antennae* represent the first set of appendages on the head and consist of three main parts: the basal part (*scape*), the middle or the second part (*pedicel*), and the remaining multisegmented part (*flagellum*). The pedicel has a sensory structure (*Johnston's organ*) in most insects, and the flagellum is the longest part in cockroaches, with approximately 88 segments in *Blattella germanica*.[1] The flagellum is also the site of various sensory receptors. Each antenna has an antennal sulcus (*circumantennal sulcus*) around its base.

Each *compound eye* is composed of a group of separate units, or *ommatidia* (2000 in *Periplaneta americana*); each ommatidium consists of an outer *cornea* or corneal lens, a *crystalline cone* (produced by four cone cells) behind the cornea, and several elongated *retinal or retinula cells* which are grouped together to form an optic rod (*rhabdom*). In addition, some pigment cells are present. A nerve connects the rhabdom to the optic nerve. For anatomical and ultrastructural details of the compound eye in *P. americana*, see the papers of Butler[2,3] and Chapter 23 in this work.

The *ocelli* (simple eyes) of cockroaches also often have a corneal lens and a crystalline cone, but are not involved in image formation. They principally act as photoreceptor cells that are capable of little more than distinguishing between light and darkness. They are also regarded as having a stimulatory effect on the compound eyes. For details, see the article by Guthrie and Tindall[4] and Chapter 23 of this text.

The *mouthparts* (Figure 2) are a set of structures that surround the mouth opening in such a manner that they enclose an extraoral or preoral space (*cibarium*). They consist of the following parts:

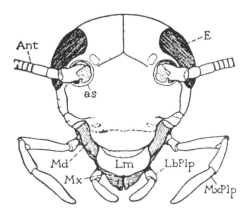

FIGURE 1. Frontal view of head of *Blatta orientalis*.[15] Ant = antenna; as = antennal sulcus; LbPlp = labial palp; Lm = labrum; Md = mandible; Mx = maxilla; MxPlp = maxillary palp; E = eye.

1. *Mandibles* (the main pair of jaws) are located on each side and are used for chewing.
2. *Maxillae* (the second pair of jaws) are located just behind the mandibles and are used chiefly for picking up and holding food; each maxilla consists of a cardo (basal part), a stipes (middle part), two distal pieces, an inner lacinia and an outer galea, and a maxillary palp.
3. The *labrum* is the upper lip; a median lobe on its internal or ventral surface is called the *epipharynx*, which serves as a taste organ.
4. The *labium*, or lower lip, represents the two fused second maxillae and closes the rear of the circle of structures around the mouth; it helps to hold food in the cibarium and serves as a tasting organ. The labium consists of a basal part (submentum), followed by a middle portion (mentum) and a distal prementum; from the latter arise a pair of inner glossae and a pair of outer paraglossae. Like the maxilla, the labium has a labial palp. Internally, there is another structure (*hypopharynx*) on the upper surface of the labium; it also serves as a taste organ or tongue.

Such primitive and generalized type of mouthparts are also found in centipedes and symphylans; the latter are considered the progenitors of insects.

B. NECK

The neck (*cervix*) in insects is considered a reduced body region (*"microthorax"*) and is strengthened on each side by lateral sclerites (cervical sclerites).

C. THORAX

A characteristic part of the insect body, the thorax (Figure 3) is the locomotor center, with three pairs of *legs* and two pairs of *wings* in the winged (pterygote) insects. It has three segments, the *pro-*, *meso-*, and *metathorax*, each one bearing a pair of legs. The wings, when present, are located only on the meso- and metathorax (pterothorax).

The thorax is like a box, consisting of a dorsal side (*notum* or *tergum*), two lateral sides (*pleura*), and a ventral or underside (*sternum*). Any hardened part or subdivision of any of the four sides of the thorax is called a *sclerite*, those of the notum, pleuron, and sternum being termed *tergite*, *pleurite*, and *sternite*, respectively. By the same token, a tergite belonging to the metathorax may be called a *metatergite* and its sternite, the *metasternite*, and so on. Each notum, pleuron, and sternum is often subdivided into parts, each part

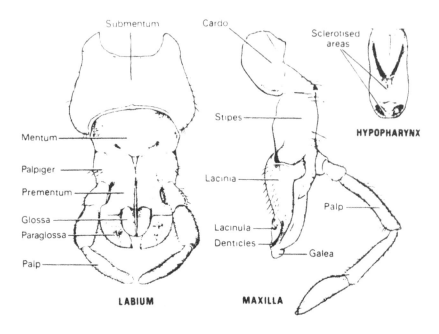

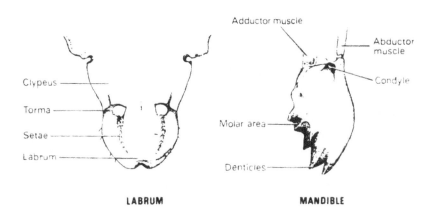

FIGURE 2.   Mouthparts of *Periplaneta americana.* (From Bell, W. S. and Adiyodi, K. G., Eds., *The American Cockroach,* Chapman and Hall, New York, 1981. With permission.)

(sclerite) bearing a different name (e.g., *scutum, scutellum,* etc.). The thorax has two pairs of spiracles (see Section V.C).

1. Legs

All three pairs of legs have a common pattern, and each leg consists of the following five parts (Figure 4):

1. The *coxa* (or limb base) acts as the ball of a ball-and-socket joint by which the leg is attached to the body.
2. The *trochanter* is a small leg segment which connects with the coxa by a dicondylic articulation and is fused with the next segment, the femur.
3. The *femur* is the largest and stoutest part of the leg; it is packed with muscles.
4. The *tibia* follows the femur; it has a dicondylic articulation with the latter and is often

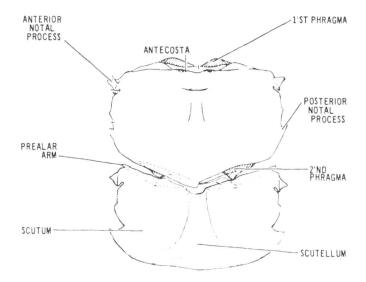

FIGURE 3. Dorsal view of the pterothorax (meso- and metathorax combined) of *Arenivaga aegyptica.*[58]

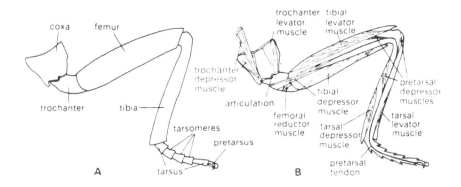

FIGURE 4. A typical insect leg and its muscles.[59]

longer and more slender than the femur. It often bears spines and (at or near its outer end) movable spurs, which provide traction during locomotion.

5.   The *tarsus,* or foot, is typically a five-jointed part of the leg.
6.   The *pretarsus* bears a pair of claws (*ungues*) at its tip and also various pads or glands.

The cockroach legs are of the running (cursorial) type.

2. Wings

Insects are the only animals that have wings (Figure 5) that evolved for flight; bird and bat wings are modified, preexisting limbs. Many believe that there were originally three pairs of wings, one on each of the three thoracic segments. However, during the course of evolution, those on the prothorax were lost.

In cockroaches, the first pair, or forewings (located on the mesothorax), are rather hardened and leathery in texture and are called *tegmina,* while the second pair (located on the metathorax) are membranous; both pairs of wings contain many veins that strengthen them.

The bases of all wings are connected to the body by membranous hinges in which are

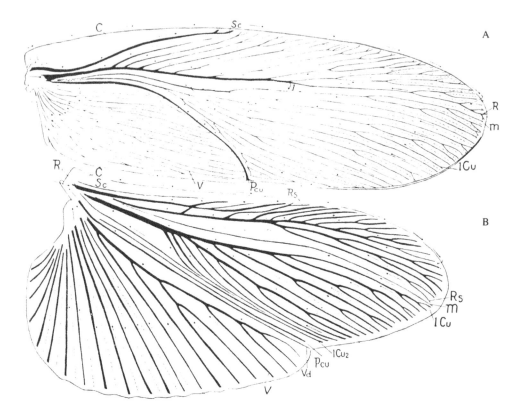

FIGURE 5. Fore- (A) and hindwings (B) of *P. americana.*[21] Letter abbreviations and arrows refer to various wing veins and directions of blood flow, respectively. Axillary sclerites at wing bases are not shown.

located a group of small sclerites (*axillary sclerites*) that articulate with the edge of the notum.

D. ABDOMEN

The abdomen typically consists of 11 segments and bears several appendages, including the *external genitalia* (see Section V.G). The jointed and prominent *cerci*, which are covered with tactile sense organs, are located on the 11th segment.

The segmental nature of the insect body is more evident in the abdomen than in the head and thorax. As in the thorax, each abdominal segment has a tergum, a sternum, and a pleuron, as well as eight pairs of spiracles (see Section V.C).

It is particularly easy to determine the sex of an adult cockroach from the terminal abdominal segments. In the male, nine sterna can be seen, while in the female, only seven are visible. In addition, females of Blattidae (e.g., *Blatta* and *Periplaneta*) have a keel-shaped subgenital plate attached to the seventh sternum (see Section V.G.1.b).

III. INTEGUMENT

The insect integument consists of a layer of epidermal cells which secrete and maintain a rigid wall, called the *cuticle*. The integument maintains the shape and rigidity of the body, acts as a barrier to toxic materials, prevents loss of body water by evaporation, and provides camouflage for the cockroach.

The integument is made up of three major layers: (1) the *epidermis*, (2) the *procuticle*, and (3) the *epicuticle*. The epicuticle is approximately 2 μm thick and can be divided into

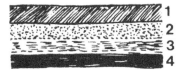

FIGURE 6. Diagram of cockroach epicuticle.[6] (1) Cement layer; (2) wax layer; (3) outer epicuticle; (4) inner epicuticle.

FIGURE 7. Cross section of the pronotum of *Blattella germanica*,[69] showing the nonsclerotized endocuticle (EN) and sclerotized exocuticle (EX). Sections were 10 μm thick and unstained.

an inner *epicuticle*, an outer *epicuticle*, a *wax layer*, and an outermost *cement layer* (Figure 6). The epicuticle is composed of proteins, lipids, and phenols.

The *procuticle*, which is the predominant region of the cuticle, can be separated into a sclerotized region (*exocuticle*) and a nonsclerotized region (*endocuticle*; see Figure 7). Sclerotization is a process by which the cuticular proteins are hardened and darkened. The process involves a variety of transformations and requires *N*-acetyldopamine.

Chemically, the cuticle is essentially made up of proteins, lipids, and carbohydrates. While the chemical characterization of individual molecules is still largely unknown, at least one important constituent, *chitin*, has been characterized. It is an *N*-acetylglucosamine polymer with a β-1,4 linkage (Figure 8). For details on the insect integument, one may refer to the works of Lipke et al.,[5] Mills,[6] and Hepburn.[7]

IV. DISSECTION INSTRUCTIONS

A. GENERAL INSTRUCTIONS

To study the *external anatomy* of any insect, one should take a particular part of the body, boil it for 10 to 15 min in 10% KOH to remove all soft parts, and then wash the material several times in water to remove all traces of the KOH. If necessary, the material to be examined may be stained in borax carmine or eosin and then studied in saline or 70% alcohol.

For an examination of the components of the *integument*, either a small portion of the body wall or the entire pronotum should be first dehydrated in an ethanol series and then embedded in a suitable medium and sectioned on a microtome. It is preferable to use unstained sections because it is easier to study the natural colorations (tanning and melanization) of

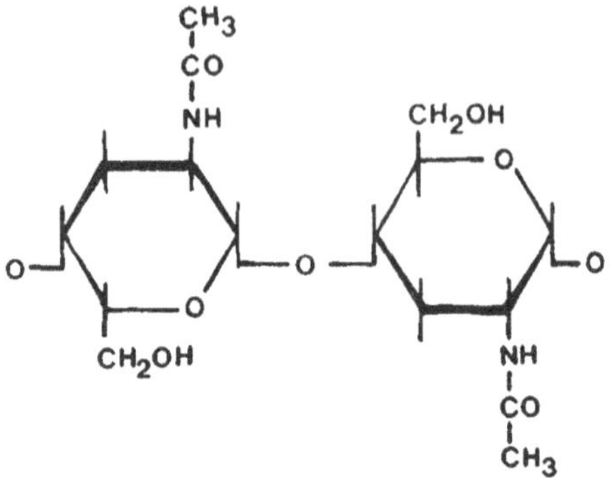

FIGURE 8.   Chemical structure of *N*-acetylglucos-amine.

the various layers. Very thin sections ($<$10 μm) are usually difficult to obtain because of the brittleness of the cuticle.

To study any of the *internal organ systems*, one has to open the body cavity (*hemocoel*), clear the fat body and the tracheae, and expose the various organs (viscera). Depending on the size of the cockroach, one would need a shallow or deep wax-layered dissection dish (DD) and physiological saline. (To prepare the latter, dissolve 9.32 g NaCl, 0.77 g KCl, 0.18 g $NaHCO_3$, and 0.01 g NaH_2PO_4 in 800 ml distilled water; after complete dissolution, add 0.5 g anhydrous $CaCl_2$, mix well, and bring the volume to the 1-l mark.) Alternately, 50 or 70% ethyl alcohol may be used.

The following protocol is recommended for cockroach dissection:

1.   First, anesthetize the cockroach with CO_2 gas or immobilize it by placing it in the freezer for 20 to 30 min. Take a pair of fine dissecting scissors and, while holding the cockroach upside down in one hand, carefully make a continuous incision along its lateral sides between the terga and the sterna. One may find it convenient to remove all legs and wings, and even the head, before making the incision, either while holding the animal or after placing it in the DD. If you prefer to place it in the DD, make sure that it remains submerged in the saline at all times, especially after exposing the viscera.

2.   After making the incision, place the cockroach on its ventral surface if you wish to look at all the internal organs except the dorsal vessel. Place it on its back for the latter structure.

3.   It is necessary at this stage to anchor the roach to the wax layer of the DD using insect pins. Insert pins, slanting away from the cockroach, all along the periphery of the terga or sterna (whichever are in contact with the wax), stretching them slightly.

4.   With fine forceps, gently peel away the unpinned body wall, exposing the viscera or the dorsal vessel. Note that all the internal organs are covered with fat body, tracheae, and tracheoles. Carefully remove them so that all the organs are clearly visible. For the study of a specific system, follow the instructions below for that particular system.

B. INSTRUCTIONS FOR VARIOUS SYSTEMS

To expose the *digestive and excretory systems* (Sections V.A and B; Figure 9), simply remove the fat body, tracheae, and tracheoles and separate the various digestive organs and the Malpighian tubules.

It is possible to study the *trachea* (Section V.C; Figure 10) in the above dissection by carefully separating the main branches of the tracheae and following some of them to the spiracles. The latter can be studied according to the methods for any part of the external anatomy.

Dissecting *muscles* (Section V.D; Figure 11) requires more care and patience, simply because most of them are very minute. Each muscle should be traced from its site of origin to that of its insertion, or vice versa.

The *dorsal vessel* (Section V.E; Figure 12) is located on the dorsal diaphragm; therefore, to study this structure, one should place the anesthetized insect on its back and remove the sterna and all internal organs to expose the heart and other associated structures (if present). The *hemocytes* can be studied and identified using Gupta's identification keys.[8,9]

The *reproductive system* (Section V.G.2; Figures 20 and 21) can be exposed by separating and removing the entire digestive system, fat body, and other tissues. Dissections of the spermatheca, its associated structures (the accessory glands), and the testes require more care and patience.

The *brain* (Section V.H; Figures 22 to 25), its associated structures, and the *ventral nerve chord* can be exposed by placing the cockroach on its ventral side and removing all overlying viscera. This will expose the entire length of the ventral nerve chord except its anterior-most part, which connects with the brain.

To dissect the brain, secure the head by inserting a small pin vertically down through the anterior-most region of the head. Starting from the prothorax, cut open the dorsal wall of the neck and the head along an arbitrary median line. To fully expose the brain, carefully separate the integument from the brain tissue and pin down the split integument into the DD. The integument may also be removed in small pieces and then discarded. Because of the small size of the brain, extreme care should be taken in removing the integument so that the brain tissue is not damaged in the process. This can be accomplished by gently pushing down the brain tissue as the integument is lifted up.

As in other animals, the nerve tissue in insects is slightly opaque, and care should be taken to decipher the location of the ganglia with proper illumination and magnification. Sometimes it is useful to stain the entire system by injecting 20 to 30 µl of 2% methylene blue in physiological saline 1 h prior to dissection.[10]

Dissection of the *endocrine system* (Section V.H; Figures 22 to 24) is more difficult than for most of the above systems. We suggest the following method:

1. After anchoring the roach to the DD, insert a small pin vertically down through the anterior-most region of the head. By holding the pin with forceps and pulling the head slightly forward (to stretch the neck region), secure the pin to the DD. A desirable posture of the head should have the mouthparts facing as far posteriorly as possible so that the neck region is fully exposed.
2. Using a pair of fine scissors, cut open the dorsal body wall along an arbitrary median line, starting from the pronotum and continuing through the head. Remove the thoracic body wall, tracheae, fat body, muscles, and digestive organs.
3. Pin down the split ends of the head integument with minute pins. The brain area, the corpora cardiaca, and the corpora allata should now be visible.
4. Under the dissecting microscope, arrange the light source so that these parts are seen clearly under minimum light intensity. In the German cockroach, the corpora allata are somewhat spindle shaped and have a maximum mean diameter of 75 µm in the male and 121 µm in the female.[10a]
5. The paired corpora cardiaca are part of the retrocerebral complex and are located dorsal to the hypocerebral ganglion. The anterior portions of the corpora cardiaca are intimately associated with the lateroventral wall of the aorta.

Since the ecdysial gland, which is juvenile tissue, persists in the adults in several cockroach species,[11] it is possible to dissect this gland in both adults and nymphs. These authors[11] have used the following procedures:

1.   Using a 100-µl Hamilton® syringe fitted with a 30-gauge needle, inject into the sixth-instar nymph or adult 20 to 30 µl of 0.25% methylene blue in double-distilled water 2 to 4 h prior to dissection.
2.   Cut open the prothorax and the neck region ventrally in ice-cold physiological saline and examine the glands.

C. DISSECTING TOOLS AND MATERIALS

Small dissecting dishes or trays can be prepared by pouring blackened (or plain) melted wax onto the bottom half of a suitable petri dish to a depth of one half to one third of the height. The thickness should be such that the dissections remain submerged without spilling the saline or alcohol.

For useful tips on microdissection, one may refer to Reference 66.

Dissecting pins, watchmaker's forceps with fine tips, iris microdissecting scissors, and other paraphernalia can be purchased from biological supply houses.

V. INTERNAL ANATOMY

A. DIGESTIVE SYSTEM

The alimentary canal (Figure 9) is divided into foregut (*stomodeum*), midgut (*mesenteron*), and hindgut (*proctodeum*). The fore- and hindgut are said to be of ectodermal origin; they have a cuticular lining on the inside, whereas the midgut has a specially synthesized peritrophic membrane that envelops the food contents.

The foregut consists of the undifferentiated *pharynx* and *esophagus*, the latter leading to a distensible *crop* that terminates in the *proventriculus* (gastric mill). Internally, the latter is lined with six hard cuticular teeth. The *stomodeal valve* is located at the end of the foregut.

The paired salivary glands, appearing as grape-like clusters (acini) with their paired reservoirs, are seen wrapped around the esophagus with delicate connective tissues. After dislodging them, the ducts can be traced passing through the neck, finally terminating at the base of the hypopharynx. The common reservoir ducts join the common salivary ducts to form the ultimate salivary duct. Salivary glands are whitish and translucent, and the reservoirs are filled with fluid.

Eight finger-like blind pouches (*gastric caeca*) are located at the junction of the fore- and midgut. The midgut is a simple tube and is the primary digestive and absorptive region. Numerous coiled Malpighian tubules are seen at the junction of the mid- and hindgut. These tubules function as excretory organs and are involved in water regulation and ionic balance.

The hindgut can be conveniently divided into a narrower *ileum*, a dilated, sac-like *colon*, and a thicker, oval-shaped *rectum*. The ileum accommodates the *pyloric valve*, consisting of six tiny spicules on the inner wall where the *peritrophic membrane* might be torn off. The rectum is provided with six longitudinal rectal pads that aid in resorption of water from the fecal matter.

Cockroaches are considered to be omnivorous, with mouthparts adapted for solid foods. However, water is an important dietary item. The ingested food passes through the alimentary canal, assisted by rhythmic peristalsis, at a specific rate, depending on the nature of the food.

The food is well moistened by the salivary secretions in the buccal cavity (cibarium) during mastication and is then passed into the esophagus via the pharynx. The crop mixes the partially digested food with the midgut digestive juices that pass forward from the midgut via antiperistaltic movement. The food is further triturated by the proventricular denticles before passing into the midgut through the stomodeal valve.

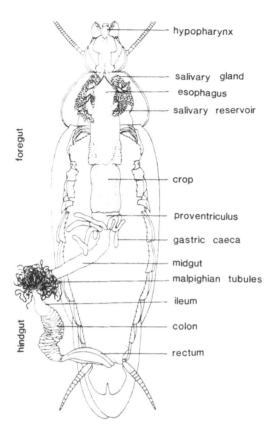

FIGURE 9.   Digestive and excretory systems of *P. amer-
icana.* (From MacLeod, E. G., *The Biology of Insects:
Laboratory Manual,* Burgess Publishing, Minneapolis, 1972.
With permission.)

The gastric caeca are thought to have secretory and absorptive functions. The midgut
is lined with a very thin peritrophic membrane which is formed by endodermal epithelial
cells and is replaced continuously. This tubular sheath, in addition to its protective function
for the delicate epithelial lining, is permeable to enzymes.

Much of the digestion takes place in the midgut. Several enzymes play a major role in
the digestion of ingested food. For example, the salivary gland in *P. americana* contains
amylase and some cellulase, whereas the midgut and caeca produce amylase, maltase,
invertase, lactase, lipase, glycosidase, and β-glucosidase. Lipase has been found in the crop,
midgut, and caeca. The enzymes detected in the crop may have been derived from either
the salivary glands or the midgut. Enzymes with a lower pH aid in carbohydrate digestion
in the crop, while trypsin and other proteinases, such as dipeptidases, aminopeptidases, and
carboxypeptidases (but not pepsin), are found in the crop and the midgut. Some midgut
enzymes are derived from intestinal microorganisms. Very little digestion takes place in the
hindgut; however, resorption of water and ions occurs in this region. For a complete list of
digestive enzymes one may refer to the chapter by Bignell in Reference 12.

B. EXCRETORY SYSTEM

The toxicity of the end product of nitrogenous metabolism appears to be correlated to
the nitrogen:hydrogen (N:H) ratio. Insects, with their high rate of metabolic activity, must
get rid of the nitrogenous waste safely and quickly. Being primarily terrestrial animals,
insects (except for aquatic species) cannot afford to have large amounts of water to excrete

this product in a soluble form. Hence, uric acid, an insoluble compound with the least toxic effect (1:1 N:H ratio), is the chief excretory product in insects.

The *Malpighian tubules* (Figure 9) are the primary excretory organs, located at the junction of mid- and hindgut in the form of coiled tubular structures. They appear to be vaguely arranged in six bundles of 12 to 15 each (or ranging from 60 to 150 in number) and are slightly yellowish in color. The tubules are of endodermal origin, with a one-cell-thick wall surrounding a lumen, and are closed distally. They are enveloped by strands of intrinsic muscles which produce the characteristic spiraling movement. A peritoneal sheath made of tracheoblasts covers the outside.

The uptake of waste material from the hemolymph can be studied by injecting dyes such as indigo carmine, neutral red, or trypan blue. For example, about 90% of the indigo carmine (2.5% aqueous) injected into the pericardial cavity of the American cockroach nymph can be excreted in about 5 h.

The probable mechanism of filtration is that soluble sodium and potassium urates, entering the tubules across the cell membrane, precipitate as insoluble uric acid and pass into the hindgut. The uric acid is then excreted along with the fecal matter. The ions and water are resorbed into the hemolymph, either at the base of the tubules or in the hindgut.

In addition to the Malpighian tubules, cockroaches are known to utilize other structures for excretion. The *fat body* can perform "storage excretion". *Uricose glands* in certain males can discharge urates by incorporating them in the *spermatophore*, which is then transferred into the female's bursa during copulation. Some waste material can be deposited in the cuticle and later expelled during the molting process. For more details on the excretory system of insects, see Reference 13.

C. RESPIRATORY SYSTEM

The high metabolic rate in insects is possible because of the availability of oxygen to the internal organs directly from the air rather than through an intermediate tissue or a fluid such as blood. This is accomplished by an elaborate network of ducts (*tracheae*) with openings on the lateral aspects of the body through which air enters and ultimately reaches the tissues and cells; this network is referred to as the *tracheal system*.

The noncellular part of the respiratory system (*ventilatory system*; see Figure 10) can be described in the following manner. Air enters the body by way of ten pairs (eight abdominal and two thoracic) of openings (*spiracles*) that regulate the flow of air by valves which are controlled by a set of muscles. Abdominal spiracles open inward and are recessed. Air then passes through the tracheal system, finally reaching the tissues such as muscles.

Tracheal structures are of ectodermal origin and are characterized by a spiral cuticular lining (*taenidia*) which prevents the collapse of the tubes. In dissection they appear as flexible silvery tubes because they contain air. Larger tubes, forming the main ducts, are called the tracheal trunks, whereas smaller to minute tubules, which are formed by repeated branchings, are known as *tracheoles*. These terminate as the tracheolar end cells that have intimate contact with the tissues.

1. Tracheal Arrangements

Using a blaberid cockroach, the following tracheal arrangement can be outlined. The dorsal view shows a pair of lateral longitudinal trunks that connect the spiracles on each side. A pair of dorsal trunks, located on either side of the middorsal heart and running parallel to the heart, are connected to each other by short commissures. These dorsal trunks are connected to the lateral trunks by stout anterior and posterior transverse connectives. In the thorax, however, the longitudinal dorsal trunks are quite enlarged, with shortened connectives.

The ventral view shows a pair of thin ventral longitudinal trunks that run on either side

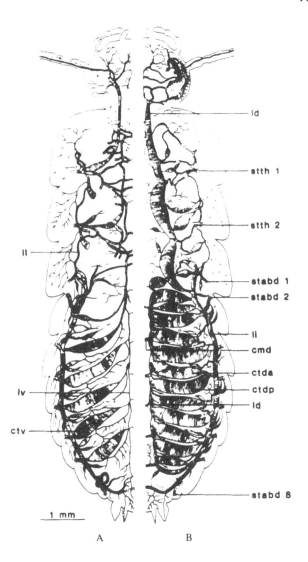

FIGURE 10.   Tracheal morphology of first-instar nymph of *Gromphadorhina portentosa*. (A) Ventral view, (B) dorsal view; cmd = dorsal commissure; ctda, ctdp = anterior and posterior dorsal transverse connectives, respectively; ctv = ventral transverse connective; ld, ll, lv = dorsal, lateral, and ventral longitudinal trunks, respectively; stabd = abdominal spiracles 1 to 8; stth = thoracic spiracles 1, 2. (From Miller, T. A., in *The American Cockroach*, Bell, W. S. and Adiyodi, K. G., Eds., Chapman and Hall, New York, 1981, chap. 8. With permission.)

of the central nervous system. These are connected to the ventral lateral trunks by single transverse connectives. The thoracic ventral longitudinals are not as enlarged on the dorsal side. In addition, there are large visceral trunks, giving out branches that anastomose into various internal structures, such as parts of the alimentary canal, Malpighian tubules, reproductive organs, and fat body. Appendicular tracheae are also seen embedded in muscles. For more details on the respiratory system of insects, refer to Reference 14.

D. MUSCULATURE

The unique feature of an insect form is its ability to accommodate large numbers of muscles (Figure 11) because of the increased surface area of the exoskeleton. This enables

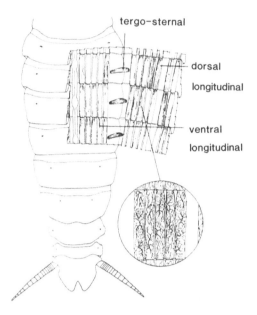

FIGURE 11.   Dorsal and ventral longitudinal and tergo-sternal muscles of the abdomen of *P. americana*. (From McLeod, E. G., *The Biology of Insects: Laboratory Manual*, Burgess Publishing, Minneapolis, 1972. With permission.)

the body to perform intricate movements despite its segmented nature. Humans with their endoskeletal system have about 700 muscles, while the American cockroach has as many, if not more. All insect muscles are of the striated type, even though functionally they can be differentiated into skeletal and visceral muscles.

The gross skeletal muscles of the American cockroach have been summarized from various sources by Guthrie and Tindall.[4] They compiled the names, numbers, and locations, along with their insertion and origin from each region of the body. For further details, refer to References 15 to 19.

Naming muscles involves functional aspects, which include many complex terms. For example, Hughes,[20] in describing leg movement, uses terminology such as protraction, promotion, retraction, remotion, adduction, abduction, levation, depression, extension, and flexion. The muscles that participate in this movement are grouped into two categories: extrinsic (arising from outside the leg) and intrinsic (arising wholly within the leg, running from one segment to the next). The coxa is moved by a set of extrinsic muscles that arise in the thorax. Intrinsic muscles are simpler and consist of pairs of antagonistic muscles on each segment. For instance, there are three levator muscles within the trochanter, arising from the coxa, and three depressor muscles (two from a coxal origin and a third from the pleural ridge). The femur is immovably attached to the trochanter, and the tibia is moved by extensor and flexor muscles arising in the femur and inserted into tendons from the membrane at the base of the tibia. The tarsal levator and depressor muscles arise from the tibia and are inserted into the top of the base of the tarsus. There are no intrinsic tarsal muscles. Characteristically, the pretarsus has only depressor muscles, whose fibers, in the form of a long tendon, are inserted into the femur and tibia and thus reach the pretarsus (Figure 4).

In addition to the large surface area created by the exoskeleton, there are modified internal portions of the skeleton, such as the *tentorium* (in the head), pleural and sternal *apophyses, phragmata* and *spinae* (in the thorax), and the *antecostal ridges* of the abdomen, which provide more areas to anchor muscles.

1. Basic Muscle Structure

Each muscle is made up of long, multinucleated fibers bounded by the *sarcolemma*, or plasma membrane. The cytoplasm and endoplasmic reticulum are referred to as *sarcoplasm* and *sarcoplasmic reticulum*, respectively. The plasma membrane is deeply invaginated into the fibers to form a regular radial canal. *Myofibrils* (filaments), a characteristic feature of muscle cells, are embedded in the sarcoplasm and extend from end to end along the fiber. The stouter fibrils are composed of myosin and the thinner ones of actin. The myosin filaments are surrounded by a ring of 10 to 12 actin filaments. Both actin and myosin filaments are linked by cross bridges, which provide structural and mechanical continuity. Units of 10 to 20 fibers are separated by tracheolated membranes. Tracheoles are in close contact with the outside of the muscle fiber, providing an adequate oxygen supply. Each fiber unit may also have its own nerve supply, or several muscle units may have common innervation. Furthermore, each unit is innervated by both a fast and a slow axon (sometimes inhibitory axons, too). The fibers are amply supplied with mitochondria.

Visceral muscles (found in the gut and heart) differ from skeletal muscles in having uninucleate cells, and the fibers are found individually rather than in groups. There are fewer mitochondria. The visceral muscles contract slowly; however, they are capable of super-contraction.

E. CIRCULATORY SYSTEM

The circulatory system of the cockroach consists of a simple tubular *dorsal vessel* (Figure 12), the blood (*hemolymph*), and the blood cells (*hemocytes*; Figure 13). Because in insects there are no real arteries or veins that carry blood directly to the tissues, the circulatory system in insects is called an *open system*, as opposed to the *closed system* in vertebrates. The body cavity (*hemocoel*) is divided longitudinally into three chambers (*sinuses*) by two longitudinal muscular partitions (*septa*; Figure 14). The dorsal septum (*diaphragm*) lies just below the heart and is attached to the terga laterally. It is made up of sheets of striated muscle fibers (*alary or aliform muscles*) and defines the *pericardial sinus* dorsally. The dorsal vessel virtually rests on it. The *ventral diaphragm*, when present, lies over the ventral nerve cord and defines the *perineural sinus*. The sinus lying between the pericardial and perineural sinuses is called the *perivisceral sinus*. The dorsal septum assists in the *diastole* of the heartbeat, while the ventral one directs blood flow (by its undulations) backward and laterally to ensure hemolymph irrigation of the nerve cord.

The *dorsal vessel* is a simple tube that runs along the dorsal midline below the terga. It is open anteriorly, but closed posteriorly, and it has two recognizable regions: the *aorta* and the *heart*, the latter being the distended, muscular posterior part. The heart is highly contractile, often confined to the abdomen, and generally consists of 10 to 12 chambers that are held in place by alary (fan-shaped) or aliform muscles. The anterior portion of the aorta, located in the head, is called the *cephalic aorta*. Blood flows into the heart through minute segmental openings (*ostia*) located on the dorsolateral sides of the heart, the cephalic aorta and the aorta being devoid of both alary muscles and ostia. The latter have valves that prevent expulsion of blood from the heart into the hemocoel during contraction (systole). Anteriorly, the cephalic aorta discharges blood near the base of the brain.

Some cockroaches (e.g., *Blaberus* and *P. americana*) also have *segmental vessels*, or "arteries" (Figure 12). In addition, some cockroaches (e.g., *P. americana*) possess the so-called *pulsatile organs*, or "hearts", located at the base of the antennae in the head and the wings.[21]

Each heartbeat consists of a *diastole* and *systole*, which together constitute a single

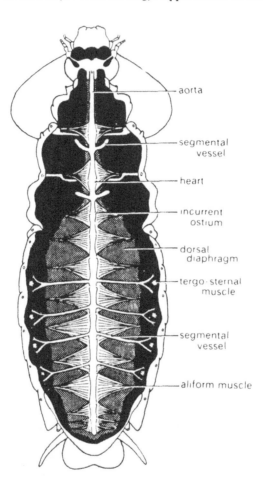

FIGURE 12.   Dissection showing the dorsal vessel, lateral arteries (segmental vessels), dorsal diaphragm, ostia, and alary (aliform) muscles of *Blaberus trapezoides*.[62]

cardiac cycle. It is controversial whether the insect heart is neurogenic or myogenic. Certainly, the heart in *Periplaneta* is heavily innervated. The heartbeat in insects originates in the posterior part of the heart and proceeds in an anterior direction. However, it is reversed under certain conditions.

The *hemocoel* is rarely full of hemolymph, which in insects is a clear liquid, serves as a medium of transport for its constituents, and constantly irrigates all the internal organs. In insects, blood does not transport oxygen to tissues and cells. It comprises about 16 to 20% of the total body weight, and it consists of *hemocytes* and all kinds of organic and inorganic ions, enzymes, and humoral immune factors (see Section V.F). Its pH ranges from 6.00 to 8.00.

Insect blood coagulates owing to hemocyte agglutination, hemolymph gelation, or a combination of both. In most insects, the granules of the granulocytes or coagulocytes cause coagulation. For details of the circulatory system and the hemolymph in insects, see the reviews of Miller[22] and Mullins,[23] respectively.

1. Hemocytes

The most commonly found hemocytes in insects are the *prohemocytes, plasmatocytes, granulocytes, spherulocytes, adipohemocytes, coagulocytes,* and *oenocytoids* (Figure 13).

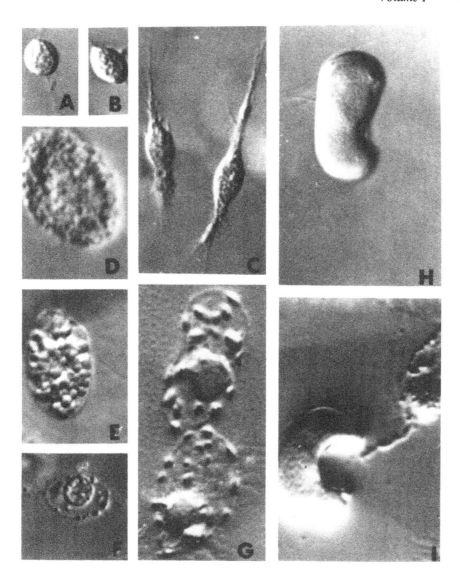

FIGURE 13. Major hemocyte types of *Gromphadorhina portentosa*. (A) Prohemocyte (magnification × 4333); (B) prohemocyte (magnification × 5633); (C) plasmatocytes (magnification × 6500); (D) granulocyte (magnification × 5687); (E) spherulocyte (magnification × 5687); (F) coagulocyte (magnification × 6314); (H) enucleate oenocytoid (magnification × 5525); (I) oenocytoid with extruded nucleus (magnification × 7150); Nomarski phase-contrast. (Reprinted with permission from Gupta, A. P., in *Comprehensive Insect Physiology, Biochemistry and Pharmacology,* Vol. 3, Kerkut, G. A. and Gilbert, L. I., Eds., Copyright 1981, Pergamon Books Ltd.)

Most of these can be identified easily using Gupta's identification key for hemocytes.[8,9] In the German cockroach, *Blattella germanica*, all seven types are present,[24,25] while in the hissing cockroach, *Gromphadorhina portentosa*, there are only six types, the adipohemocytes being absent.[26] For various details of hemocytes and their functions, see the reviews of Gupta.[8,9,27]

F. IMMUNE SYSTEM

The immune system is essentially a part of the circulatory system and consists of two types of blood cells as well as several humoral factors found in the blood. Insects have

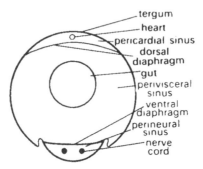

FIGURE 14.   Cross section of abdomen of a generalized insect, showing the three sinuses.[63]

several kinds of blood cells (see Section V.G), two of which (the *granulocytes* [GRs] and *plasmatocytes*; Figures 13 and 15) were called *immunocytes* by Gupta[26,27] because of their ability to discriminate between self and nonself tissues. These two immunocytes function in two important ways: (1) they serve as the first line of defense (*cell-mediated immunity*) and (2) they synthesize and secrete the following immunologic factors, some of which provide the second line of defense (analogous to *antibody-mediated immunity*): *hemagglutinins* (lectins),* *lysozymes* (also produced by the fat body), *complement-like factors, antibacterial* and *antiviral factors*, the *prophenoloxidase system, coagulogens,* and the *clotting enzyme*. For details of the insect immune system, see References 27 and 28.

GRs are very reactive to foreign antigens and even chemicals. When exposed to them, GRs become activated and are characterized by the presence of numerous plasma membrane extensions (pseudopods; see Figure 15) and pronounced development of peripheral microtubules and nuclear pores.[29,30]

G. REPRODUCTIVE SYSTEM
1. External Genitalia
External genitalia represent the greatly modified appendages of the eighth and ninth abdominal segments. In males, they are used to clasp the female during mating; in the female, they are concerned with oviposition and are therefore called *ovipositors*. The stingers of wasps, bees, and ants are ovipositors that have become adapted for defense and offense.

a. Male Genitalia
In insects, the external genitalia of the male develop from a pair of phallic lobes (*phallomeres*), which divide to form a pair of inner (*mesomeres*) and a pair of outer (*parameres*) structures. In *Blatta*, there is also an additional ventral phallomere (*hypophallus*).[31]

The *male genitalia* in insects are located on the ninth segment (Figure 16) and consist of a pair of claspers (*parameres*) and the penis (*aedeagus*). The inner chamber of the aedeagus, which continues with the *ejaculatory duct* (see below), is called the *endophallus*, and its opening at the tip or end of the aedeagus is the *phallotreme*. Note that the true *gonopore* is at the junction of the ejaculatory duct and the endophallus; hence, it is internal. However, in many insects, the endophallic duct is reversible so that the gonopore assumes a terminal position during copulation (Figure 17). A well-formed endophallus does not occur in *Blatta*.[31] The claspers and the aedeagus may be mounted on a common base, the *phallobase*. Snodgrass[15] uses the term *phallus* to include the phallobase and its various processes,

---

* Functionally analogous to antibodies; it must be mentioned that insects do not possess antibodies similar to those in vertebrates, but have molecules that perform antibody functions.

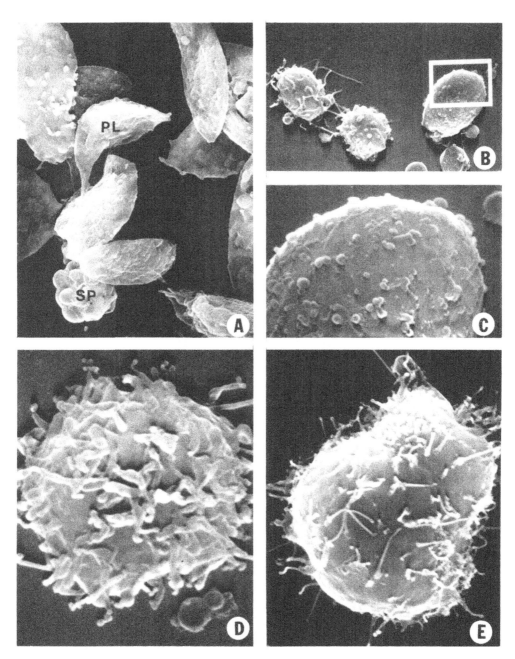

FIGURE 15. Granulocytes in various stages of activation. (A) Cluster of hemocytes (plasmatocyte, granulocytes, and spherulocyte) of *B. germanica*; PL = plasmatocyte, SP = spherulocyte. All unlabeled cells are granulocytes (GRs). Note that the GR is the major immunocyte in this cockroach. (Magnification × 4,000.) (B) Three GRs of *B. germanica*. (Magnification × 3,000.) (C) Magnified view of portion of one GR (in Figure 15B, right). (Magnification × 12,000.) (D) Activated GR of *B. germanica*. Note that this immunocyte, when reacting to foreign material (i.e., in its activated state), including chemicals in a saline solution, gives rise to numerous extensions of its plasma membrane (also called pseudopods). This type of cellular reaction of the GR is analogous to similar reaction of the vertebrate T-lymphocyte. (Magnification × 10,000.) (E) Activated GR of *G. portentosa*.

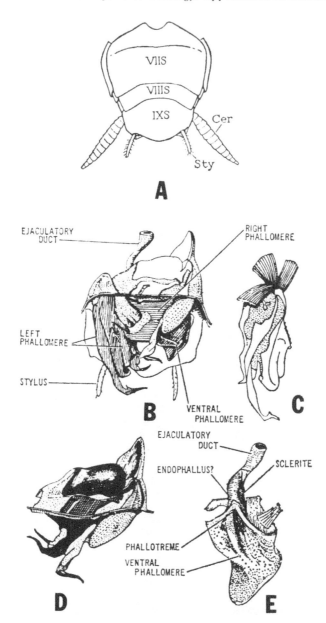

FIGURE 16. External male genitalia of *B. germanica*. (A) End of abdomen showing three abdominal segments (7 to 9), the cerci, and stylus;[15] (B) genitalia within the genital chamber; (C) left phallomere; (D) right phallomere; (E) ventral phallomere, with phallotreme from endophallic pouch continuous with ejaculatory duct.[31]

the aedeagus, and the endophallus. Note that in Dictyoptera not only are the typical claspers and aedeagus absent (because the phallomeres do not form these typical structures; see article by Khalifa[32]), but there is bilateral asymmetry in the genitalia (e.g., in *Blattella*). According to Matsuda,[31] the male genitalia in Blattaria are concealed by the ninth sternum (*subgenital plate*). For more details of the external genitalia in Blattaria see References 31, 33, and 34, and for further reading on genitalic muscles and nerves see References 35 and 36.

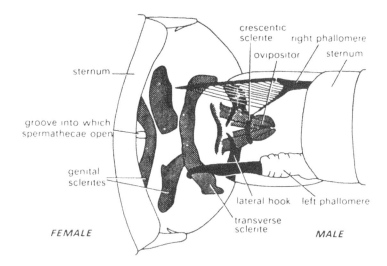

FIGURE 17. Ventral view of male and female genitalia of *Blattella* in copula (end-to-end position).[32] Subgenital plate and endophallus have been removed.

FIGURE 18. Ootheca of *Blatta orientalis,* split open to show arrangement of eggs. (From *The Oriental Cockroach,* Vol. 4, Neil A. Maclean Co., Belmont, CA, 1969.)

b. Female Genitalia

When present, the *female genitalia* in insects are located on the eighth and ninth segments. In Blattaria, the subgenital plate (seventh sternum) together with the dorsal sclerites of the eighth and the ninth sterna form a large chamber, which has a smaller anterior chamber and a larger posterior one (vestibule); the egg case (ootheca; see Figure 18) is formed in the latter chamber, and the anterior chamber supports the opening of the common oviduct (Figure 19). The floor of this chamber forms the *genital chamber,* which in cockroaches represents the forerunner of the highly sclerotized *bursa copulatrix* found in other insects.[31]

The ovipositor is often hidden by the subgenital plate in most cockroaches, and the stylus of the ninth segment is absent. Typically, an ovipositor (Figure 19B) has a base (*gonocoxa* or *valvifer*) that is articulated to an elongated, slender process (*gonapophysis* or *valvula*) that forms the shaft of the ovipositor. At the base of the ovipositor is a small sclerite (*gonangulum*) which, in generalized insects, is attached to the base of the first gonapophysis

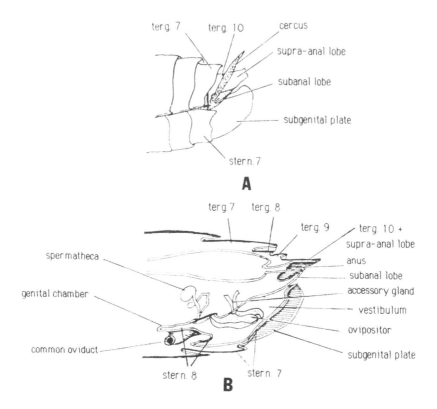

FIGURE 19.   Female genitalia of *Blatta orientalis*.[31] (A) Lateral view of posterior abdominal segments (eighth and ninth segments concealed). (B) Median section through posterior abdominal segments.

and articulates with the second gonocoxa (of the ninth segment) and the tergum of this segment. It probably represents the coxa of the ninth segment. In some insects, the second gonocoxa has a process (the *gonoplac* or third valvula).

In Blattaria there are three pairs of reduced valvulae that form the ovipositors, and both the second (inner) and third (lateral) valvulae are supported by the second gonocoxa (valvifer).[31]

2. Internal Reproductive Organs

a. Male Organs

The male reproductive organs (Figure 20) in insects consist of a pair of *testes* (made up of sperm tubes) that are connected to a median duct (*ejaculatory duct*) by a pair of lateral ducts (*vasa deferentia*). The posterior part of each vas deferens is distended and is called a *seminal vesicle*. In addition, there are accessory glands whose secretion (together with the semen) produces capsules containing sperm (*spermatophores*) in the ejaculatory duct. This duct opens to the exterior through an opening (*gonopore*).

According to Matsuda,[31] in Blattaria the number of sperm tubes in the testes of various cockroaches varies: *Blattella* spp. have four, while *Cryptocercus* and *Leucophaea* have many. Furthermore, their shapes also vary: globular or fusiform (*Blatta*), rounded or oval (*Blattella*), rounded (*Cryptocercus*), and mulberry like (*Blaberus*).

The *vasa deferentia* in cockroaches terminate in an *ampulla*, which is an anterior mesodermal part of the definitive ejaculatory gland.[31]

The *accessory glands* in cockroaches may be of several types. Louis and Kumar[37] found three types in *P. americana* and *Blaberus*, on the basis of their lengths. However, such

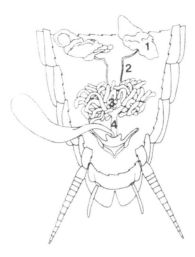

FIGURE 20. Internal reproductive organs of male *P. americana*; 1 = testis, 2 = vas deferens, 3 = accessory gland, 4 = ejaculatory duct. (From MacLeod, E. G., *The Biology of Insects: Laboratory Manual*, Burgess Publishing, Minneapolis, 1972. With permission.)

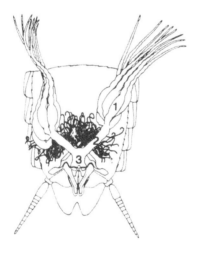

FIGURE 21. Internal reproductive organs of female *P. americana*; 1 = ovary, 2 = lateral oviduct, 3 = common oviduct, 4 = colleterial or accessory gland. (From MacLeod, E. G., *The Biology of Insects: Laboratory Manual*, Burgess Publishing, Minneapolis, 1972. With permission.)

differences do not occur in *Leucophaea, Nauphoeta,* and *Blattella*. Large amounts of uric acid are often found in the accessory glands in Blattellidae and Blaberidae; most of this is poured over the spermatophore during copulation. For details of spermatophore formation in cockroaches, consult References 4, 32, and 38 to 43.

b. *Female Organs*

The female internal reproductive organs (Figure 21) in insects consist of a pair of *ovaries* that are connected to a pair of *lateral oviducts*. These, in turn, open posteriorly into a common oviduct, which opens into the *genital chamber* formed by an inflection of the body wall behind the eighth sternum. Because the genital chamber serves as a copulatory pouch during mating, it is also called the *bursa copulatrix*.[15] When the genital chamber becomes elongated and tubular and is continuous with the median oviduct, it is called the *vagina*. In addition, there are accessory or *colleterial glands, spermathecae,* and a *spermathecal gland*.

Each ovary is composed of several egg tubes (*ovarioles*), which are of the primitive *panoistic* type (i.e., without specialized nurse cells) in cockroaches. The number of ovarioles varies: *Diploptera* have 6; *Blatta*, 6; *Leucophaea*, 15 to 20; and *Periplaneta*, 8. The paired lateral oviducts are short. The number of spermathecae also varies. In *Blatta* and *Periplaneta* (Blattidae), it is a two-branched structure. *Blattella* and *Diploptera* have two pairs and *Leucophaea*, one pair. Whenever there are branched spermathecae, one branch is considered to function as a spermathecal gland.[31] For more details on spermathecae, see References 33 and 44.

The accessory or colleterial glands in cockroaches also open into the genital chamber via a single basal tube, and they consist of a number of long tubules whose secretion forms the egg case (ootheca). For details on accessory glands and the formation of the egg case, see References 4, 34, 43, and 45 to 51.

H. NEUROENDOCRINE SYSTEM

Essentially, the insect pars intercerebralis-corpora cardiaca complex is functionally analogous to the vertebrate hypothalamic-hypophyseal complex in terms of coordinating the activities of the nervous and endocrine systems.

Historically, insects have fascinated scientists for their ability to metamorphose, and they have provided some basic insight into the science of endocrinology. As early as 1922, Kopeć[52] found that a chemical factor released from the brain was necessary for pupation in the gypsy moth larva, thus implicating, for the first time in any animal, the nervous system in endocrine control of growth and development. Subsequently, Clever and Karlson[53] provided the first evidence on the mode of action of steroid hormones at the cellular and molecular levels in a fly.

There are four main components in the insect neuroendocrine system: (1) neurosecretory cells in the brain and ganglia of the ventral nerve cord (Figure 22); (2) the corpora cardiaca (paired neurohemal organs that store and release neurosecretions; Figure 23); (3) the ecdysial, molting, or prothoracic glands (Figure 24); and (4) the corpora allata (paired lateral glands; Figure 23). The latter two are nonneural epithelial endocrine glands.[54]

Growth, differentiation, and morphogenesis are regulated by three classes of developmental hormones, viz., the neurohormones, the ecdysones, and the juvenile hormones, produced by the brain and/or the neurohemal organs, the ecdysial glands, and the corpora allata, respectively. In a typical situation, the pars intercerebralis neurosecretory cells of the protocerebrum of the brain secrete a tropic hormone (*ecdysiotropin*) which passes into the hemolymph through the corpora cardiaca. The ecdysial glands, upon stimulation by ecdysiotropin, secrete a morphogenetic hormone (*β-ecdysone*;[54,64] Figure 25) which initiates the molting process. The other morphogenetic hormone, *juvenile hormone* (Figure 26), determines the characteristics of the molt (nymphal or adult). During the early nymphal stages the titer of the juvenile hormone is high, and its gradual decline and eventual "absence" results in the nymphal-adult molt.

The juvenile hormone, however, reappears in the adult cockroach for a variety of different roles (i.e., egg maturation, activity of the reproductive accessory glands, and sex pheromone production and release). The chapters in Part V of this work consider some of these topics in more detail. For details of neurohemal organs and insect endocrinology, one may refer to References 55 to 57.

VI. SUMMARY

This chapter consists of basic information on the external anatomy of the cockroach

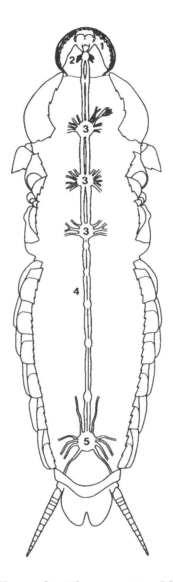

FIGURE 22.   Diagram of central nervous system of *P. americana*;[61] 1 = head (brain) region, 2 = subesophageal ganglion, 3 = thoracic ganglia, 4 = ventral nerve chord, 5 = terminal abdominal ganglion.

(including the integument), general and specific instructions for dissecting cockroaches and their various systems, and the internal anatomy, including the digestive, excretory, respiratory, musculature, circulatory, immune, reproductive, and endocrine systems. It is mainly intended for scientists in the biomedical and veterinary sciences who may not be familiar with insect anatomy.

ACKNOWLEDGMENTS

This is the New Jersey Agricultural Experiment Station Publication No. F08125-02-88, supported by state funds and U.S. Hatch funds.

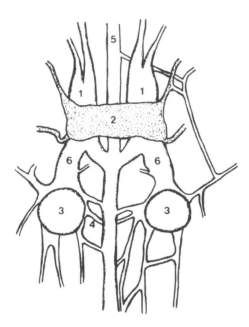

FIGURE 23.   Diagram of partial retrocerebral complex of *P. americana*;[12] 1 = corpus cardiacum, 2 = cardiaca-commissural organ (posterior portion removed), 3 = corpus allatum, 4 = allatal commissure, 5 = recurrent nerve, 6 = nervi corporis cardiaci.

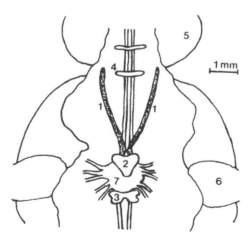

FIGURE 24.   Ventral view of partial dissection of cervical and prothoracic region of *P. americana* nymph;[12] 1 = ecdysial glands, 2 = sternum, 3 = poststernum, 4 = cervical sclerite, 5 = head, 6 = coxa, 7 = thoracic ganglion.

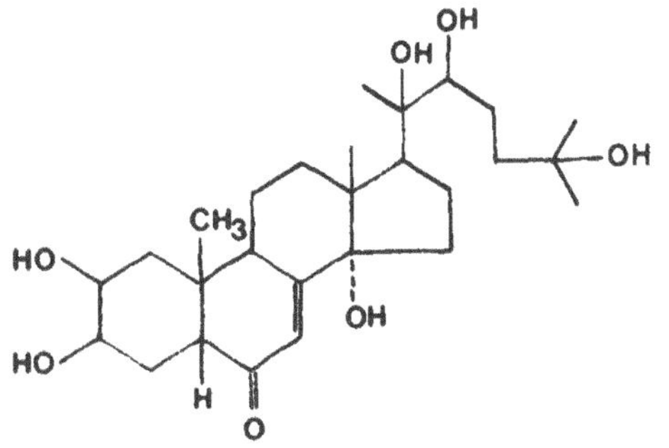

FIGURE 25. Structure of β-ecdysone.

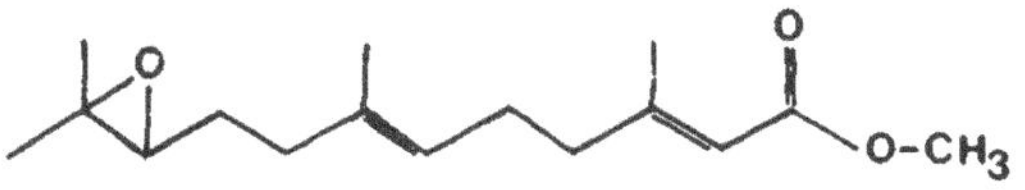

FIGURE 26. Structure of cockroach juvenile hormone: methyl *cis*-10,11-epoxy-3,7,11-trimethyl-*trans*-2,*trans*-6-dodecadienoate.[65]

REFERENCES

1. **Ramaswamy, S. B. and Gupta, A. P.**, Sensilla of the antennae and the labial and maxillary palps of *Blattella germanica* (L.) (Dictyoptera: Blattellidae): their classification, *J. Morphol.*, 168, 269, 1981.
2. **Butler, R.**, The anatomy of the compound eye of *Periplaneta americana* (L.). I. General features, *J. Comp. Physiol.*, 83, 223, 1973.
3. **Butler, R.**, The anatomy of the compound eye of *Periplaneta americana* (L.). II. Fine structure, *J. Comp. Physiol.*, 83, 239, 1973.
4. **Guthrie, D. M. and Tindall, A. P.**, *The Biology of the Cockroach*, Edward Arnold, London, 1968.
5. **Lipke, H., Graves, B., and Leto, S.**, Polysaccharides and glycoprotein formation in the cockroach. II. Incorporation of D-glucose-^{14}C into bound carbohydrate, *J. Biol. Chem.*, 240, 601, 1965.
6. **Mills, R. R.**, Integument, in *The American Cockroach*, Bell, W. S. and Adiyodi, K. G., Eds., Chapman and Hall, New York, 1981, chap. 2.
7. **Hepburn, H. R.**, Structure of the integument, in *Comprehensive Insect Physiology, Biochemistry and Pharmacology*, Vol. 3, Kerkut, G. A. and Gilbert, L. I., Eds., Pergamon Press, Oxford, 1985, chap. 1.
8. **Gupta, A. P.**, Identification key for hemocyte types in hanging drop preparations, in *Insect Hemocytes*, Gupta, A. P., Ed., Cambridge University Press, London, 1979, chap. 17.
9. **Gupta, A. P.**, Cellular elements in the hemolymph, in *Comprehensive Insect Physiology, Biochemistry and Pharmacology*, Vol. 3, Kerkut, G. A. and Gilbert, L. I., Eds., Pergamon Press, Oxford, 1985, chap. 10.
10. **Hazarika, L. K. and Gupta, A. P.**, Anatomy of the retrocerebral complex of *Blattella germanica* (L.) (Dictyoptera: Blattellidae), *Zool. Anz.*, 219(5/6), 257, 1987.
10a. **Das, Y. T. and Gupta, A. P.**, unpublished data, 1988.
11. **Hazarika, L. K. and Gupta, A. P.**, Structure, innervation, persistence, and effects of juvenile hormone on the prothoracic glands in adult *Blattella germanica* (L.) (Dictyoptera: Blattellidae), *Zool. Sci.*, 3, 145, 1986.
12. **Bell, W. S. and Adiyodi, K. G., Eds.**, *The American Cockroach*, Chapman and Hall, New York, 1981.
13. **Bradley, T. J.**, The excretory system: structure and physiology, in *Comprehensive Insect Physiology, Biochemistry and Pharmacology*, Vol. 4, Kerkut, G. A. and Gilbert, L. I., Eds., Pergamon Press, Oxford, 1985, chap. 10.
14. **Mill, P. J.**, Structure and physiology of the respiratory system, in *Comprehensive Insect Physiology, Biochemistry and Pharmacology*, Vol. 3, Kerkut, G. A. and Gilbert, L. I., Eds., Pergamon Press, Oxford, 1985, chap. 13.

15. **Snodgrass, R. E.,** *Principles of Insect Morphology,* McGraw-Hill, New York, 1935.

16. **Carbonell, C. S.,** The thoracic muscles of the cockroach *Periplaneta americana, Smithson. Misc. Collect.,* 107, 1, 1947.

17. **Chadwick, L. E.,** The ventral intersegmental muscles of cockroaches, *Smithson. Misc. Collect.,* 131, 1, 1957.

18. **Shankland, D. L.,** Nerves and muscles of the pregenital abdominal segments of the American cockroach, *Periplaneta americana* (L.), *J. Morphol.,* 117, 353, 1965.

19. **Alsop, D. W.,** Comparative analysis of the intrinsic leg musculature of the American cockroach, *Periplaneta americana, J. Morphol.,* 158, 199, 1978.

20. **Hughes, G. M.,** The coordination of insect movements. I. The walking movements of insects, *J. Exp. Biol.,* 29, 267, 1952.

21. **Yeager, J. F. and Hendrickson, G. O.,** Circulation of blood cells in wings and wing pads of the cockroach, *Periplaneta americana* (Linn.), *Ann. Entomol. Soc. Am.,* 27, 739, 1934.

22. **Miller, T. A.,** Structure and physiology of the circulatory system, in *Comprehensive Insect Physiology, Biochemistry and Pharmacology,* Vol. 3, Kerkut, G. A. and Gilbert, L. I., Eds., Pergamon Press, Oxford, 1985, chap. 8.

23. **Mullins, D. E.,** Chemistry and physiology of the hemolymph, in *Comprehensive Insect Physiology, Biochemistry and Pharmacology,* Vol. 3, Kerkut, G. A. and Gilbert, L. I., Eds., Pergamon Press, Oxford, 1985, chap. 9.

24. **Hazarika, L. K. and Gupta, A. P.,** Variations in hemocyte populations during various developmental stages of *Blattella germanica* (L.) (Dictyoptera: Blattellidae), *Zool. Sci.,* 4, 307, 1987.

25. **Chiang, A. S., Gupta, A. P., and Han, S. S.,** Arthropod immune system. I. Comparative light and electron microscopic accounts of immunocytes and other hemocytes of *Blattella germanica* (Dictyoptera: Blattellidae), *J. Morphol.,* 198, 257, 1988.

26. **Gupta, A. P.,** The identity of the so-called crescent cell in the hemolymph of the cockroach, *Gromphadorhina portentosa* (Schaum) (Dictyoptera: Blaberidae), *Cytologia,* 50, 739, 1985.

27. **Gupta, A. P.,** Arthropod immunocytes: identification, structure, functions, and analogies to the functions of vertebrate B- and T-lymphocytes, in *Hemocytic and Humoral Immunity in Arthropods,* Gupta, A. P., Ed., John Wiley & Sons, New York, 1986, chap. 1.

28. **Gupta, A. P., Ed.,** *Hemocytic and Humoral Immunity in Arthropods,* John Wiley & Sons, New York, 1986.

29. **Han, S. S. and Gupta, A. P.,** Arthropod immune system. V. Activated immunocytes (granulocytes) of the German cockroach, *Blattella germanica* (L.) (Dictyoptera: Blattellidae) show increased number of microtubules and nuclear pores during immune reaction to foreign tissue, *Cell Struct. Funct.,* 13, 333, 1988.

30. **Han, S. S. and Gupta, A. P.,** Arthropod immune system. II. Encapsulation of implanted nerve cord and "plain gut" surgical suture by granulocytes of *Blattella germanica* (L.) (Dictyoptera: Blattellidae), *Zool. Sci.,* 6, 303, 1989.

31. **Matsuda, R.,** *Morphology and Evolution of the Insect Abdomen,* Pergamon Press, Oxford, 1976.

32. **Khalifa, A.,** Spermatophore production in *Blattella germanica* (L.) (Orthoptera: Blattidae), *Proc. R. Entomol. Soc. London,* A25, 53, 1950.

33. **McKittrick, F. A.,** Evolutionary Studies of Cockroaches, Cornell Univ. Exp. Stn. Mem. No. 389, Cornell University, Ithaca, NY, 1964.

34. **Roth, L. M.,** Evolution and taxonomic significance of reproduction in Blattaria, *Annu. Rev. Entomol.,* 15, 75, 1970.

35. **Pipa, R. L.,** Muscles and innervation of the genital and postgenital abdominal segments of the female American cockroach *Periplaneta americana* (L.) (Dictyoptera: Blattidae), *Int. J. Insect Morphol. Embryol.,* 16, 17, 1987.

36. **Pipa, R. L.,** Muscles and nerves of the posterior abdomen and genitalia of the male *Periplaneta americana* (L.) (Dictyoptera: Blattidae), *Int. J. Insect Morphol. Embryol.,* 17(6), 455, 1988.

37. **Louis, D. and Kumar, R.,** Morphology and histology of the mushroom-shaped gland in some Dictyoptera, *Ann. Entomol. Soc. Am.,* 64, 977, 1971.

38. **Ito, H.,** Contribution histologique et physiologique à l'étude des annexes des organes genitaux des Orthoptères, *Arch. Anat. Microsc.,* 20, 343, 1924.

39. **Zabinski, J.,** Copulations extérieure chez les Blattes, *C.R. Soc. Biol.,* 112, 596, 1933.

40. **Gupta, B. L.,** On the structure and function of spermatophore in the cockroach, *Periplaneta americana* (Linn.), *Indian J. Entomol.,* 8, 79, 1947.

41. **Van Wyk, L. E.,** The morphology and histology of the genital organs of *Leucophaea maderae* (L.), *Entomol. Soc. S. Afr.,* 15, 3, 1952.

42. **Davey, K. G.,** The evolution of the spermatophore in insects, *Proc. R. Entomol. Soc. London,* A35, 107, 1960.

43. **Cornwell, P. B.,** *The Cockroach,* Vol. 1, Hutchinson Press, London, 1968.
44. **Gupta, B. L. and Smith, D. S.,** Fine structural organization of the spermatheca in the cockroach, *Periplaneta americana, Tissue Cell,* 1(2), 295, 1969.
45. **Brunet, P. C. J.,** The formation of the ootheca by *Periplaneta americana, Q. J. Microsc. Sci.,* 92, 113, 1951.
46. **Hagan, H. R.,** *Embryology of the Viviparous Insect,* Ronald Press, New York, 1951.
47. **Roth, L. M.,** The evolutionary significance of rotation of the ootheca in Blattaria, *Psyche,* 74, 85, 1967.
48. **Roth, L. M.,** Water changes in cockroach oothecae in relation to the evolution of ovoviviparity and viviparity, *Ann. Entomol. Soc. Am.,* 60, 828, 1967.
49. **Roth, L. M.,** Uricose glands in accessory sex gland complex of male Blattaria, *Ann. Entomol. Soc. Am.,* 60, 1203, 1967.
50. **Roth, L. M.,** Oothecae of Blattaria, *Ann. Entomol. Soc. Am.,* 61(1), 83, 1968.
51. **Roth, L. M.,** Additions to the oothecae, uricose glands of Blattaria, *Ann. Entomol. Soc. Am.,* 64, 127, 1971.
52. **Kopeć, S.,** Studies on the necessity of the brain for the inception of insect metamorphosis, *Biol. Bull.,* 42, 322, 1922.
53. **Clever, U. and Karlson, P.,** Induktion von Puff-veranderungen in den Speicheldrusen Chromosomen von *Chironomus tentans* durch Ecdyson, *Exp. Cell Res.,* 20, 623, 1960.
54. **Doane, W. W.,** Role of hormones in insect development, in *Developmental Systems: Insects,* Vol. 2, Counce, S. J. and Waddington, C. H., Eds., Academic Press, New York, 1973, chap. 4.
55. **Gupta, A. P., Ed.,** *Neurohemal Organs of Arthropods,* Charles C Thomas, Springfield, IL, 1983.
56. **Kerkut, G. A. and Gilbert, L. A., Eds.,** *Comprehensive Insect Physiology, Biochemistry and Pharmacology,* Vol. 7 and 8, Pergamon Press, Oxford, 1985.
57. **Gupta, A. P., Ed.,** *Morphogenetic Hormones in Arthropods,* Vol. 1 to 3, Rutgers University Press, New Brunswick, NJ, 1989.
58. **Matsuda, R.,** Morphology and evolution of the insect thorax, *Mem. Entomol. Soc. Can.,* No. 76, 1, 1970.
59. **Snodgrass, R. E.,** Morphology and mechanism of the insect thorax, *Smithson. Misc. Collect.,* 80(1), 1, 1927.
60. **Das, Y. T. and Gupta, A. P.,** Nature and precursors of juvenile hormone-induced excessive cuticular melanization in German cockroach, *Nature (London),* 268(5616), 139, 1977.
61. **MacLeod, E. G.,** *The Biology of Insects: Laboratory Manual,* Burgess Publishing, Minneapolis, 1972.
62. **Nutting, W. L.,** A comparative anatomical study of the heart and accessory structures of the orthopteroid insects, *J. Morphol.,* 89, 501, 1951.
63. **Richards, A. G.,** Ventral diaphragms of insects, *J. Morphol.,* 113, 17, 1963.
64. **King, D. S. and Marks, E. P.,** The secretion and metabolism of α-ecdysone by cockroach (*Leucophaea maderae*) tissue *in vitro, Life Sci.,* 15, 147, 1974.
65. **Muller, P. J., Masner, P., Trautman, K. H., Suchy, M., and Wipf, H. K.,** The isolation and identification of juvenile hormone from cockroach corpora allata *in vitro, Life Sci.,* 15, 915, 1974.
66. **Wyman, R. J.,** A beginners guide to microdissection: *Drosophila* style, *Fine Sci. Points,* 1, 1, 1987.

Section II. Nervous System

Chapter 4

ANATOMY OF THE CENTRAL NERVOUS SYSTEM AND ITS USEFULNESS AS A MODEL FOR NEUROBIOLOGY

Charles R. Fourtner and Charles J. Kaars

TABLE OF CONTENTS

I. INTRODUCTION

The nervous system of the cockroach has been and continues to be an excellent experimental model for a variety of physiological, pharmacological, toxicological, and behavioral investigations. The chapters of this work provide ample testimony to the usefulness of the cockroach nervous system as a model for the discovery and verification of phenomena of general biological significance. It is our intent to provide a brief introduction to the overall gross morphology of the nervous system in the cockroach, to discuss some of the neuronal components of the cockroach nervous system, to describe the neuronal circuits underlying some behavioral models, and, finally, to describe the glial sheath around the central nervous system (CNS) and some of the interesting changes that occur in the glial elements during peripheral regeneration. Where gross and cellular morphologies are presented elsewhere in this work, we will direct the reader to the appropriate figures.

II. GROSS MORPHOLOGY OF THE CENTRAL NERVOUS SYSTEM

The cockroach CNS is comprised of two basic gross morphological elements: ganglia and connectives. Ganglia are aggregations of neurons. The neuronal cell bodies form a peripheral rind (cortex). The local processes of these neurons and the processes of neurons whose somata are located in other ganglia and in the periphery form a central neuropil. Neuronal processes are found within two morphologically distinct regions: tracts and neuropil. Tracts are regions in which all the neuronal processes have the same general orientation; they are virtually free of synaptic contacts. Neuropil is characterized by processes which have no general orientation, but in which synaptic contacts are made between processes. Connectives are the central neural structures which join adjacent ganglia, and tracts are the neural pathways within a ganglion which are continuous with anterior and posterior connectives. The laterally running neural pathways within a ganglion are known as commissures and are the major communications link between the segmental (right and left) hemiganglia. Except for regions of synaptic contacts, neurons and their processes are invested with conneural glial cells. The entire CNS is enveloped in a sheath composed of glia and noncellular elements.

Embryologically, each segment develops a protoganglion. During embryonic development there occurs considerable fusion of protoganglia, with the result that the postembryonic nervous system has 11 principal ganglia: 2 in the head (the supra- and subesophageal ganglia, which are joined by the circumesophageal connectives), 3 in the thorax (the pro-, meso-, and metathoracic ganglia), and 6 in the abdomen (for review see References 1 to 3).

A. THE CEPHALIC NERVOUS SYSTEM

The supraesophageal ganglion is primarily a region for integration of sensory information, as well as sensory-to-motor synaptic interactions, initiation of locomotory commands, and the integration and initiation of neuroendocrine reflexes. The supraesophageal ganglion is the product of fusion of an uncertain number of segmental ganglia (most likely three) during embryonic development. Three distinct regions of the supraesophageal ganglion can be recognized. Figure 1 illustrates the gross morphology and the nerves arising from the cephalic nervous system.

1. Protocerebrum

The protocerebrum, the largest and most anterior region of the supraesophageal ganglion, receives input from the ocelli and compound eyes. The neural pathways and projections from the ocelli and compound eyes to this region are discussed in more detail in Chapter

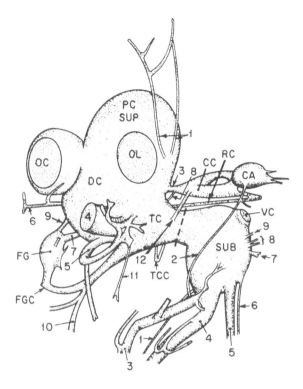

FIGURE 1. A schematic illustration of the left lateral aspect of the cephalic region of the nervous system of *Periplaneta americana,* with the peripheral projections of the optic lobe (OL) and the ocellar nerve (OC) severed. SUP, supraesophageal ganglion; SUB, subesophageal ganglion; CA, corpus allatum; CC, corpus cardiacum; FG, frontal ganglion; FGC, frontal ganglion connective; RC, recurrent nerve; TCC, tritocerebral commissure; VC, ventral connective; PC, protocerebral region; DC, deutocerebral region; TC, tritocerebral region. The dashed line designates the circumesophageal connectives joining the supra- and subesophageal ganglia. For a description of the distribution of the peripheral nerves, see Tables 1 and 2. (From Pipa, R. and Delcomyn, F., in *The American Cockroach,* Bell, W. J. and Adiyodi, K. G., Eds., Chapman and Hall, London, 1981, 175. With permission.)

23 of this work. Also within the protocerebrum are the pars intercerebralis and the pars lateralis, which play an extremely important role in neuroendocrine function; from these regions emerge the major neuronal processes that innervate the corpora cardiaca (CC), an organ which serves as a neurosecretory store analogous to the anterior pituitary in vertebrates.

Using a variety of histological stains and cobalt sulfide backfilling procedures it is clear that a number of cells along the medial border of the pars intercerebralis send axonal processes into the CC nerves 1 and 2. For further discussion on this literature see the review of Pipa and Delcomyn,[2] and for a general description of the gross morphology of the supraesophageal/neuroendocrine pathways see Chapter 5 in this work. Krauthamer,[4] using electrophysiological procedures and injection of horseradish peroxidase (HRP), recently demonstrated that there are two types of cells within the pars intercerebralis. Type 1 is clearly a neurosecretory type of neuron and sends axonal processes to the CC nerves. Type 2, a nonneurosecretory type cell, is larger in diameter, with its axonal process projecting to the circumesophageal connectives. Since these cells can be clearly defined by their physiological properties, they represent a good model for the study of neurosecretory cells having their somata within the

CNS. Thompson and Tobe (see Chapter 5 of this work) have reviewed the cellular morphology of the neural elements associated with the supraesophageal/neuroendocrine pathways in the cockroach *Diploptera punctata.*

The largest structures within the supraesophageal ganglion are the mushroom bodies (the corpora pedunculata). Each mushroom body is subdivided into two lobes (a and b) and two calyces, a lateral and a medial (see Figure 3 in Chapter 26 of this work). Although there have been a substantial number of studies regarding the structure of the mushroom bodies, of the neuronal projections into and out of these highly organized structures, and of the neuronal types within the pedunculata, there is still considerable debate as to the function of these dominant structures. Numerous hypotheses regarding their function have been proposed, from serving a purely olfactory purpose (e.g., olfactory memory[5]) to being a highly complex integrating region for parallel processing of visual and olfactory information.[6]

There have been, as yet, few studies on the physiology of thse structures in the cockroach. Due to the projections from the antennal afferents, Weiss[7] has argued that the mushroom bodies may be a region of olfactory integration.

2. Deutocerebrum

Posterior to the protocerebrum is the deutocerebrum, the principal center for the antennal olfactory input which is processed in the many neuropilar regions or glomeruli. Output from the glomeruli projects to the mushroom bodies of the protocerebrum. The morphology of the olfactory pathways and the neuronal projections is summarized and discussed by Seelinger in Chapter 26 of this work.

3. Tritocerebrum

The most posterior region of the supraesophageal ganglion is the tritocerebrum, which is intimately associated with the frontal ganglion, the primary ganglion of the stomatogastric (also termed the stomodaeal or vegetative) nervous system. In addition to receiving input directly from the frontal ganglion, it also receives input from the hypocerebral ganglion via the frontal ganglion. The hypocerebral ganglion is a small ganglion of the stomatogastric system through which other, more caudal stomodaeal inputs/outputs pass and is connected to the frontal ganglion by the recurrent nerve of the stomodaeal nervous system.[8] Stefano et al., in Chapter 16 of this work, further consider some of the pharmacological studies involving the stomatogastric nervous system, particularly the posterior recurrent nerve.

There is a large neuropilar area of the tritocerebrum which may function as an area for integration of sensory input from the mouth parts[9] and the antennal neuropil of the deutocerebrum. It also appears to be another important neuroendocrine pathway, since the third nerve (NCC III) projecting to the CC arises from the tritocerebrum; however, it is not clear if any of the axons exiting NCC III originate in the tritocerebrum. For a general discussion of the anatomical arrangements of the neuroendocrine and stomodaeal projections to the supraesophageal ganglion, see Chapter 5 in this work and the review of Pipa and Delcomyn.[2]

4. Subesophageal Ganglion

The other principal component of the cephalic nervous system is the subesophageal ganglion, the primary motor control system for the muscles operating the mouth parts (the mandibles, maxillae, and labium). While it most likely results from the fusion of three embryonic segments, its origins remain uncertain due to the complexity of its neuropilar structure and the few detailed studies of its structure. The subesophageal ganglion may also function as a site of origin for command elements responsible for controlling locomotory behavior, since there is behavioral evidence that some inhibitory influences on reflexes are removed when the thoracic ganglia are isolated from the subesophageal ganglion.[10] The gross anatomy of the subesophageal ganglion is depicted in Figure 1.

The subesophageal ganglion of the locust has been studied extensively during the past decade. In a recent review of that literature, Altman and Kien[11] describe the gross and cellular morphology and discuss the role of the subesophageal ganglion in maintaining motor excitability, but more interestingly as a cross (left-right) comparator aiding motor coordination. Although the cockroach CNS would be amenable to extensive studies of the subesophageal ganglion, only a few of the cellular components of this structure have been identified.[12]

Figure 1 also illustrates the major nerve branches arising from the supra- and subesophageal ganglia, and the legend provides information concerning the primary sites innervated by the cephalic nerves.

A very recent compendium of papers on the arthropod brain[13] should serve as a useful description for those wishing to compare the cockroach supra- and subesophageal ganglia to those of other arthropods.

B. THE THORACIC NERVOUS SYSTEM

The thoracic CNS of the cockroach is composed of three ganglia (pro-, meso-, and metathoracic) joined by their paired connectives. Each of these ganglia is responsible for processing and integrating sensory information from and controlling motor output to their appropriate thoracic segments. The pro- and mesothoracic ganglia each gives rise to eight bilaterally symmetrical pairs of nerves. The metathoracic ganglion, a composite of the embryonic ganglia of the third thoracic and first abdominal segments, gives rise to eight pairs of segmental nerves, with nerve 8 innervating the spiracle of the first abdominal segment. Figure 2 illustrates the locations and branching patterns of the nerves arising from the mesothoracic ganglia.

The nomenclature for the peripheral nervous system is a simple alphanumeric system in which each branch (ramus) arising from a segmental nerve is numbered in sequential order, ramus number 1 being the most centrally located. For example, the first ramus branching from segmental nerve 5 is labeled 5r1. In those cases in which a nerve decussates into two or three branches of similar diameter an alpha designation is assigned; for example, segmental nerve 3 decussates into two nerves of similar size which are labeled nerve 3A (anterior) and nerve 3B (posterior). Branching patterns of the individual rami in the periphery are noted by lower-case alpha designations. For example, the nerve innervating one of the muscles located in the coxa of a walking leg is designated 5r1a.[14,15]

For a brief description of the segmental nerves see Tables 1 to 3, and for a more complete description of the peripheral distribution of the nerves and the approximate number of axonal profiles contained in each nerve and subsequent branch see the review by Guthrie and Tindall[1] and the series of papers by Dresden and Nijenhuis[14-16] and by Pipa et al.[17,18]

C. THE ABDOMINAL NERVOUS SYSTEM

Within the abdomen there are six ganglia, identified as abdominal ganglia 1 through 6. Abdominal ganglion 1 is the ganglion of the second embryonic segment, and ganglion 6 results from the fusion of ganglia originating in embryonic segments 7 to 11. The gross anatomy of the five anterior ganglia is similar in a given species of cockroach, with two bilateral segmental pairs, termed Nerve 1 (dorsal) and Nerve 2 (ventral). In some species there is a sharp distinction between Nerves 1 and 2, but in others fusion can occur at the ganglion or at various distances from the ganglion. In addition, in some species there is a very small transverse nerve which projects to the periphery and primarily contains neurosecretory axons. Nerve 1 innervates the spiracles, with Nerve 2 innervating the abdominal musculature.[18a]

The description of the gross anatomy of the sixth abdominal ganglion is presented by Hue and Callec in Chapter 8 of this work, which is on synaptic mechanisms in the CNS.

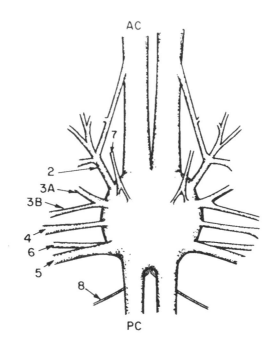

FIGURE 2.   A schematic illustration of the mesothoracic ganglion of the cockroach *Periplaneta americana*, in ventral view. AC, anterior connective; PC, posterior connective. The numbers 2 through 8 designate the segmental nerves from anterior to posterior consecutively. The anterior connective is considered the first nerve emanating from the thoracic ganglia. For a distribution of the peripheral nerves, see Table 3.

TABLE 1
Peripheral Nerves Arising from Supraesophageal Ganglion

Nerve #	Name	Type[a]	Region
1	Dorsal integumentary	Sen	Vertex and frons
2	Ocellar	Sen	Ocelli
3	Dorsal cardiac	Mot	Corpus cardiacum
4	Antennal	Sen/mot	Antenna
5	Dorsal deutocerebral	Mot/sen	Extrinsic dorsal muscles
6	Ventral	Mot/sen	Antenna
7	Median frontal	Connective	Frontal ganglion
8	Main cardiac (NCC II)	Mot	Corpus cardiacum
9	Lateral frontal	Connective	Frontal ganglion
10	Labral	Sen/mot	Clypeus/labrum
11	Lateral integumentary	Sen	Genal region
12	Ventral cardiac (NCC III)	Mot	Corpus cardiacum

[a] Sen = sensory, mot = motor.

III. MICROSCOPIC ORGANIZATION OF THE CENTRAL NERVOUS SYSTEM

We will limit our discussion here primarily to the thoracic ganglia. A comprehensive study on the microscopic anatomy of the brain, comparable to that of the honeybee,[6] has

TABLE 2
Peripheral Nerves Arising from Subesophageal Ganglion

Nerve #	Name	Type[a]	Region
1	Hypopharyngeal	Mot/sen	Hypopharynx
2	Corpus allatum	Mot	Corpus allatum
3	Mandibular	Mot/sen	Mandibular muscles, hypopharynx
4	Maxillary	Mot/sen	Maxilla
5	Labial	Mot/sen	Labrum
6	Cervical-1	Mot	Ligula neck muscles
7	Cervical-2	Mot/sen	Ligula neck muscles
8	Cervical-3	Mot	Ligula neck muscles
9	Prothoracic gland	Mot	Prothoracic gland

[a] Mot = motor, sen = sensory.

TABLE 3
Peripheral Distribution of Nerves Arising from Thoracic Ganglia

Nerve #	Type[a]	Innervation
2A	Sen	Ventral thoracic musculature
2B	Sen/mot	Wings, meso and meta; neck muscles, pro
2C	Mot	Dorsal thoracic and neck musculature
3A	Mot	Coxal and thoracic musculature
3B	Sen/mot	Primary sensory nerve to coxa; coxal and femoral musculature
4	Mot	Bifunctional muscles which originate in thorax and insert on trochantin
5	Sen/mot	Coxal, femoral, and distal leg musculature and sense organs
5r8	Sen	Large sensory nerve to distal leg segments
6A	Mot	Tergal and coxal musculature
6B	Mot/sen	Coxal; muscles/meron
7	Sen	Basisternum
8	Sen/mot	Spiracle

Note: For a more complete description of specific rami, see References 12, 14—18, 35, and 78.

[a] Sen = sensory, mot = motor.

not been published. There are descriptions of a few cellular pathways in the supraesophageal ganglion and some of the neurons associated with those pathways elsewhere in this work (see Chapter 23). One recent study on the morphology of third-order ocellar interneurons in the protocerebrum is a particularly excellent example of this type of neuronal-anatomical research in the cockroach.[19,20] Figure 3 illustrates one of those interneurons (OL-II) which projects to the optic lobes on both sides of the brain and is included as an example of the morphological complexity and of the distance over which an interneuron in the supraesophageal ganglion can project.

The pathways in the thoracic ganglia originally described by Pipa et al.[18] and recently in more detail by the superb studies of Gregory[21,22] have proven to be exceptionally useful for those studying the central neural components responsible for a variety of behaviors.

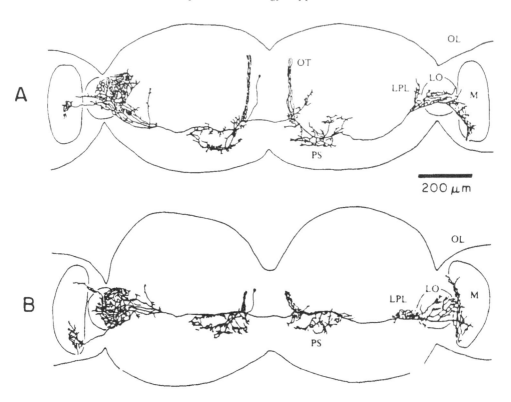

FIGURE 3. The microscopic anatomy of an interneuron (OL-II, or optic lobe neuron II) located in the protocerebrum. The cell body is located in the pars intercerebralis, and its processes form an extensively branching arborization in both optic lobes and in the area of both ocellar tracts (OT). PS, posterior slope of protocerebrum; M, medulla; LO, lobula; LPL, lateral protocerebral lobe. (A) Dorsal view; (B) posterior view. (From Mizunami, M. and Tateda, H., *J. Exp. Biol.*, 125, 57, 1986. With permission.)

Using a variety of silver staining procedures, Gregory has clearly defined the regions of longitudinal tracts, commissures, and synaptic interaction within the neuropil of the thoracic ganglia. Thoracic ganglia consist of a rind two to seven cells in thickness surrounding the neuropilar areas, along with the tracts and commissures connecting the neuropilar areas.

The thoracic ganglia contain seven longitudinal tracts connecting each anterior connective to its corresponding posterior connective. These are illustrated in Figure 4 and are labeled as follows: (1) the dorsal tract, which consists of medial and lateral regions (the MDT and LDT, respectively); (2) the dorsal intermediate tract (DIT); (3) the dorsal medial tract (DMT); (4) the ventral lateral tract (VLT); (5) the ventral medial tract (VMT); (6) the ventral intermediate tract (VIT); and (7) the ventral tract, which consists of five (inner and outer) lateral bundles and three medial bundles (o.LVT, i.LVT, MVT1, MVT2, and MVT3).

There are also ten commissures which connect the right and left hemiganglia; six are located dorsally and are labeled DCI to DCVI (anterior to posterior), while four are ventral: (1) ventral commissure I (VCI), (2) ventral commissural loop II (VCLII), (3) posterior ventral commissure (PVC), and (4) supramedian commissure (SMC).

Gregory[21] also charted the projections of the segmental nerves as they enter the ganglion and invade the appropriate neuropilar and commissural regions. For example, the fibers in nerve 2 divide into two roots, dorsal and ventral, after they enter the ganglion; nerve 3 divides into seven roots, 4 into two, 5 (which is the largest) into eleven identifiable roots, 6 into four, and 7 into three. Nerve 8 is formed from a single root. From these data he was able to identify the specific projections for a number of identified motoneurons (see below). Whereas the commissures and the connective tracts are primarily areas of nonsynaptic

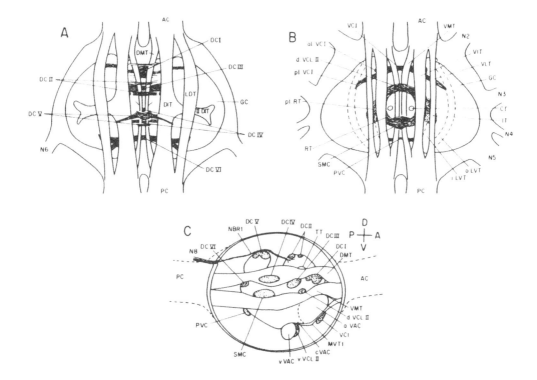

FIGURE 4. Schematic illustration of the tracts and commissures in the mesothoracic ganglion. (A) frontal section, dorsal view; (B) frontal section, ventral view; (C) midsaggital view. See text (Section III) for description of the tracts and commissures and the corresponding abbreviations not listed here. AC, anterior connective; PC, posterior connective; GC, ganglionic core; RT, ring tract; pl.RT, posterior limb of ring tract; IT, I-tract; CT, C-tract; TT, T-shaped tract; ll.DIT, lateral limb of dorsal intermediate tract; al.VCI and pl.VCI, anterior and posterior limbs of ventral commissure I, respectively; d.VCLII and a.VCII, dorsal and anterior regions of ventral commissural loop II, respectively; N8R1, nerve 8 root 1; a.VAC, c.VAC, and v.VAC, anterior, cylindrical, and posterior region of the ventral association centers, respectively. (From Gregory, G. E., *Philos. Trans. R. Soc. London*, 267, 421, 1974. With permission.)

interaction, the nerve roots are very probably the regions of significant synaptic input to motoneurons.[21]

Gregory[21] has also extended the previous morphological work on the location and distribution of somata. By using tracts, commissures, and glial invaginations into the somatal rind, he has grouped somata by their location in the ganglion. He has arranged the somata of the mesothoracic ganglion into approximately 25 groups and numerous subgroups. There are two anterior lateral groups (with a possible third), four anterior ventral, one anterior dorsolateral, four anterior ventrolateral, two posterior lateral, two posterior ventrolateral, and three posterior ventral groups. The remaining somata are grouped in regions on the dorsal and ventral midline and are described in detail with their neuritic projections into the various tracts and commissures of the ganglion.

IV. STRUCTURE AND FUNCTION OF IDENTIFIED NEURONS IN VENTRAL NERVE CORD

A. TECHNICAL CONSTRAINTS IN PHYSIOLOGICAL STUDIES OF THE NERVOUS SYSTEM

Anderson et al. (Chapter 10 of this work) have described the various techniques for electrophysiological recording that are currently in use for investigating neural activity in the central and peripheral nervous systems of the cockroach. Some additional procedures which are generally appropriate follow.

Animals should first be anesthetized with cold or carbon dioxide (a 5-s exposure to CO_2 is usually sufficient to keep animals quiet for a minute or two). Regardless of whether a relatively intact or highly dissected preparation is used as an experimental model, the dorsal surface of the animal should first be opened and the gut removed. If the gut is left in place, any perforation of the midgut will lead to the release of enzymes which will cause rapid deterioration of the centrally generated neural responses. The gut can be extracted by gently pulling it with two pairs of forceps, thus separating the gut at both the oral and anal ends. When the gut is removed carefully, the entire reproductive system will also be extracted.

Removal of the exoskeleton must be done with care, particularly when exposing the ventral surface of the animal, since it is extremely important to keep the tracheal system and oxygen flow intact. In fact, Krauthamer[23] devised an interesting experimental system in his investigations of the neurosecretory cells of the pars intercerebralis. To study the cells *in situ*, he used an isolated head in which the main neck tracheae were cannulated and the cephalic tracheal system was perfused with humidified O_2. This procedure maintained the supra- and subesophageal ganglia in viable condition for up to 30 h.

One further constraint on the experimental protocols associated with investigation of the CNS *in situ* is use of the proper saline. For experiments in which the neurolemma surrounding the nervous system is left intact, any one of a variety of ionic compositions can be utilized. For example, for *Periplaneta americana* we suggest a MOPS (morpholinopropane-sulfonic acid)-buffered saline (124 m*M* NaCl, 10 m*M* KCl, 5 m*M* $CaCl_2$, 1 m*M* $MgCl_2$, 3 m*M* MOPS, 3.9 m*M* MOPS sodium salt [pH 7.21], and 40 m*M* sucrose[27]). The MOPS buffer allows a higher calcium concentration than the standard phosphate buffers and also permits oxygenation of the saline with little change in pH. The sucrose is added for proper osmotic balance; it should be added immediately preceding the experiment to reduce the possibility of the growth of microorganisms, which often produce toxins.

For experiments in which the neurolemma must be removed, we suggest a saline with a composition similar to the ionic environment surrounding the neural elements of the CNS;[25-27] for example, 124 m*M* NaCl, 10 m*M* KCl, 8 m*M* $MgCl_2$, 20 m*M* $CaCl_2$, 3 m*M* MOPS, 3.9 m*M* MOPS sodium salt, and 40 m*M* sucrose.

B. MOTONEURONS

The cockroach nervous system has proven to be a superb experimental model for analysis of the morphology of single, physiologically identified neurons. The original attempts at mapping the motoneurons of the meso- and metathoracic ganglia in cockroaches were performed using various staining procedures,[28,29] the most useful being that of Young,[29] which involved staining the perinuclear RNA. Figure 5 is an example of such a map for the mesothoracic ganglia. With the advent of the use of Procion dyes,[30] either pressure injected or iontophoresed into somata, several investigators began to study not just location of the motoneuronal somata, but the three-dimensional structure of the neuropilar arborization of the physiologically identified motoneurons. The study of the morphology of the processes of identified neurons was advanced tremendously through the work of Pitman et al.,[31] who demonstrated the utility of iontophoresing cobaltous ions directly into the cell body of a physiologically identified motoneuron, D_f, the motoneuron which produces the very rapid twitch of the femur. This procedure proved more useful because the cobaltous ion moved throughout the cell much more quickly than the less mobile Procion dyes. Standard microscopy could then be utilized for viewing the cellular morphology, thus permitting a more complete description of the entire neuronal arbor of the injected motoneurons (see also Chapter 11 in this work).

A typical motoneuron consists of a rather large unipolar cell body, generally between

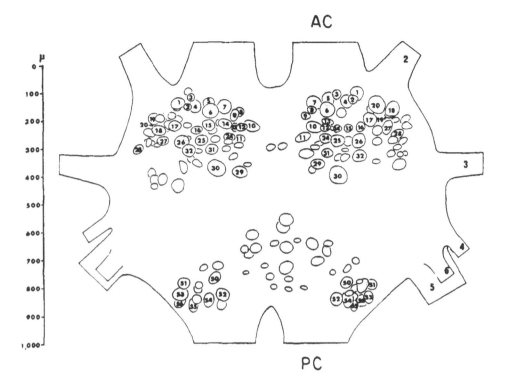

FIGURE 5.   A map of the position of neural somata in the mesothoracic ganglion of the cockroach. The somata placement was sufficiently consistent from one preparation to another to enable some of the profiles to be identified and numbered. This type of map may prove beneficial for further studies of identified neurons, particularly motoneurons, in the CNS. AC, anterior connective; PC, posterior connective. (From Young, D., *J. Insect Physiol.*, 15, 1175, 1969. With permission.)

20 and 50 μm in diameter. Arising from the cell body is a very thin fiber which may run for several tens of microns and may expand to as large as 5 μm in diameter. This expanded fiber, the neuropilar segment,[32] is generally located in one of the nerve roots described by Gregory;[21] from it originate a large number of smaller processes which may extend to many regions of the neuropil ipsilateral to the soma. For an exquisite view of the complexity of the processes of one motoneuron, D_s, see Figure 9A in Chapter 14 of this work. The peripheral axon also arises from this expanded process. Near the edge of the ganglion there is generally a slight reduction in the diameter of the axon where the axon exits the neuropilar region; this constricture may be the low-threshold initial segment of the motor axon.

The expanded region (neuropilar segment) of the motoneuronal neuritic field is a neurophysiologist's dream, since it is possible to penetrate this region of an identified neuron and record synaptic events, particularly those which influence the nearby spike-initiating zone. In original experiments on insect central nervous elements, recordings were obtained primarily from the somata which were spatially and electrically isolated from the extensive neuronal arbor.[33]

The cell bodies of the motoneurons are generally located on the ventral and ventral lateral surfaces of the ganglion, ipsilateral to the nerve root of origin for the motor axon. There are two notable exceptions to this general rule. First, the cell body of motoneuron D_s, innervating the slower muscles which extend the femur, is located on the ipsilateral dorsoposterior surface of the thoracic ganglia[34] in the posterior lateral group. Second, the cell body of the motoneuron innervating the dorsal longitudinal flight muscle is located on the contralateral ventroanterior surface of the ganglion.[12,35]

In addition to serving as a model for the analysis of the morphology of identified motoneurons, the cockroach CNS has been an experimental model for the study of populations of motoneurons innervating individual muscles and even subgroups of muscles of known fiber type.[36] Cobalt backfilling is the technique most usefully applied for such studies. This technique was originally used to identify the number and distribution of the common inhibitory motoneurons which innervate several of the muscles of the cockroach leg.[37] Two types of inhibitory neurons were found, one termed the widespread common inhibitory neuron since its axon projects into the segmental nerves 3, 4, 5 (via 4), and 6. Its soma is located in the posterior ventral medial group. The other type of inhibitor is the local inhibitory neuron, the axon of which exits through nerve 5, but bifurcates almost immediately after leaving the ganglion, sending an axonal branch to nerve root 4 through the small anastomosis connecting segmental nerves 4 and 5. The somata of the local common inhibitors are located within the posterior ventral groups.

Several studies have revealed the location and structure of the somata and neuronal processes of several motoneurons innervating the thoracic musculature of the cockroach.[35-36,38] Pearson and Fourtner[37] mapped the location of several motoneuronal somata which innervate the musculature of the walking legs. Using the dye Lucifer yellow both to backfill motor nerves and to iontophorese into physiologically identified motoneurons, Ritzmann et al.[35] and Davis[12] have described the locations and the major neuropilar projections of 16 motoneurons which innervate flight muscles and are active during flight behavior. As in the case of limb motoneurons, the neuronal processes of the flight motoneurons are extensive and radiate into many of the neuropilar regions of the thoracic ganglia.

Using the cobalt backfilling techniques, Iles[38] studied the motoneurons of the prothoracic ganglion and found with few exceptions that the location of somata and the distribution of neurites were similar to those previously described in the meso- and metathoracic ganglia.

Backfilling motoneuronal axons in concert with recording from individual motoneurons within the CNS has allowed the physiological characterization of over 50 identified motoneurons within the meso- and metathoracic ganglia of the cockroach which are associated with various leg and wing movements. Although the neuronal maps of the meso- and metathoracic ganglia described above would certainly contain the somata of these identified motoneurons, the specific locations of most of the physiologically identified motoneuron somata have not been charted on those maps. Although the ability to identify a specific motoneuronal soma from preparation to preparation would allow analysis comparable to that on identified neurons in molluscs,[39] with currently available information it appears that it is only possible to identify a given soma as belonging to one of Gregory's groups prior to penetration of the neuron.

The cockroach nervous system has also proven to be an excellent model for studying the role of motoneurons in a variety of behavior patterns. Since the gross morphology and the action of most thoracic skeletal muscles have been well described and the distribution of motor axons to individual muscles and muscle groups has been studied, the neural basis of behaviors ranging from simple reflexes to complex motor patterns such as locomotion, walking, or flight can be and has been investigated extensively. Using the type of electrophysiological techniques described by Anderson et al. in Chapter 10 of this work, several investigators have correlated muscle activity occurring during well-described behavioral patterns with centrally generated motor output recorded extracellularly from peripheral motor axons or intracellularly from motoneurons or other central neurons (for review, see References 2, 31, and 40). In addition, the cockroach has served as an excellent model for the investigation of sensory-to-motor reflexes and of sensory inputs to interneurons. For a comprehensive discussion of sensory-motor interactions controlling reflexes of the limb, see Chapter 25 of this work.

For a number of years, the cockroach has been the insect of choice for investigating walking behavior and interlimb coordination (for review of the earlier literature in this field, see Reference 41). During the past 20 years, several investigators, using high-speed cinematography, electrophysiological recording techniques, and the staining procedures described above, have refined and advanced models proposed by earlier workers. It is now clear that insect walking is controlled by a circuit of neurons in the CNS, the central pattern generator (CPG) of which is activated by some type of central command and can be modulated by a variety of sensory inputs. This literature has been reviewed on numerous occasions during the past decade[42-44] and will not be discussed except in reference to specific interneurons.

C. DORSAL UNPAIRED MEDIAL (DUM) NEURONS

A group of neurons having somata located along the dorsal midline were originally described and were found, using electrophysiological data, to send axonal branches to both sides of the body.[45,46] In addition, Crossman et al.[45] suggested that members of this group might send axonal projections out several of these bilateral nerve roots. The physiological importance of these cells has been established only recently, with most of the research concentrated on the locust rather than the cockroach.[47-49] One DUM neuron which innervates the extensor tibiae muscle in the locust releases octopamine in the vicinity of the muscle and can modulate muscular contraction.[50] For a review of this literature, see the article by Evans and Meyers[50] and Chapter 17 of this work. Studies on the DUM cells in the cockroach and the effects of transmitters such as octopamine and proctolin (see Chapter 14 of this work) eventually may explain the manner in which cockroaches maintain and modulate the high level of resting tension in skeletal muscles.

Gregory[51] has identified three groups of dorsal median cells, with the DUM cells probably being located in the posterior medial group. It is evident that not all medial cells have the characteristic bilateral symmetry of DUM cells. Gregory has identified one neuron, located in the middorsal medial group, which appears to innervate only one side of the animal and which has also been identified by Davis[12] as being a common neuron innervating all three of the dorsal longitudinal flight muscles. This strongly suggests that not all dorsal medial neurons are unpaired.

D. INTRAGANGLIONIC NEURONS

Investigations of the cockroach nervous system and its role in the control of locomotion led to the discovery of another group of neurons, each member of which is confined entirely within a single thoracic ganglion. Pearson and colleagues,[52,53] using cobalt iontophoresis, described a pair of interneurons that played a significant role in the neuronal circuit, the CPG responsible for producing rhythmic leg movements. These intraganglionic interneurons had small cell bodies (15 to 20 μm in diameter) located on the ventral surface contralateral to their extensively branching processes (Figure 6). Experimental manipulation of the membrane potential of these interneurons demonstrated that they influenced a broad pool of motoneurons innervating the leg contralateral to their somata, particularly those motoneurons responsible for movement of the femur.

One other important physiological characteristic of these local intraganglionic interneurons is that they are not capable of producing all-or-nothing action potentials or even the graded spikes which have been observed in other neuronal types (e.g., amacrine cells). Therefore, these cells apparently produce their postsynaptic effects through graded release of transmitter. The studies on these intraganglionic neurons established a new class of physiologically and morphologically identified, local, nonspiking neurons which (1) were clearly part of a CPG and (2) could control activity in a specific motoneuronal pool through the graded release of neurotransmitters. This model system has led to the discovery of similar neuronal types in other arthropods[44,54-58] and may prove to be a useful model for nonspiking neurons in vertebrate CNSs.

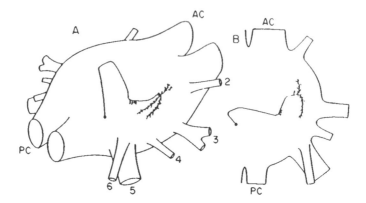

FIGURE 6.   Drawing of a cobalt sulfide-stained interneuron (Interneuron I) in the metathoracic ganglion of the cockroach.[37] This neuron is an essential element of the CPG which produces rhythmic leg movements. It has no obvious axon and does not produce action potentials. (A) A dorsal three-dimensional view; (B) a dorsal two-dimensional view. AC, anterior connective; PC, posterior connective.

E. INTERGANGLIONIC NEURONS

In addition to the intraganglionic interneurons, a large number of interganglionic interneurons have also been characterized morphologically and physiologically in the thoracic nervous system of the cockroach. The most obvious of these are the giant axons to be discussed in Section IV.F below. However, Ritzmann et al.[59,60] have recently described a series of interganglionic interneurons which are intercalated between the giant fibers and the motoneurons controlling the behavioral escape responses in the cockroach. The processes of these interganglionic interneurons branch extensively in the neuropil of both sides of the ganglion, with the cell bodies located contralateral to the connectives through which their axons exit the ganglion. Five of these interneurons were identified and named on the basis of the characteristic shape of their neuritic processes: "Lambda", "Cross", "J", "Reversed J", and "T". Figure 7 illustrates several examples of these identified interganglionic interneurons.

One population of these interganglionic interneurons, the dorsal posterior group (DPG) neurons, is found in the dorsal posterior region of the meso- and metathoracic ganglia. The morphology of these cells has been elucidated using iontophoresis of the dye Lucifer yellow,[61] a powerful staining technique, since even after injection of the dye the cell continues its normal physiological activity unless selectively inactivated by directing UV light at the region containing the dye.[62] Using the mapping nomenclature of Gregory,[21] Ritzmann and Pollack[59] have described the intraganglionic projections of each of these interganglionic neurons and their axonal projections from the ganglion. For example, "Lambda" projects anteriorly and posteriorly via contralateral DIT; "J", anteriorly via LDT; "Reversed J", posteriorly via LDT; and "Cross", anteriorly and posteriorly via VIT (see Figure 7). Another population of neurons called "T" cells have their somata in the ventral region at the level of nerve 3 and have their axonal projection through the contralateral anterior DIT.

DPG neurons excite the motoneuron D_s, which innervates a slow extensor of the femur. All of the DPGs except the "Cross" cell initially excite the widespread common inhibitor. "Lambda" activates it with a very short latency. This explains the rapid activation of the common inhibitor observed following direct stimulation of the abdominal ventral nerve cord.[63] These motor responses always occur in the motor units contralateral to the DPG somata. One of the fascinating riddles yet to be answered is the purpose of the common inhibitors, particularly with reference to their early activation by the DPG interneurons.

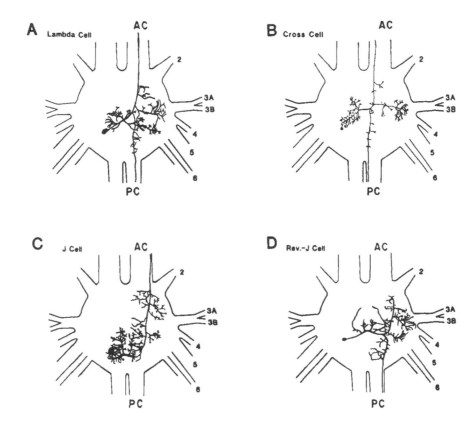

FIGURE 7. Drawings of four identified interganglionic interneurons associated with specific motoneuronal activity in the meta- and mesothoracic ganglia of the cockroach. The somata of these neurons are all located within the dorsal posterior group of somata. (A) "Lambda"; (B) "Cross"; (C) "J"; (D) "Reversed J". AC, anterior connective; PC, posterior connective. See Section IV.E of text for a complete description of the physiology of these cells. (From Ritzmann, R. E. and Pollack, A. J., *J. Comp. Physiol. A*, 159, 639, 1986. With permission.)

The "T" neuron evoked a more complex motor response, with an initial contralateral excitation of both D_s and the widespread common inhibitor followed by a long-lasting burst of activity in the leg flexor (levator) motor neuronal pool.

F. GIANT AXONS

One of the most extensively studied systems in the cockroach CNS is the system of giant axons, which run the length of the ventral nerve cord.[64-66] Ever since the description of the rapid escape response in the cockroach, there has been continuing discussion on the role which these giant axons might play in escape behavior. The giant axons have been divided into two morphological groups: a ventral group of four, three extremely large axons (50 to 60 μm in diameter) and a smaller one (30 μm), and a dorsal group of three axons ranging from 20 to 30 μm in diameter.[67]

Each of the giant axons has a soma which is located in the sixth abdominal ganglion. The soma is contralateral to the connective containing its axon and its major dendritic arbor. There are few small processes ipsilateral to the cell body. There are also processes arising from the giant axons as they pass through the meta- and mesothoracic ganglia. These are small and are generally located in the ventral neuropilar regions. Figure 8 illustrates the morphology of the seven giant axons (see also Figure 1 in Chapter 8 of this work for a photograph of the morphology of a cobalt-stained ventral giant 2).

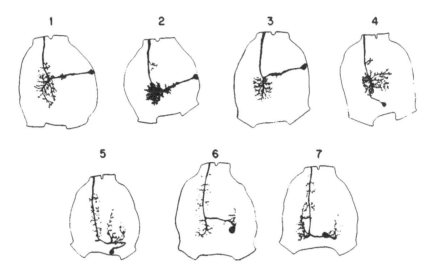

FIGURE 8.   Drawings of cobalt-filled giant fiber processes and somata in the sixth abdominal ganglion. The numbers above each drawing correspond to the number of the giant cell depicted; 1 through 4 represent the ventral giants and 5 through 7 the dorsal giants. The soma is always contralateral to the process projecting to the anterior connective. (From Daley, D. L., Vardi, N., Appignani, B., and Camhi, J. M., *J. Comp. Neurol.*, 196, 41, 1981. With permission.)

Each of the seven giant axons is relatively easy to penetrate with microelectrodes in the abdominal nerve cord without disrupting sensory input via the cercal afferents and motor output from the thoracic ganglia. Westin et al.[68,69] demonstrated that each of the giants is responsive to air streams directed at the cerci and has unique receptive fields, thus permitting a selective activation of giants with well-controlled noninvasive inputs.

Studies of the motor output from thoracic ganglia resulting from stimulation of giant fibers (either singly or in pairs) have revealed that the giants, other than ventral giants 2 and 3, can influence the motor output to the population of motoneurons used in walking.[70-72] Recent physiological studies have demonstrated separate functions for the dorsal and ventral giants. It is now fairly well established that the giant axons (at least the dorsal ones) can be utilized to initiate escape by activating the motoneuronal pool used for running and may directly activate the CPG for running. The ventral giants may be extremely important in determining the directionality of escape.[71,71a] Daley and Delcomyn,[73-75] using a tightly tethered cockroach with the abdominal ventral nerve cord held in place on an immovable platform, were able to record from the giant axons intracellularly. While their data strongly suggested that during walking there was simultaneous activity in the dorsal giants, but little or no activity in the ventral giants, there may in fact be inhibition of the ventral giants.[74]

Ritzmann et al.[76,77] have demonstrated that activation of an individual dorsal giant axon can evoke activity in the leg motoneurons associated with running in order to avoid a predator, or, depending on the sensory input arising from the legs, the dorsal giants can initiate flight behavior by turning on the CPG for flight, driving the wing motoneurons instead of the limb motoneurons for walking or running. This is an example of the importance of sensory input for "gating" the activity arriving in a single giant axon to selectively activate one or the other CPG.

This work on giant axon activation of flight has led to some recent investigations of the flight system of the cockroach as a model for interneuronal activation of behavior. As in the case of the walking system, the flight musculature and its activity during bouts of flight have been described.[78] Some extremely large muscles which function as flight muscles are also active during rapid walking bouts.[78] Another interesting finding is that some of the flight muscles appear to be active continuously during flight rather than being active during only one phase of the flight pattern. The innervation of the flight musculature is also well known, allowing the precise description of motoneurons given above. It is clear that as a model system for studies of flight the locust currently offers a better model system. Nevertheless, the cockroach flight system is extremely useful for investigating the neural mechanisms responsible for the onset of flight activity.

Another extremely important aspect of the giant axons as an experimental model is that this system permits simultaneous microelectrode recordings from a single giant axon in the abdominal cord and from neuropilar processes of neurons within a thoracic ganglion. This arrangement allows the study of the synaptic interaction between giant axons and other neurons within the meso- and metathoracic ganglia. Ritzmann and co-workers[59,60] have recently completed an extensive investigation using this preparation, demonstrating synaptic interactions between giant fibers and other interganglionic interneurons. The DPG neurons receive excitatory input from the ventral giant axons and apparently are unaffected by dorsal giant stimulation. There is, however, a sidedness to the input, since "Lambda" responds to contralateral ventral giants 1 and 2 and ipsilateral 3, whereas "Reverse-J" and "Cross" are driven by ipsilateral 1 and 2. "J" neuron received input from ipsi- and contralateral ventral giant 2. The "T" cell, on the other hand, fires action potentials following activation of the ipsilateral dorsal giant axons 5 and 7 and most likely is unaffected by ventral giant input.

More recently, the effects of sensory input on at least the DPGs have been studied.[79] For a review of these data, as well as the role of sensory input on limb reflexes and on the control of leg movements, see Chapter 25 of this work.

V. GLIA

In addition to the neural elements of the CNS functioning as model systems for studies of basic physiological processes, the glia cells and their associated noncellular matrix, the neurolemma, surrounding the CNS of the cockroach have recently been studied thoroughly for their possible role in controlling the chemical and, more particularly, the ionic environment in and around the nerve cells and neuronal processes of the nervous system.[80-89] Lane[90] recently has written a review on the ultrastructure of the glial components associated with the CNS. Also presented in that review are some previously unpublished observations on the ultrastructure of unidentified synaptic connections in the neuropil of the thoracic ganglia. In addition, the glial elements play the primary role in the repair and regeneration of the sheath surrounding the CNS and the neural processes, respectively.

There are generally conceived to be two classes of glial cells surrounding the CNS: the perineural glial cells forming the outer sheath around the CNS and the neuroglia sending processes around the axons and into the neuropil of the CNS. The perineurium, generally assumed to be the blood-brain barrier,[85-87] consists of cells which are somewhat similar to epithelial cells in that there are extensive gap junctions between adjacent perineurium glial cells. Figure 9 illustrates the morphological relationships among the glial elements and neural elements of the CNS. For a thorough discussion of the role of the neural sheath and glial elements in control of the chemical milieu surrounding the neural elements and the repair function of glia, see Chapter 11 in this work and References 91 to 93.

In the peripheral nervous system, perineural glia cells are also found directly beneath the neurolemma, and subperineural glia is found surrounding individual motor axons. There may be as many as six or seven layers of glial sheath around individual motor axons, whereas there will be a glial sheath around pools of sensory axons rather than around single ones.

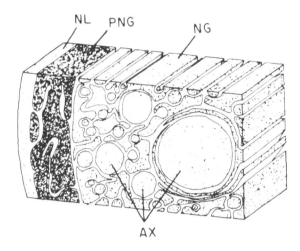

FIGURE 9.   Three-dimensional schematic drawing of the nerve sheath surrounding the abdominal connectives. AX, axon; NL, neurolemma; PNG, perineuronal glia; NG, neuroglia. (From Treherne, J. E. and Schofield, J. E., *J. Exp. Biol.*, 95, 61, 1979. With permission.)

Figure 10A illustrates the morphological arrangement in a purely motor nerve containing axons innervating the extensor tibiae muscle of the cockroach leg. In addition, the glial membrane surrounding the motor axons envelops many regions of electron-dense material. In the normal glia there are occasional occurrences of microtubules, but these are usually located near an area where a nerve is splitting, sending a ramus to a group of muscle fibers.

Although the role of glia in repair and regeneration has been studied extensively in the CNS, studies on the role of glia in peripheral regeneration are lacking. In recent studies in our laboratory, we have shown that there is a remarkable transformation in the organization of the glial elements within 5 days of nerve damage.[95] The model for these studies is a small motor nerve innervating the muscle producing extension of the tibia (as seen in Figure 10B). The nerve contains four axonal profiles: the fast extensor tibiae (the largest); a somewhat smaller diameter axon, the slow extensor tibiae; and two small common inhibitors. The two excitors exit the thoracic ganglia via nerve root 3B, while the inhibitors exit via nerve 5, but "cross over" to 3B in an anastomosis near the end of the coxa.

The utility of this model system is that the two excitors or the two inhibitors can be damaged selectively and the effect of degenerating or regenerating fibers on intact fibers can be studied. One of our findings is that, following lesioning of the excitors, within 1 week there is hypertrophy in the cross-sectional diameter of the intact inhibitory axons. However, as mentioned above, the most remarkable occurrence is the transformation occurring in the subperineural glia. By day 5 following lesioning, the glia surrounding the degenerating excitors have invaded that axonal space and have begun to synthesize large numbers of microtubules. By 7 to 10 days after lesioning, the microtubules have begun to align axially along the nerve, and after 2 to 3 weeks the perineural glial has become an exquisite network of axially arranged microtubules (Figure 10C) which may serve in guiding the regenerating excitatory motor axons.[95]

VI. SUMMARY

With the development of the technical expertise to record simultaneously from various areas in the nervous system, the cockroach CNS has been and will continue to be a superb

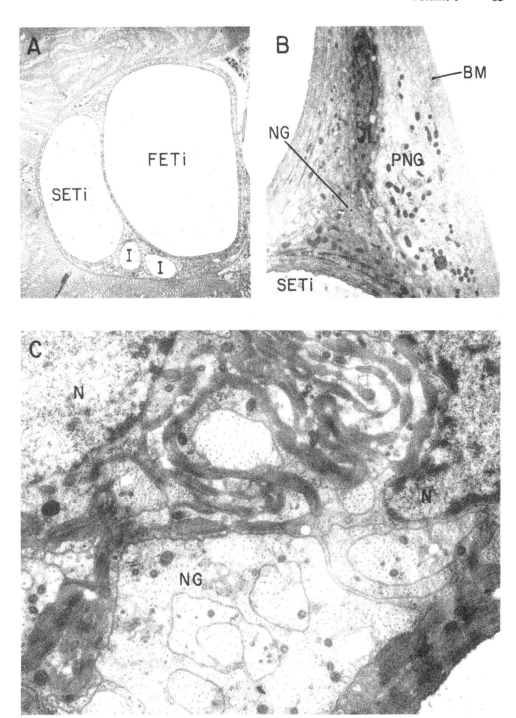

FIGURE 10. Electron microscopic illustration of peripheral motor nerve 3B11. (A) Cross section of the nerve, revealing the four motor units; FETi and SETi are the fast and slow extensor tibiae motor axons respectively; the two widespread common inhibitors = I. (Magnification × 1200.) (B) Cross section of the neuroglia surrounding SETi and FETi, revealing the basement membrane (BM), the perineural glia (PNG), lighter staining neural glial elements, the neural glia (NG), with denser staining cytoplasm and layered extracellular material circumferentially arranged around the axons. (Magnification × 7250.) (C) Cross section of transformed glia 14 d after nerve section, revealing glial processes with axially arranged microtubules; N = nucleus. (Magnification × 23,000.)

model for investigating neural events modulating behavior, neural mechanisms at the neuronal level, and regeneration of neural tissue. As a summary example, the case of the escape response of the cockroach offers just one instance of a complex neural response in which most of the neural mechanisms from sensory reception to motor activation are amenable to anatomical and physiological investigations. These studies will continue to serve as a basis for neurobiological mechanisms that span the animal kingdom.

REFERENCES

1. **Guthrie, D. M. and Tindall, A. R.,** *The Biology of the Cockroach,* Edward Arnold, London, 1968.
2. **Pipa, R. and Delcomyn, F.,** Nervous system, in *The American Cockroach,* Bell, W. J. and Adiyodi, K. G., Eds., Chapman and Hall, London, 1981, 175.
3. **Weevers, R. deG.,** The insect ganglia, in *Comprehensive Insect Physiology, Biochemistry and Pharmacology,* Vol. 5, Kerkut, G. A. and Gilbert, L. I., Eds., Pergamon Press, Oxford, 1985, 213.
4. **Krauthamer, V.,** Morphology of identified neurosecretory and nonneurosecretory cells in the cockroach pars intercerebralis, *J. Exp. Zool.,* 234, 221, 1985.
5. **Strausfeld, N. J.,** *Atlas of an Insect Brain,* Springer-Verlag, New York, 1976.
6. **Mobbs, P. G.,** Brain structure, in *Comprehensive Insect Physiology, Biochemistry and Pharmacology,* Vol. 5, Kerkut, G. A. and Gilbert, L. I., Eds., Pergamon Press, Oxford, 1985, 309.
7. **Weiss, M. J.,** Neuronal connections and the function of the corpora pedunculata in the brain of the American cockroach, *Periplaneta americana* (L.), *J. Morphol.,* 142, 21, 1974.
8. **Willey, R. B.,** The morphology of the stomodeal nervous system in *Periplaneta americana* (L.) and other Blattaria, *J. Morphol.,* 108, 219, 1962.
9. **Ernst, K.-D., Boeckh, J., and Boeckh, V.,** A neuroanatomical study on the organization of the central antennal pathway in insects. II. Deuterocerebral connections in *Locusta migratoria* and *Periplaneta americana, Cell Tissue Res.,* 176, 285, 1977.
10. **Bodenstein, D.,** Studies on nerve regeneration in *Periplaneta americana, J. Exp. Zool.,* 136, 89, 1957.
11. **Altman, J. S. and Kien, J.,** Functional organization of the subesophageal ganglion in arthropods, in *Arthropod Brain,* Gupta, A. P., Ed., John Wiley & Sons, New York, 1987, 265.
12. **Davis, N. T.,** Serial homologies of the motor neurons of the dorsal intersegmental muscles of the cockroach, *Periplaneta americana* (L.), *J. Morphol.,* 176, 197, 1983.
13. **Gupta, A. P., Ed.,** *Arthropod Brain,* John Wiley & Sons, New York, 1987.
14. **Dresden, D. A. and Nijenhuis, E. D.,** Fibre analysis of the nerves of the second thoracic leg in *Periplaneta americana, Proc. K. Ned. Akad. Wet.,* C61, 213, 1958.
15. **Nijenhuis, E. D. and Dresden, D.,** A micro-morphological study on the sensory supply of the mesothoracic leg of the American cockroach, *Periplaneta americana, Proc. K. Ned. Akad. Wet.,* C55, 300, 1955.
16. **Nijenhuis, E. D. and Dresden, D.,** On the topographical anatomy of the nervous system of the mesothoracic leg of the American cockroach *(Periplaneta americana), Proc. K. Ned. Akad. Wet.,* C58, 121, 1955.
17. **Pipa, R. L. and Cook, E. F.,** Studies on the hexapod nervous system. I. The peripheral distribution of the thoracic nerves of the adult cockroach, *Periplaneta americana, Ann. Entomol. Soc. Am.,* 52, 695, 1959.
18. **Pipa, R. L., Cook, E. F., and Richards, A. G.,** Studies on the hexapod nervous system. II. The histology of the thoracic ganglia of the adult cockroach, *Periplaneta americana* (L.), *J. Comp. Neurol.,* 113, 401, 1959.
18a. **Kaars, C. J.,** Innervation and control of tension in abdominal muscles of *Blaberus discoidalis* and *Gromphadorhina portentosa, J. Insect Physiol.,* 25, 209, 1983.
19. **Mizunami, M., Yamashita, S., and Tateda, H.,** Intracellular stainings of the large ocellar second order neurons in the cockroach, *J. Comp. Physiol.,* 149, 215, 1982.
20. **Mizunami, M. and Tateda, H.,** Classification of ocellar interneurones in the cockroach brain, *J. Exp. Biol.,* 125, 57, 1986.
21. **Gregory, G. E.,** Neuroanatomy of the mesothoracic ganglion of the cockroach, *Periplaneta americana* (L.). I. The roots of the peripheral nerves, *Philos. Trans. R. Soc. London,* 267, 421, 1974.
22. **Gregory, G. E.,** Silver staining of insect central nervous system by the Bodian protargal method, *Acta Zool. (Stockholm),* 51, 169, 1970.
23. **Krauthamer, V.,** Electrophysiology of identified neurosecretory and non-neurosecretory cells in the cockroach pars intercerebralis, *J. Exp. Zool.,* 234, 234, 1985.

24. **Chesler, M. and Fourtner, C. R.**, Mechanical properties of a slow muscle in the cockroach, *J. Neurobiol.*, 12, 391, 1981.
25. **Carr, C. E. and Fourtner, C. R.**, Pharmacological analysis of a monosynaptic reflex in the cockroach, *Periplaneta americana, J. Exp. Biol.*, 86, 259, 1980.
26. **Thomas, M. V. and Treherne, J. E.**, An electrophysiological analysis of extra-axonal sodium and potassium concentrations in the central nervous system of the cockroach *(Periplaneta americana* L.), *J. Exp. Biol.*, 63, 801, 1975.
27. **Treherne, J. E., Scofield, P. K., and Lane, N. J.**, Physiological and ultrastructural evidence for an extracellular anion matrix in the central nervous system of an insect *(Periplaneta americana), Brain Res.*, 247, 255, 1982.
28. **Cohen, M. J. and Jacklet, J. W.**, The functional organization of motor neurons in an insect ganglion, *Philos. Trans. R. Soc. London*, 252, 561, 1967.
29. **Young, D.**, The motor neurons of the mesothoracic ganglion of *Periplaneta americana, J. Insect Physiol.*, 15, 1175, 1969.
30. **Stretton, A. O. W. and Kravitz, E. A.**, Neuronal geometry: determination with a technique of intracellular dye injection, *Science*, 162, 132, 1968.
31. **Pitman, R. M., Tweedle, C. D., and Cohen, M. J.**, Branching of central neurons: intracellular cobalt injection for light and electron microscopy, *Science*, 176, 412, 1972.
32. **Tyrer, N. M. and Altman, J. S.**, Motor and sensory flight neurones in a locust demonstrated using cobalt chloride, *J. Comp. Neurol.*, 157, 117, 1974.
33. **Hoyle, G.**, Cellular mechanisms underlying behavior — neuroethology, *Adv. Insect Physiol.*, 7, 349, 1970.
34. **Pearson, K. G. and Bradley, A. B.**, Specific regeneration of excitatory motoneurons to leg muscles in the cockroach, *Brain Res.*, 47, 492, 1972.
35. **Ritzmann, R. E., Fourtner, C. R., and Pollack, A. J.**, Morphological and physiological identification of motor neurons innervating flight musculature in the cockroach, *Periplaneta americana, J. Exp. Zool.*, 225, 347, 1983.
36. **Fourtner, C. R. and Pearson, K. G.**, Morphological and physiological properties of motor neurons innervating insect leg muscles, in *Identified Neurons and Behavior of Arthropods*, Hoyle, G., Ed., Plenum Press, New York, 1977, 87.
37. **Pearson, K. G. and Fourtner, C. R.**, Identification of the somata of common inhibitory motoneurones in the metathoracic ganglion of the cockroach, *Can. J. Zool.*, 51, 859, 1973.
38. **Iles, J. F.**, Organization of motoneurones in the prothoracic ganglion of the cockroach, *Periplaneta americana* (L.), *Philos. Trans. R. Soc. London Ser. B*, 276, 205, 1976.
39. **Kandel, E. R. and Schwartz, J. H.**, *Principles of Neural Science*, Elsevier, New York, 1985.
40. **Delcomyn, F.**, Walking and running, in *Comprehensive Insect Physiology, Biochemistry and Pharmacology*, Vol. 5, Kerkut, G. A. and Gilbert, L. I., Eds., Pergamon Press, Oxford, 1985, 439.
41. **Wilson, D. M.**, Central nervous mechanism for the generation of rhythmic behavior in arthropods, *Symp. Soc. Exp. Biol.*, 20, 199, 1966.
42. **Delcomyn, F.**, Neural basis of rhythmic behavior in animals, *Science*, 210, 492, 1980.
43. **Fourtner, C. R.**, Central nervous control of cockroach walking, in *Neural Control of Locomotion*, Herman, R. M., Grillner, S., Stein, P. S. G., and Stuart, D. G., Eds., Plenum Press, New York, 1977, 401.
44. **Pearson, K. G.**, Interneurons in the ventral cord of insects, in *Identified Neurons and Behavior of Arthropods*, Hoyle, G., Ed., Plenum Press, New York, 1977, 329.
45. **Crossman, A. R., Kerkut, G. A., Pitman, R. M., and Walker, R. J.**, Electrically excitable nerve cell bodies in the central ganglia of two insect species, *Periplaneta americana* and *Schistocerca gregaria*. Investigation of cell geometry and morphology by intracellular dye injection, *Comp. Biochem. Physiol.*, 40A, 579, 1971.
46. **Crossman, A. R., Kerkut, G. A., and Walker, R. J.**, Electrophysiological studies on the axon pathways of specified nerve cells in the central ganglia of two insect species, *Periplaneta americana* and *Schistocerca gregaria, Comp. Biochem. Physiol.*, 43A, 393, 1972.
47. **Hoyle, G.**, A function for neurons (DUM) neurosecretory on skeletal muscle of insects, *J. Exp. Zool.*, 189, 401, 1974.
48. **Hoyle, G.**, The dorsal, unpaired median neurons of the locust metathoracic ganglion, *J. Neurobiol.*, 9, 43, 1978.
49. **Hoyle, G., Dagan, D., Moberly, B., and Colquhoun, W.**, Dorsal unpaired median insect neurons make neurosecretory endings on skeletal muscle, *J. Exp. Zool.*, 187, 159, 1974.
50. **Evans, P. D. and Meyers, C. M.**, Peptidergic and aminergic modulation of insect skeletal muscle, *J. Exp. Biol.*, 124, 143, 1986.
51. **Gregory, G. E.**, Neuroanatomy of the mesothoracic ganglion of the cockroach *Periplaneta americana*. II. Median neuron cell body groups, *Philos. Trans. R. Soc. London*, 306, 191, 1984.
52. **Pearson, K. G., Fourtner, C. R., and Wong, R. K.**, Nervous control of walking in the cockroach, in *Control of Posture and Locomotion*, Stein, R. B., Pearson, K. G., Smith, R. S., and Redford, J. B., Eds., Plenum Press, New York, 1973, 495.

53. **Pearson, K. G. and Fourtner, C. R.,** Nonspiking interneurons in walking system of the cockroach, *J. Neurophysiol.*, 38, 33, 1975.
54. **Watkins, B. L., Burrows, M., and Siegler, M. V. S.,** The structure of locust non-spiking interneurones in relation to the anatomy of a segmental ganglion, *J. Comp. Neurol.*, 240, 233, 1985.
55. **Burrows, M.,** The control of sets of motoneurones by local interneurones, *J. Physiol.*, 298, 213, 1980.
56. **Burrows, M. and Siegler, M. V. S.,** Graded synaptic transmission between local interneurons and motoneurones in the metathoracic ganglion of the locust, *J. Physiol.*, 285, 231, 1978.
57. **Roberts, A. and Bush, B. M. H.,** *Neurons without Impulses,* Cambridge University Press, London, 1981.
58. **Pearson, K. G.,** Nerve cells without action potentials, in *Simpler Networks and Behavior,* Fentress, J. C., Sinauer Associates, Sunderland, MA, 1976, 99.
59. **Ritzmann, R. E. and Pollack, A. J.,** Identification of thoracic interneurons that mediate giant interneuron-to-motor pathways in the cockroach, *J. Comp. Physiol. A,* 159, 639, 1986.
60. **Ritzmann, R. E., Pollack, A. J., and Westin, J.,** Integration of directional wind field information in thoracic interneurons of the cockroach, *Neurosci. Abstr.*, 13, 140, 1987.
61. **Stewart, W. W.,** Functional connections between cells as revealed by dye-coupling with a highly fluorescent napthalimide tracer, *Cell,* 14, 741, 1968.
62. **Miller, J. P. and Selverston, A. I.,** Rapid killing of single neurons by irradiation of intracellularly injected dye, *Science,* 206, 702, 1979.
63. **Fourtner, C. R. and Drewes, C. D.,** Excitation of the common inhibitory motor neuron: a possible role in the startle reflex of the cockroach, *Periplaneta americana, J. Neurobiol.,* 8, 477, 1977.
64. **Ritzmann, R. E.,** The cockroach escape response, in *Neural Mechanisms in Startle Behavior,* Eaton, R. C., Ed., Plenum Press, New York, 1984, 93.
65. **Roeder, K. D.,** Organization of the ascending giant fiber system in the cockroach *(Periplaneta americana),* *J. Exp. Zool.,* 108, 243, 1948.
66. **Parnas, I. and Dagan, D.,** Functional organizations of giant axons in the central nervous system of insects: new aspects, *Adv. Insect Physiol.,* 9, 95, 1971.
67. **Daley, D. L., Vardi, N., Appignani, B., and Camhi, J. M.,** Morphology of the giant interneurons and cercal nerve projections of the American cockroach, *J. Comp. Neurol.,* 196, 41, 1981.
68. **Westin, J., Langberg, J. L., and Camhi, J. M.,** Responses of giant interneurons of the cockroach, *Periplaneta americana* to wind puffs of different directions and velocities, *J. Comp. Physiol.,* 121, 307, 1977.
69. **Westin, J. and Ritzmann, R. E.,** The effect of single giant lesions on wind-evoked motor responses in the cockroach, *Periplaneta americana, J. Neurobiol.,* 13, 127, 1982.
70. **Ritzmann, R. E. and Camhi, J. M.,** Excitation of leg motor neurons by giant interneurons in the cockroach, *Periplaneta americana, J. Comp. Physiol.,* 125, 305, 1978.
71. **Ritzmann, R. E. and Pollack, A. J.,** Motor responses to paired stimulation of giant interneurons in the cockroach, *Periplaneta americana.* I. The dorsal interneurons, *J. Comp. Physiol.,* 143, 61, 1981.
71a. **Ritzmann, R. E.,** personal communication.
72. **Ritzmann, R. E.,** Motor responses to paired stimulation of giant interneurons in the cockroach, *Periplaneta americana.* II. The ventral interneurons, *J. Comp. Physiol.,* 143, 71, 1981.
73. **Daley, D. L. and Delcomyn, F.,** Modulation of excitability of cockroach giant interneurons during walking. I. Simultaneous excitation and inhibition, *J. Comp. Physiol.,* 138, 231, 1980.
74. **Daley, D. L. and Delcomyn, F.,** Modulation of excitability of cockroach giant interneurons during walking. II. Central and peripheral components, *J. Comp. Physiol.,* 138, 241, 1980.
75. **Delcomyn, F.,** Activity and structure of movement-signalling (corollary discharge) interneurons in a cockroach, *J. Comp. Physiol.,* 150, 185, 1983.
76. **Ritzmann, R. E., Tobias, M. L., and Fourtner, C. R.,** Flight activity initiated via giant interneurons of the cockroach: evidence for bifunctional trigger interneurons, *Science,* 210, 443, 1980.
77. **Ritzmann, R. E., Pollack, A. J., and Tobias, M. L.,** Flight activity mediated by intracellular stimulation of dorsal giant interneurons of the cockroach, *Periplaneta americana, J. Comp. Physiol.,* 147, 313, 1982.
78. **Fourtner, C. R. and Randall, J. B.,** Studies on cockroach flight: the role of continuous neural activation of non-flight muscles, *J. Exp. Zool.,* 221, 143, 1982.
79. **Murrain, M. P.,** Morphological and Physiological Analysis of Proprioceptive Inputs to DPG Interneurons in the Cockroach, Ph.D. dissertation, Case Western Reserve University, Cleveland, OH, 1987, 1.
80. **Ashhurst, D. E.,** The connective tissue of insects, *Annu. Rev. Entomol.,* 13, 45, 1968.
81. **Radojcic, T. and Pentreath, V. W.,** Invertebrate glia, *Prog. Neurobiol.,* 12, 115, 1979.
82. **Lane, N. J.** Invertebrate neuroglia — junctional structure and development, *J. Exp. Biol.,* 95, 7, 1981.
83. **Lane, N. J. and Treherne, J. E.,** Studies on perineurial junctional complexes and sites of uptake of microperoxidase and lanthanum in the cockroach central nervous system, *Tissue Cell,* 4, 427, 1972.
84. **Maddrell, S. H. P. and Treherne, J. E.,** The ultrastructure of the perineurium in two insect species, *Carausius morosus* and *Periplaneta americana, J. Cell Sci.,* 2, 119, 1967.

85. **Schofield, P. K. and Treherne, J. E.**, Localization of the blood-brain barrier of an insect: electrical model and analyses, *J. Exp. Biol.*, 109, 319, 1984.
86. **Schofield, P. K., Swales, L. S., and Treherne, J. E.**, Potentials associated with the blood-brain barrier of an insect: recordings from identified neuroglia, *J. Exp. Biol.*, 109, 307, 1984.
87. **Lane, N. J. and Treherne, J. E.**, Uptake of peroxidase by the cockroach central nervous system, *Tissue Cell*, 2, 413, 1970.
88. **Treherne, J. E. and Schofield, P. K.**, Mechanisms of ionic homeostasis in the central nervous system of an insect, *J. Exp. Biol.*, 95, 61, 1981.
89. **Treherne, J. E. and Pichon, Y.**, The insect blood-brain barrier, *Adv. Insect Physiol.*, 9, 257, 1972.
90. **Lane, N. J.**, Structure of components of the nervous system, in *Comprehensive Insect Physiology, Biochemistry and Pharmacology*, Vol. 5, Kerkut, G. A. and Gilbert, L. I., Eds., Pergamon Press, Oxford, 1985, 1.
91. **Treherne, J. E., Harrison, J. B., Treherne, J. M., and Lane, N. J.**, Glial repair in an insect central nervous system: effects of surgical lesioning, *J. Neurosci.*, 4, 2689, 1984.
92. **Smith, P. J. S., Leech, C. A., and Treherne, J. E.**, Glial repair in an insect central nervous system: effects of selective glial disruption, *J. Neurosci.*, 4, 2698, 1984.
93. **Meiri, H., Dormann, A., and Spira, M. E.**, Comparison of ultrastructural changes in proximal and distal segments of transected giant fibres in the cockroach, *Periplaneta americana*, *Brain Res.*, 263, 1, 1983.
94. **Treherne, J. E. and Schofield, J. E.**, Ionic homeostasis of the brain microenvironment in insects, *J. Exp. Biol.*, 95, 61, 1979.
95. **Fourtner, C. R. and Kaars, C. J.**, in preparation.

Chapter 5

INNERVATION AND ELECTROPHYSIOLOGY OF THE CORPUS ALLATUM

C. S. Thompson and S. S. Tobe

TABLE OF CONTENTS

I. INTRODUCTION

The corpora allata (CA) of insects are endocrine glands closely associated with and innervated by axons originating in the central nervous system (CNS). They are a major component of the retrocerebral system, but are also anatomically associated with the stomatogastric nervous system.[1] Indeed, these glands were incorrectly thought to be part of the nervous system when they were first described in the 19th century.[2] Numerous studies conducted since that time have established that the CA are composed of parenchymal cells which arise from the embryonic ectoderm and that they are the site of synthesis and release of juvenile hormone (JH).[3] JH plays a vital role in regulating the processes of metamorphosis and reproduction. The importance of JH in the regulation of these processes is reflected in the enormous literature dealing with JH and with the CA (see, for example, References 3 to 5).

An interesting feature of the brain-corpus cardiacum-corpus allatum system of insects is the superficial anatomical resemblance to the hypothalamic-pituitary system of vertebrates.[6] This analogy between the CA and the adenohypophysis suggests that the CA may be regulated at least in part by hormones produced by neurosecretory cells in the brain. This idea is supported by the observations that there are neurosecretory cells in the insect brain which have terminals in the CA and that neural connections between the brain and the CA are necessary for the normal functioning of these endocrine glands. However, there are important differences between the CA and the adenohypophysis. As far as is known, the CA produce only a single hormone rather than several different hormones, and JH is a lipid-soluble sesquiterpenoid. Thus, JH is not stored in membrane-bound vesicles as are the peptide hormones of the adenohypophysis. Rather, JH is released into the hemolymph as it is synthesized and appreciable quantities are not stored in the glands.[7,8] Thus, mechanisms acting to control JH biosynthesis may provide a model for mechanisms acting to control the rate of biosynthesis of the steroid hormones of vertebrates.[9]

The Pacific beetle cockroach, *Diploptera punctata* (Eschscholtz), has provided a useful model system in which to study factors controlling the biosynthesis and release of JH. In this species, the CA are relatively large glands (200 to 300 μm in adults)[10] and, when active, produce large amounts of JH in comparison to other species.[3] Furthermore, in adult females, JH is synthesized and released in a well-defined cycle associated with the vitellogenic growth and maturation of oocytes.[11] As reviewed in Chapter 21 of this work, several factors involved in the regulation of biosynthesis of JH by the CA of adult female *D. punctata* have been studied. Among these are neural connections with the brain, humoral factors from other sources, and the second messengers calcium and cyclic adenosine monophosphate (cAMP).

II. INNERVATION OF THE CORPORA ALLATA

The innervation of the CA is complex and has not been determined precisely for any species; i.e., the precise location and identity of neurons in the CNS remain to be described. Both conventional and neurosecretory neurons from many parts of the brain and various ganglia contribute axons to the nerves of the retrocerebral complex, and there is a great deal of species variability in the size, number, and location of these neurons as well as in the neuroanatomy of the retrocerebral complex itself. Whereas conventional staining techniques have revealed several groups of neurons which send axons to the CA in a variety of species, the functional significance of these neurons has been difficult to assess. Some axons may bypass the CA without giving off terminal branches, whereas some neurosecretory neurons may have terminal branches in the CA, but release a neurohormone which is not involved in the regulation of the CA. This is apparently the case in Lepidoptera in which cerebral neurosecretory neurons apparently containing the prothoracicotropic hormone have terminals in or on the CA.[12] The CA also may be regulated by neurosecretory neurons that release neurohormones at sites some distance from the CA.

In cockroaches, as in most other insect species, the CA appear to be innervated by axons from both the brain and the subesophageal ganglion. In *D. punctata*, the nickel chloride backfilling technique has been used to show that some neurons in both the medial and the lateral groups of neurosecretory cells in the protocerebrum send axons to the CA by way of the nervi corporis cardiaci I (NCC I) and the nervi corporis cardiaci II (NCC II), respectively. A few neurons in the tritocerebrum also send axons to the CA by way of the nervi corporis cardiaci III (NCC III), and about 20 neurons in the subesophageal ganglion send axons by way of the nervi corporis allati II (NCA II).[13] This is summarized in Figure 1. Similar groups of neurons have been shown to send axons to the CA in a variety of other species.[9] A comparison of brains backfilled from NCA I with brains backfilled from the postallatal nerves indicates that about 40 neurons in the contralateral pars intercerebralis (PI) — medial neurosecretory cells (MNSC) — and about 20 in the ipsilateral pars lateralis (PL) — lateral neurosecretory cells (LNSC) — actually terminate in each corpus allatum.[10,14] There is some evidence that neurons in these two groups are involved in regulation of the CA. Radiocautery of either group or of the axon tracts leading from them will stimulate JH biosynthesis by the CA of virgin females.[15] Neurons in the subesophageal ganglion may not be directly involved in the regulation of JH biosynthesis and release. In *Periplaneta americana* the function of the CA was not altered by severance of NCA II,[16] and in *D. punctata* this procedure did not prevent activation of the CA.[17] No information is available concerning the role of tritocerebral neurons in the regulation of CA activity.

More recently, intracellular recording and dye injection techniques have been employed to identify neurons with terminals in the CA of *D. punctata* and to study their physiology.[18] Neurosecretory neurons in the protocerebrum of cockroaches are located easily with microelectrodes because their somata are close to the surface of the brain and because they display long-duration somatic action potentials. Krauthamer[19,20] has shown that neurons in the PI of *P. americana* which generate somatic action potentials contain numerous large, electron-opaque neurosecretory granules, whereas those neurons which do not show somatic action potentials lack neurosecretory granules. Thompson et al.[18] injected the dye Lucifer yellow into electrically active somata in the protocerebrum of *D. punctata* to locate their terminal branches. Most of the neurons of the PI labeled in this manner showed extensive terminal arborizations in the contralateral corpus cardiacum (CC) (Figure 2B). In a few cases, axons of neurons passed through the contralateral CC and hypocerebral ganglion to the recurrent nerve and lacked any terminal branches, and a few neurons showed terminals in both the contralateral CC and CA. In no case were the terminals of the neurons originating in the PI restricted only to the contralateral CA. Some neurons of the PL had extensive terminal arborizations in the ipsilateral CA and, in some cases, the terminals appeared to be restricted to the CA (Figure 2A). It was also found that intrinsic neurons of the CC had terminal branches in the cap region of the CA as well as throughout the CC[18] (Figure 2C). Because of their close proximity to the CA, these neurons previously had been undetected by other methods of visualization.

One interesting feature of the protocerebral neurosecretory system of *D. punctata* is that small groups of neurons are coupled by low-resistance pathways. Thompson et al.[18] observed that injecting dye into one cell body labeled one or more other cell bodies about 50% of the time. Also, when electrical current was injected into one neuron, membrane responses could be recorded from other neurons nearby. This provides another parallel with the vertebrate hypothalamic-pituitary system, in which small groups of neurosecretory cells in the hypothalamus are coupled by low-resistance pathways.[22] Although the functional significance of this coupling is unknown, it has been proposed that such coupling may be a means of coordinating metabolic and electrical activity among neurons that secrete the same hormone.[22]

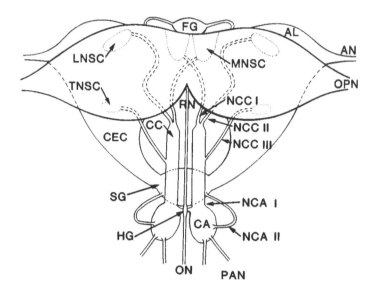

FIGURE 1.   Diagramatic representation of brain-retrocerebral complex of *D. punctata* (dorsal view). AL: antennal lobe; AN: antennal nerve; CA: corpora allata; CC: corpora cardiaca; CEC: circumoesophageal connective; FG: frontal ganglion; HG: hypocerebral ganglion; LNSC: lateral neurosecretory cells located in the pars lateralis (PL); MNSC: medial neurosecretory cells located in the pars intercerebralis (PI); NCA: nervi corporis allati; NCC: nervi corporis cardiaci; ON: esophageal nerve; OPN: optic nerve; PAN: postallatal nerves; RN: recurrent nerve; SG: subesophageal ganglion; TNSC: tritocerebral neurosecretory cells.

III. CALCIUM AND CYCLIC AMP

A major advance in the study of JH production in insect species came with the development of an *in vitro* radiochemical assay sensitive enough to accurately measure the quantity of JH produced by a single pair of CA[7,23,24] over a short interval. This has permitted the precise determination of the rates of JH biosynthesis at different stages in the life cycle of various species, as well as the examination of factors which may be involved in the control of JH biosynthesis and release (see, for example, References 3 and 25). Recently, the role of the putative second messengers calcium and cAMP in these processes has been examined in *D. punctata*.

Kikukawa et al.[26] have shown that release of JH by the CA is almost totally inhibited in the absence of extracellular calcium ions and increases in a dose-dependent manner, with maximal release occurring between 3 and 5 m*M* calcium. Furthermore, when CA are incubated in medium containing either the calcium channel blocker lanthanum or the calcium ionophore A23187, JH release is strongly inhibited. Other agents known to alter calcium flux across the cell membrane are also able to modulate JH release *in vitro*. In calcium-free medium, rates of JH release fall to very low levels and only small quantities of JH are retained within the glands. However, stimulation of JH biosynthesis by the addition of the JH precursor farnesoic acid[23,24] in calcium-free medium results in substantial release of JH and in a significant accumulation of both JH and its immediate precursor, methyl farnesoate, within the glands. Thus, calcium ions play a role in the regulation of both the biosynthesis of JH and its release from the CA *in vitro*. Whether changes in intracellular calcium ion concentration regulate biosynthesis or release of JH *in vivo* remains to be determined. Provided that extracellular calcium is present *in vitro*, the CA of *D. punctata* store little JH (~10% of that released per hour) and the rate of release is precisely correlated with the rate of biosynthesis over the entire range of CA activity.[7,24]

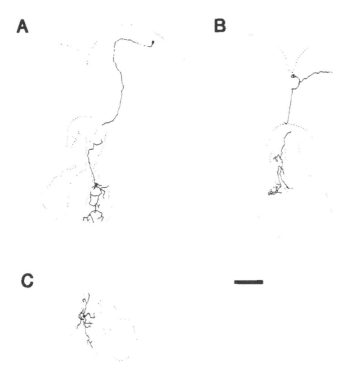

FIGURE 2. Neurons labeled by injecting Lucifer yellow into nerve terminal varicosities on the surface of the CA. (A) A lateral neurosecretory cell; (B) a medial neurosecretory cell; (C) an intrinsic neuron of the CC. Because these neurons were labeled by injecting terminal varicosities, the dendrites in the brain (A and B) were not labeled extensively. (Scale bar = 200 μm.)

The CA of *D. punctata* contain high levels of cAMP (up to 0.4 pmol per pair) and show a dose-dependent accumulation of cAMP following incubation in medium containing the adenylate cyclase activator forskolin and the phosphodiesterase inhibitor 3-isobutyl-1-methylxanthine (IBMX).[27] A number of observations indicate that cAMP may be involved in the regulation of JH biosynthesis and release by the CA. Incubation of CA of adult females with forskolin, IBMX, or 8-bromo-cAMP causes a rapid, dose-dependent inhibition of JH biosynthesis.[27] Incubation of CA in medium with an elevated potassium ion concentration also results in an increase in the cAMP content of the glands[28] and an inhibition of JH biosynthesis.[29] In addition, treatment of CA with aqueous extracts of the protocerebral lobes of the brain results in a dose-dependent elevation of cAMP content and an inhibition of JH biosynthesis.[30] These observations support the hypothesis that cAMP acts as a second messenger for an allatostatin that is released from neurosecretory cell terminals within the CA.

Additional experimental results indicate that calcium ions may play a role in the control of JH biosynthesis by cAMP. Incubation of CA in calcium-free medium causes a dramatic increase in cAMP content. On the other hand, elevated extracellular calcium levels substantially reduce the ability of brain extracts to inhibit JH biosynthesis *in vitro*, although such elevated levels of calcium do not block the ability of brain extracts to elevate the cAMP content of the glands.[30] These results suggest the existence of at least two separate regulatory pathways for JH biosynthesis, one dependent on and the other independent of calcium ions.

IV. ELECTROPHYSIOLOGY OF CORPORA ALLATA CELLS

The CA of adult female *D. punctata* are composed of several thousand irregularly shaped parenchymal cells. Most of these are up to 10 μm in length, although a few spindle-shaped cells are up to 20 μm in length. The parenchymal cells are coupled extensively by low-resistance intercellular pathways which permit the spread of electrical current and the fluorescent dye Lucifer yellow from cell to cell throughout the glands.[31] It is likely that such intercellular communication is mediated by gap junctions, since both transmission and freeze-fracture electron microscopy have revealed the presence of numerous gap junctions between CA cells.[31,32]

When electrical current is injected into a single, electrically isolated cell, a change in the membrane potential of the cell is generated as the current flows across the cell membrane to ground. The amplitude of this passive membrane response, or electrotonic potential, is the product of the current injected and the total membrane resistance (input resistance) of the cell (Ohm's law). When current is injected into a cell which is part of a network of electrically coupled cells, some current will flow through gap junctions into contiguous cells and across their cell membranes to ground, thus eliciting an electrotonic potential. The amplitude of this response will depend on both the gap junctional conductance and the conductance of the nonjunctional membrane in a manner which varies with the topology of the network of coupled cells.[33] Because of the small size of the CA cells of *D. punctata*, it has not been possible to insert two microelectrodes into the same cell without causing significant damage to the cell membrane. Thus, the input resistance of single CA cells and the coupling coefficient between cells (the ratio of the membrane response of a coupled cell to the membrane response to the injected cell during a current pulse) have not been determined accurately. It is possible to inject current into one cell and record membrane responses from other cells in the gland, however. In the CA of *D. punctata*, when current is injected into one cell and electrotonic potentials from other cells are recorded with a second microelectrode, there is little decrement in amplitude as the recording electrode is moved farther away from the current-passing electrode. The distance over which the electrotonic potential would show a significant decrease in amplitude is greater than the length of the gland itself.[31] Thus, it is likely that gap junctional conductance is large relative to nonjunctional conductance and that the coupling coefficient is close to unity. All of the cells of a gland appear to be coupled together, forming an electrical syncytium, since not a single instance of an uncoupled cell has been observed in thousands of penetrations.[44] However, the possibility that some cells in the center of a gland are uncoupled from the rest has not been eliminated. Electrical coupling was observed in every CA preparation from animals of the fourth stadium through day 13 of adulthood. The presence of such cell-to-cell coupling is a common feature of endocrine glands.[34] In many cases, however, electrical coupling is not as extensive as that found in the CA of *D. punctata*.[35]

When current pulses of differing polarity and amplitude are injected into one CA cell and electrotonic potentials from another cell are recorded, the resulting current/voltage (I/V) plot appears similar to that of many nonexcitable cell types (Figure 3). Current-voltage relationships are linear when negative current (hyperpolarizing) is injected, but show a pronounced decrease in slope with increasing amplitude of positive current (depolarizing) injection. This could be due to either an increase in the nonjunctional conductance of the cell membranes of the CA or a decrease in the gap junctional conductance, or both. The observation that the time constant of decay of the electrotonic potential (the time required for the electrotonic potential to decay to 1/e of its steady-state value) decreases with increasing amplitude of depolarizing current pulses (Figure 3) indicates that an increase in membrane conductance accounts for much of the decrease in slope. The time constant is the product of membrane resistance and capacitance and should not be influenced by gap junctional conductance, nor should membrane capacitance change. However, the possibility that a decrease in gap junctional conductance occurs during positive current injection cannot be excluded. In many tissues an elevation of intracellular calcium ion concentration has been

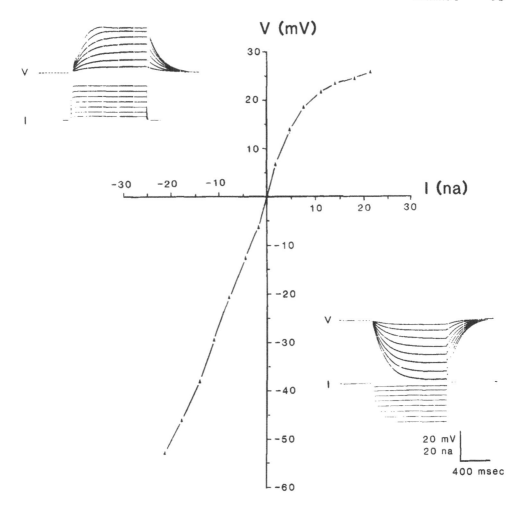

FIGURE 3. Typical current/voltage (I/V) relationships of the network of electrically coupled CA cells. I/V relationships are linear in the hyperpolarizing direction, but show a decrease in slope in the depolarizing direction. These data are from a corpus allatum of a day 11 mated female. Examples of the data used to construct such curves are shown in the upper left (depolarizing) and lower right (hyperpolarizing) quadrants; these were obtained from another preparation. Resting potential is set at 0 on the y axis.

shown to decrease gap junctional conductance, although very high levels of calcium typically are required to effect a significant change.[36] Thus, if calcium ions were to enter the CA cells during depolarizing current pulses (see below), a reduction in gap junctional conductance might occur. Also, although not a common phenomenon, several cases of rectifying gap junctions have been described.[36] If the gap junctions between CA cells were to pass negative current more easily than positive current, then a decrease in slope of the I/V curve might occur in the depolarizing direction.

In normal cockroach saline,[19] the "input resistance" (ratio of electrotonic potential to current injected) of the network of electrically coupled CA cells ranges from 1 to 3 MΩ.[31] Values of up to 10 MΩ have been observed following the addition of the potassium channel blockers tetraethylammonium (TEA) ions and 4-aminopyridine (4-AP) to the saline.[44] Under these conditions, positive current pulses may elicit active responses, or action potentials, from the cell membranes of the CA (Figure 4A).[37] In most other excitable cells, action potentials are attributable to the opening of voltage-sensitive cation channels followed by an influx of either sodium ions or calcium ions, or both, down their electrochemical gradients

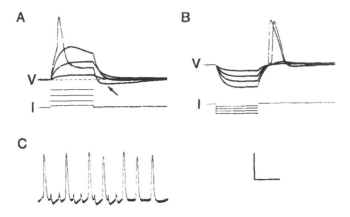

FIGURE 4.   Examples of action potentials generated by CA cell membranes. (A) In the presence of TEA (40 mM) and 4-AP (200 μM), depolarizing current pulses elicit action potentials above a threshold depolarization of about 25 mV. These action potentials are followed by a large decrease in the electronic potential and an after-hyperpolarization at the end of the current pulse (arrow). (B) In the presence of TEA, 4-AP, and barium ions (15 mM), action potentials may follow hyperpolarizing pulses. (C) In some preparations, spontaneous action potentials occur following perfusion with saline containing TEA, 4-AP, and barium ions. Scale bars: (A,B) vertical — upper trace, 20 mV; lower trace, 20 nA; horizontal — 1.0 s. (C) vertical — 10 mV; horizontal — 4.0 s.

into the cells.[38] In the CA cells of *D. punctata*, action potentials occur at a threshold depolarization of 15 to 30 mV and are followed by a large decrease in electrotonic potential and a pronounced after-hyperpolarization (Figure 4A). Several observations indicate that these action potentials are mediated by both sodium ions and calcium ions flowing through voltage-sensitive channels. The spikes are blocked by the sodium channel blocker tetrodotoxin and sodium-free saline. They are also blocked by the calcium channel blockers cobalt and lanthanum ions.[37] Barium ions, on the other hand, are known to pass through calcium channels easily and to block potassium channels.[38] In the presence of barium ions, the CA cells initially become hyperexcitable. Action potentials increase in amplitude and duration, and spontaneous spikes may occur (Figure 4B). Spikes also may occur following hyperpolarizing current pulses (Figure 4C). The decrease in electronic potential and the after-hyperpolarization which follows action potentials could be the result of the entry of calcium into the CA cells followed by the activation of a calcium-sensitive potassium conductance,[39] although this remains to be proved.

The above experimental findings are preliminary in nature, but they indicate that the CA cell membranes of *D. punctata* do contain a variety of ion channels and that some of them are voltage sensitive. Whether or not action potentials occur under normal circumstances *in vivo* is not known.

V. RESPONSES OF CORPORA ALLATA CELLS TO NERVE STIMULATION

Electrical stimulation of axons innervating the CA of *D. punctata* produces two distinct types of response. Stimulation of NCA I, containing axons of the medial and lateral protocerebral neurosecretory cells, elicits a hyperpolarization of the CA cells and an increase in the electrotonic potential (Figure 5A). These responses occur at a sharp threshold as stimulating voltage is increased from zero, and they have a relatively long latency. On the

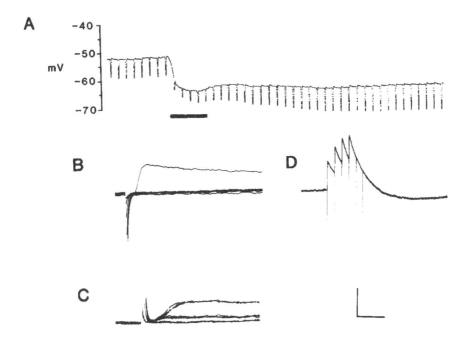

FIGURE 5. Responses of CA cell membranes to electrical stimulation of axons innervating the CA. (A) Current pulses (-5.0 nA, 1700-ms duration) were injected into one CA cell every 15 s, and electronic potentials (downward deflections) were recorded from another CA cell with a second microelectrode. Electrical stimulation of NCA I (0.5-ms pulses at 5 Hz; bar) elicited a hyperpolarization and a gradual increase in the size of the electrotonic potential. (B,C,D) stimulation of unidentified axons elicited apparent EPSPs in CA cells. (B) Six superimposed sweeps of the oscilloscope are shown. Stimulating voltage was increased slightly after each sweep. The final (largest) stimulus pulse elicited a short-latency depolarization of the CA cells. (C) Several sweeps of the oscilloscope are shown. Stimulus voltage was varied between each sweep. Two distinct amplitudes of EPSPs are apparent. (D) The all-or-none nature of the EPSPs is shown. Six stimulus pulses were delivered, but only the first four elicited an EPSP. Scale bars: (A) vertical — 10 mV; horizontal — 1.5 min; (B) 4 mV, 20 ms; (C) 4 mV, 10 ms; (D) 4 mV, 400 ms.

other hand, short-latency depolarizing responses also may be elicited by electrical stimulation (Figure 5B,C). These also occur at a sharp threshold and resemble conventional excitatory postsynaptic potentials (EPSPs).[40] The location of the axons eliciting the latter responses is uncertain. Higher stimulating voltages are typically required to elicit such responses, and fine axon branches within the CA may be activated by current spread from the nearby stimulating electrode.

These observations indicate that activity in neurons innervating the CA of *D. punctata* can produce both depolarizing and hyperpolarizing responses. The neurotransmitters involved and the ionic basis of these responses remain unknown, as does their role in the control of JH biosynthesis and release.

Changes in membrane potential of CA cells have been correlated with various conditions which alter JH release *in vitro*. High-potassium saline produces a marked inhibition of JH biosynthesis and release as well as an increase in the cAMP content of the glands, as reviewed above. High-potassium saline depolarizes CA cells and reduces the electrotonic potential (Figure 6B). On the other hand, aqueous extracts of the brain, which inhibit JH release and elevate cAMP content *in vitro*,[29] elicit a hyperpolarization of the CA cells and an increase in the electrotonic potential (Figure 6A). The application of forskolin and IBMX to the CA also causes a hyperpolarization of the CA cells and an increase in electrotonic potential, as well as an elevation of cAMP and an inhibition of JH release.[31]

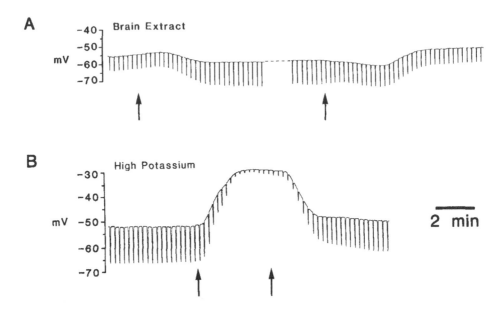

FIGURE 6.   Responses of CA cells to brain extract and high-potassium saline. (A) Current pulses (-5.0 nA, 1700 ms) were injected into one CA cell and electrotonic potentials recorded from another cell with a second microelectrode. The preparation initially was perfused with normal saline. Brain extract (0.5 brain equivalents per 100 μl saline) entered the preparation dish at the first arrow. At the second arrow, perfusion with normal saline resumed. A total of 8.0 min was deleted from the continuous record. (B) Current pulses were injected into one CA cell, as in A. The preparation was perfused continually, first with normal saline (12 m*M* K$^+$), then with high-potassium saline (56 mM K$^+$) (first arrow), and, finally, with normal saline (second arrow).

Thus, it appears that JH biosynthesis and release bear no simple relationship to the membrane potential of the CA cells *in vitro. In vivo,* however, membrane potential may act to regulate calcium flux across CA cell membranes. It is likely that the level of intracellular calcium ions plays an important role in controlling JH biosynthesis and release as well as other cellular functions.

VI. SUMMARY AND CONCLUSIONS

Our understanding of the neural control of JH biosynthesis by the CA of *D. punctata* is far from complete; however, a general picture is beginning to emerge. Anatomical studies have revealed groups of neurons in three distinct locations within the nervous system which have terminal arborizations on or in the CA. Electrical stimulation of axon tracts near the CA indicates that at least some of these neurons make functional contact with CA cells: both long-term hyperpolarizing responses and brief depolarizing responses may be elicited by such stimulation. The ability to selectively stimulate these groups of neurons should aid in identifying which neurons make functional contact with CA cells and which neurotransmitters are involved. Also, measurements of CA cell membrane potentials and electrotonic potentials provide a relatively simple assay which can be used in attempting to purify neurotransmitters or neurohormones involved in control of the CA.

While the precise effects of electrical stimulation on JH biosynthesis remain unknown, there is some evidence for the involvement of Ca^{2+} and cAMP in the transduction of neural signals. The application of potassium channel blockers to CA cells reveals that they are not electrically unexcitable, as they initially appear to be, which is also the case with several other cell types.[41] This is consistent with the idea that CA cell membranes have voltage-sensitive calcium channels and that membrane potential is an important determinant of the level of ionized calcium in the cytoplasm. It also appears that CA cell membranes have receptors which are coupled to adenylate cyclase and that neurotransmitters or neurohormones act in the control of intracellular levels of cAMP.

The usefulness of the CA of cockroaches as potential biomedical models is somewhat limited by the fact that so much more is known about vertebrate endocrine glands than about insect CA. Also, CA cells are relatively small and have not yet been grown successfully in tissue culture. However, there appear to be many fundamental similarities between steroid-secreting endocrine cells of vertebrates and the CA cells of insects. The biosynthetic pathway for JH appears to be identical to that for sterol production up to the branchpoint of farnesyl pyrophosphate. Also, there is increasing evidence that intracellular signal molecules and control mechanisms have been highly conserved during the course of evolution. Thus, the CA may provide a system in which to develop alternate strategies to regulate sterol biosynthesis.

The parenchymal cells of the CA are intercoupled by gap junctions. This is a feature common to many endocrine glands, but coupling in the CA appears to be more extensive than in most vertebrate endocrine glands. It has recently been demonstrated that antibodies to gap junction proteins can alter the properties of gap junctions *in vivo* and produce developmental abnormalities in several invertebrate and vertebrate species.[42] The CA may provide a good model system in which to study the role of gap junction communication in the control of hormone biosynthesis by the nervous system, and other cellular processes as well. For example, when the CA of adult female cockroaches undergo a cycle of JH biosynthesis, there is an increase of about 50% in both cell size and cell number.[43] Thus, the nervous system is acting to control the size of the CA and, again, gap junction communication may be involved.

REFERENCES

1. **Willey, R. B.**, The morphology of the stomodeal nervous system in *Periplaneta americana* (L.) and other Blattaria, *J. Morphol.*, 13, 219, 1961.
2. **Cassier, P.**, The corpora allata of insects, *Int. Rev. Cytol.*, 57, 1, 1979.
3. **Tobe, S. S. and Stay, B.**, Structure and regulation of the corpus allatum, *Adv. Insect Physiol.*, 18, 305, 1985.
4. **Engelmann, F.**, *The Physiology of Insect Reproduction*, Pergamon Press, Oxford, 1970.
5. **de Kort, C. A. D. and Granger, N. A.**, Regulation of juvenile hormone titer, *Annu. Rev. Entomol.*, 26, 1, 1981.
6. **Wigglesworth, V. B.**, Historical perspectives, in *Comprehensive Insect Physiology, Biochemistry and Pharmacology*, Vol. 7, Kerkut, G. A. and Gilbert, L. I., Eds., Pergamon Press, Oxford, 1985, 1.
7. **Tobe, S. S. and Pratt, G. E.**, Dependence of juvenile hormone release from corpus allatum on intraglandular content, *Nature (London)*, 252, 474, 1974.
8. **Tobe, S. S. and Pratt, G. E.**, The influence of substrate concentrations on the rate of insect juvenile hormone biosynthesis by corpora allata of the desert locus *in vitro*, *Biochem. J.*, 114, 107, 1974.
9. **Feyereisen, R.**, Regulation of juvenile hormone titer: synthesis, in *Comprehensive Insect Physiology, Biochemistry and Pharmacology*, Vol. 7, Kerkut, G. A. and Gilbert, L. I., Eds., Pergamon Press, Oxford, 1985, 391.
10. **Lococo, D. J. and Tobe, S. S.**, Neuroanatomy of the retrocerebral complex, in particular the pars intercerebralis and partes laterales in the cockroach *Diploptera punctata* Eschscholtz (Dictyoptera: Blaberidae), *Int. J. Insect Morphol. Embryol.*, 13, 65, 1984.
11. **Tobe, S. S and Stay, B.**, Corpus allatum activity *in vitro* during the reproductive cycle of the viviparous cockroach *Diploptera punctata* (Eschscholtz), *Gen. Comp. Endocrinol.*, 31, 138, 1977.
12. **Carrow, G. M., Calabrese, R. L., and Williams, C. M.**, Spontaneous and evoked release of prothoracicotropin from multiple neurohemal organs of the tobacco hornworm, *Proc. Natl. Acad. Sci. U.S.A.*, 78, 5866, 1981.

13. **Lococo, D. J.**, Neuronal and Intercellular Communication in the Corpus Allatum of *Diploptera punctata*, Ph.D. thesis, University of Toronto, Toronto, Ontario, Canada, 1985.

14. **Lococo, D. J. and Tobe, S. S.**, Retrograde and orthograde axon transport by brief exposure to nickel chloride: methodology and parameters for success in the brain-retrocerebral complex of the cockroach *Diploptera punctata*, *J. Insect Physiol.*, 30, 635, 1984.

15. **Ruegg, R. P., Lococo, D. J., and Tobe, S. S.**, Control of corpus allatum activity in *Diploptera punctata*: roles of the pars intercerebralis and pars lateralis, *Experientia*, 39, 1329, 1983.

16. **Pipa, R. L.**, Neural influences on corpus allatum activity and egg maturation in starved virgin *Periplaneta americana*, *Physiol. Entomol.*, 7, 449, 1982.

17. **Stay, B. and Tobe, S. S.**, Control of juvenile hormone biosynthesis during the reproductive cycle of a viviparous cockroach. I. Activation and inhibition of corpora allata, *Gen. Comp. Endocrinol.*, 33, 531, 1977.

18. **Thompson, C. S., Lococo, D. J., and Tobe, S. S.**, Anatomy and physiology of neurons terminating in the corpora allata of the cockroach *Diploptera punctata*, *J. Comp. Neurol.*, 261, 120, 1987.

19. **Krauthamer, V.**, Electrophysiology of identified neurosecretory and non-neurosecretory cells in the cockroach pars intercerebralis, *J. Exp. Zool.*, 234, 207, 1985.

20. **Krauthamer, V.**, Morphology of identified neurosecretory and non-neurosecretory cells in the cockroach pars intercerebralis, *J. Exp. Zool.*, 234, 221, 1985.

21. **Scharrer, B.**, Neurosecretion. XIII. The ultrastructure of the corpus cardiacum of the insect *Leucophaea maderae*, *Z. Zellforsch.*, 60, 761, 1963.

22. **Renaud, L. P., Bourgne, C. W., Day, T. A., Ferguson, A. V., and Randle, J. C. R.**, Electrophysiology of mammalian hypothalamic supraoptic and paraventricular neurosecretory cells, in *The Electrophysiology of the Secretory Cell*, Poisner, A. M. and Trifaro, J. M., Eds., Elsevier, New York, 1985, 165.

23. **Pratt, G. E. and Tobe, S. S.**, Juvenile hormones radiobiosynthesized by corpora allata of adult female locusts *in vitro*, *Life Sci.*, 14, 575, 1974.

24. **Feyereisen, R., Tobe, S. S., and Friedel, T.**, Farnesoic acid stimulation of C_{16} juvenile hormone biosynthesis by corpora allata of adult female *Diploptera punctata*, *Insect Biochem.*, 11, 401, 1981.

25. **Kikukawa, S. and Tobe, S. S.**, Juvenile hormone biosynthesis in female larvae of *Diploptera punctata* and the effect of allatectomy on haemolymph ecdysteroid titre, *J. Insect Physiol.*, 32, 981, 1986.

26. **Kikukawa, S., Tobe, S. S., Solowiej, S., Rankin, S. M., and Stay, B.**, Calcium as a regulator of juvenile hormone biosynthesis and release in the cockroach *Diploptera punctata*, *Insect Biochem.*, 17, 179, 1987.

27. **Meller, V. H., Aucoin, R. R., Tobe, S. S., and Feyereisen, R.**, Evidence for an inhibitory role of cyclic AMP in the control of juvenile hormone biosynthesis by cockroach corpora allata, *Mol. Cell. Endocrinol.*, 43, 155, 1985.

28. **Aucoin, R. R.**, The Role of Cyclic AMP and Calcium in the Regulation of JH Biosynthesis in *Diploptera punctata*, Master's thesis, University of Toronto, Toronto, Ontario, Canada, 1986.

29. **Rankin, S. M., Stay, B., Aucoin, R. R., and Tobe, S. S.**, *In vitro* inhibition of juvenile hormone synthesis by corpora allata of the viviparous cockroach, *Diploptera punctata*, *J. Insect Physiol.*, 32, 151, 1986.

30. **Aucoin, R. R., Rankin, S. M., Stay, B., and Tobe, S. S.**, Calcium and cyclic AMP involvement in the regulation of juvenile hormone biosynthesis in *Diploptera punctata*, *Insect Biochem.*, 17, 965, 1987.

31. **Lococo, D. J., Thompson, C. S., and Tobe, S. S.**, Intercellular communication in an insect endocrine gland, *J. Exp. Biol.*, 121, 407, 1986.

32. **Johnson, G. D., Stay, B., and Rankin, S. M.**, Ultrastructure of corpora allata of known activity during the vitellogenic cycle in the cockroach *Diploptera punctata*, *Cell Tissue Res.*, 239, 317, 1985.

33. **Socolar, S. J.**, The coupling coefficient as an index of junctional conductance, *J. Membr. Biol.*, 34, 29, 1977.

34. **Meda, P., Perrelet, A., and Orci, C.**, Gap junctions and cell-to-cell coupling in endocrine glands, *Mod. Cell Biol.*, 3, 131, 1984.

35. **Petersen, O. H.**, *The Electrophysiology of Gland Cells*, Monogr. Physiol. Soc. No. 36, Academic Press, London, 1980, 253.

36. **Spray, D. C. and Bennett, M. V. L.**, Physiology and pharmacology of gap junctions, *Annu. Rev. Physiol.*, 47, 281, 1985.

37. **Thompson, C. S. and Tobe, S. S.**, Electrical properties of membranes of cells of the corpora allata of the cockroach *Diploptera punctata*: evidence for the presence of voltage-sensitive calcium channels, in *Insect Neurochemistry and Neurophysiology*, Borkovec, A. B. and Gelman, D. B., Eds., Humana Press, Clifton, NJ, 1986, 375.

38. **Hille, B.**, *Ionic Channels of Excitable Membranes*, Sinauer Associates, Sunderland, MA, 1984.

39. **Petersen, O. H. and Maruyama, Y.**, Calcium-activated potassium channels and their role in secretion, *Nature (London)*, 307, 693, 1984.

40. **Thompson, C. S., Lococo, D. J., and Tobe, S. S.**, Cell coupling and electrical activity in the corpora allata of the cockroach *Diploptera punctata*, *Neurosci. Abstr.*, 10, 480, 1984.

41. **O'Donnell, M. J.,** Potassium channel blockers unmask electrical excitability of insect follicles, *J. Exp. Zool.*, 245, 137, 1988.
42. **Marx, J. L.,** Symposium focuses on genes in development: gap junctions needed for development, *Science*, 239, 1493, 1988.
43. **Szibbo, C. M. and Tobe, S. S.,** Cellular and volumetric changes in relation to the activity cycle in the corpora allata of *Diploptera punctata*, *J. Insect Physiol.*, 27, 655, 1981.
44. **Thompson, C. S.,** unpublished data.

Chapter 6

NEUROANATOMICAL METHODS

Jonathan M. Blagburn and David B. Sattelle

TABLE OF CONTENTS

I. INTRODUCTION

With limited space available to cover an area to which several volumes have already been devoted, we will mainly focus on techniques with which we have personal experience and which we know are applicable to cockroaches. In view of the need for brevity, we apologize to those workers who would like to see more detailed methods included and refer them to more comprehensive works — in particular, two excellent volumes devoted to insect neuroanatomy, edited by Strausfeld and Miller[1] and by Strausfeld.[2]

In this chapter, we examine optical sectioning of unstained nervous tissue and general staining methods, together with more specific histochemical techniques for localizing cell types and cell components. Recent advances in immunocytochemistry offer the prospect of detailed mapping of cell constituents, especially following the successful application of monoclonal antibodies to particular peptides and even amino acids. Autoradiography and *in situ* hybridization techniques also offer complementary approaches for the localization of specific cell components. Approaches to neuronal morphology include the Golgi method, degeneration studies, and a wide range of intracellular marking techniques, most notably cobalt, horseradish peroxidase (HRP), and fluorescent dye injection. While advances in electron microscopy continue to add to our understanding of insect neuroanatomy, they are only discussed briefly in the present chapter.

The majority of the techniques described have been applied to vertebrate and invertebrate preparations; some, such as cobalt staining or fluorescent dye injection, were first attempted in insects or molluscs and subsequently acquired importance in the biomedical field. For studies which utilize intracellular microelectrodes to combine electrophysiological recording with dye injection, insect ganglia are particularly attractive because of the relatively small number of nerve cells, many of which are of large size and unique identity, organized with their cell bodies in a peripheral rind or cortex.

II. OPTICAL SECTIONING OF NERVOUS TISSUE WITH NOMARSKI OPTICS

This technique, although limited in its usefulness to tissues of favorable dimensions and degree of transparency, is the most rapid of the anatomical techniques since it is performed on living material. Optical sectioning with Nomarski optics is ideally suited to identifying cell bodies or other regions of neurons prior to impalement with intracellular microelectrodes[3-5] (Figure 1). A tissue which is amenable to being examined in this way is also likely to be a suitable candidate for studies utilizing voltage- or calcium-sensitive dyes and optical recording techniques.

Insect ganglia and those of many other invertebrates are organized so that the inner neuropil contains the axons and dendritic branches and their synaptic connections. The neuronal cell bodies are generally devoid of synaptic contacts and are concentrated in an outer layer, in contrast to the cell bodies of vertebrate neurons. This arrangement gives the advantage that the cell body forms a convenient and relatively easy site in which to place intracellular microelectrodes. However, it is the axons and dendrites which are the sites of functional interactions. The particular advantage of Nomarski optics is that the technique allows the experimenter to see structures which are inside the tissue; this is particularly useful when it is necessary to place microelectrodes in them. Vertebrate central nervous system (CNS) tissues are generally too large to be suitable for this technique, although there are notable exceptions in the cases of larval fish and amphibians. However, Nomarski optics can be useful in visualizing the neurons within brain slices.

It is necessary to use a microscope equipped with Nomarski (differential interference contrast) optics. Zeiss® Nomarski is recommended. A 40× water-immersion objective lens of approximately 0.6 to 0.7 numerical aperture (N.A.) gives a suitable magnification, allowing axons of 5 μm diameter to be visualized easily, but also giving a good overview of the tissue. In order to obtain good visibility within the tissue, it should be no thicker than 200 μm and should not contain pigment. It may be necessary to adjust the osmolarity of the saline slightly in order to improve the visibility of cells within the tissue. Under good conditions it is possible to visualize all neuronal cell bodies in a living ganglion from a first- or second-instar cockroach nymph and, by focusing through the ganglion, neurites, axons, and major dendritic branches are revealed.

If it is only necessary to examine the tissue, any type of microscope will suffice; however, if examination is to be coupled with intracellular recording or dye injection, then two important conditions should be fulfilled. First, the working distance of the objective should be at least 1 to 2 mm. This allows microelectrodes to be mounted at an angle of 30 to 45° such that they readily clear the walls of the experimental chamber. Suitable objectives include the Zeiss® Achromat 40×/0.75 N.A. water immersion. Second, if more than one cell is to be impaled with microelectrodes, the micromanipulators must be fixed relative to the stage. Most modern upright microscopes are designed to focus by moving the stage, an arrangement

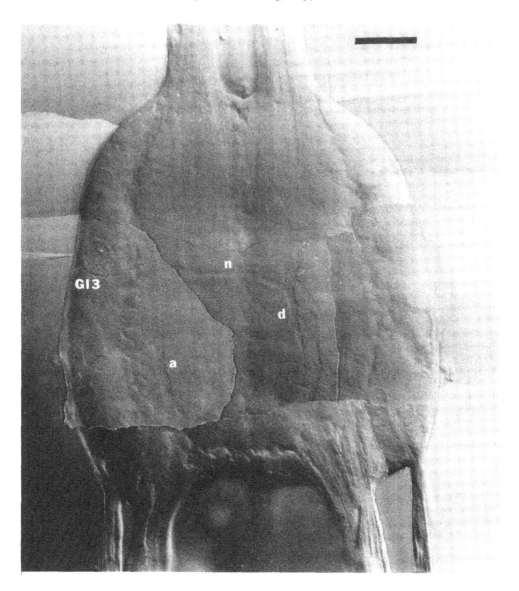

FIGURE 1.   Montage of photographs of the terminal abdominal ganglion of first instar *Periplaneta americana*, viewed with Nomarski optics. Some of the different parts of the picture were taken at slightly different planes of focus. The anterior of the ganglion gives rise to the interganglionic connectives (top), while the cercal nerves arise from the posterior. The tip of a microelectrode can been seen on the left, touching the cell body of giant interneuron 3, in which can be seen the nucleus and nucleolus. The cell body, located in the outer cortex of neuronal somata, sends a neurite (n) into the neuropil, where it crosses to the other side of the ganglion and forms dendritic branches. The primary dendrite (d) is visible. Large sensory axons (a) are also visible within the neuropil. (Scale bar: 50 μm.)

which precludes their use in multiple-electrode electrophysiological studies unless the manipulators holding the electrodes are mounted directly onto the stage. The Microtec® M2 manipulation microscope (Micro Instruments, Oxford, England) is designed specifically for such studies, while its counter-balanced head also allows the mounting of heavy camera equipment. Alternatively, an existing microscope can be modified in a workshop facility. An inverted microscope equipped with Nomarski optics may be used if the tissue is thin, but this type of microscope will not be suitable for visualizing cells on the upper surface of tissue more than about 100 μm thick.

III. GENERAL STAINING TECHNIQUES

A. METHYLENE BLUE

Intra vitam methylene blue staining can be used to stain nerve fibers in the insect nervous system, thereby providing an overview of the structures in the ganglionic neuropil.[6,7] Precise information about the shapes of individual nerve cells is best obtained using intracellular tracers (see Section IV.C).

Methylene blue (1 to 2.5 ml of a 0.08 to 1% solution of methylene blue in double-distilled water containing 0.75 to 2.25% NaCl or physiological saline) is injected into the hemocoel of the insect so that the body wall is distended. For peripheral elements, the insect is left at room temperature for 0.5 to 1 h before dissection (for central neurons, 1 to 5 h is sufficient). The best results are obtained with newly molted animals. After the staining period, the animal is rapidly dissected and flooded with fixative (12% ammonium molybdate). The tissues are transferred into fixative at 4°C for 7 to 16 h. The addition of two to four drops of 0.25% OsO_4 solution added to 50 ml ammonium molybdate improves fixation, possibly at the expense of staining density. After washing in distilled water, the tissue is dehydrated in three changes of absolute ethanol (the first change being no longer than 30 s because nonabsolute ethanol extracts the stain). After clearing in xylene the preparation is mounted in dammar-xylol, as Canada balsam, a more convenient mounting medium, also leaches out the stain.

B. ACRIDINE ORANGE (FOR LIVING CELLS)

Acridine orange (0.1 μg/ml in insect saline or culture medium) can be used to assess neuronal viability. Under UV illumination the nucleus appears orange and the cytoplasm green in living cells. This technique has been used to assess viability of dissociated cockroach neurons maintained in short-term culture[8] and is applicable to all types of dissociated neurons.

C. TOLUIDINE BLUE

Neuronal cell bodies can be localized in whole insect ganglia by staining with toluidine blue.[9,10] This technique serves as a useful preliminary to intracellular recording from identified somata and enables the morphology of single neurons (as delineated, for example, by cobalt staining [Section IV.C.1]) to be related to the overall distribution of cell bodies.

Freshly dissected ganglia are placed in staining solution at 50°C and left for 0.25 to 1 h (the optimal staining time should be determined for the particular preparation and staining conditions). The staining solution consists of 1 g toluidine blue, 6 g borax, and 1 g boric acid in 100 ml distilled water. The tissue is transferred into several changes (5 min each) of Bodian's fixative, which differentiates the stain. This should be continued under visual control until the cell bodies are seen clearly and the neuropil and nerves are white. The fixative consists of 5 ml formalin, 5 ml glacial acetic acid, and 90 ml 80% ethanol. The fixed tissue is dehydrated in two changes of 90% ethanol (5 min each) and two changes of 100% ethanol (10 min each), then cleared in methyl benzoate; the preparations are then mounted in neutral Canada balsam or a synthetic substitute.

D. ETHYL GALLATE

This staining technique is useful for revealing the tracts and commissures of insect ganglia.[11,12] The tissue is prepared as for electron microscopy (see Section V.A). After osmication, the tissue is washed in buffer and then stained for 2.5 h in ethyl trihydrobenzoate (ethyl gallate) before dehydrating and embedding in soft Araldite® and sectioning at 10 μm.

E. REDUCED SILVER IMPREGNATION TECHNIQUES

These somewhat complex methods provide a detailed picture of neuronal cell bodies, large fibers, fiber tracts, and neuropilar regions. A major disadvantage of these techniques is that they are capricious. The Bodian protargol technique is perhaps the simplest; a protocol is given in brief here. The Holmes and Ungewitter reduced silver techniques give similar results and have been dealt with in detail elsewhere.[13,14]

1. Bodian Protargol Technique

This technique has been described in detail by Gregory,[15,16] who has worked extensively with the nervous system of *Periplaneta americana*. A brief description of the technique is given here. The ganglia are fixed for 16 to 24 h at room temperature in "aged" alcoholic Bouin, which consists of 5 ml formaldehyde, 25 ml ethanol, 5 ml acetic acid, 5 ml ethyl acetate, 15 ml diethoxymethane, and 0.5 g picric acid, brought up to 100 ml with distilled water. The tissue is dehydrated, cleared, and embedded in paraffin wax; 20-μm sections are then cut, mounted on slides, and dewaxed.

The sections are impregnated for 18 to 24 h at 37°C in 2% protargol solution, adjusted to pH 8.4 with 10% ammonium hydroxide and containing 1 to 4 g copper; they are then washed briefly in distilled water. The stain is developed for 10 to 15 min in freshly made developer solution (containing 3 or 10% sodium sulfite in 1% hydroquinone, depending on the amount of copper used). After three 3-min washes in distilled water, all the previous stages are repeated. Intensification in 1% gold chloride solution for 15 min is followed by a rinse in distilled water, then reduction in 2% oxalic acid solution for 10 min. Finally, the sections are washed in water, then treated with 5% sodium thiosulfate solution for 10 to 15 min to remove residual silver. Three 3-min washes in distilled water are followed by de-hydration, clearing, and mounting in the usual way.

IV. LOCALIZATION OF NEURONAL CONSTITUENTS

An important class of neuroanatomical methods reveal constituents of neurons. In most cases the molecule of interest is a neurotransmitter, neuromodulator, or neurohormone (or the synthetic or degradative enzyme of the above). However, it is also possible to localize the functional activity of neurons anatomically, using the deoxyglucose technique. In a few cases the substance of interest is spread throughout the neuron, and if the technique employed to localize it has high resolution, details of neuronal shape are obtained. More often, the localization technique must be coupled with other neuroanatomical methods if the anatomy of the neurons is to be studied.

A. *IN VIVO* OBSERVATIONS

Using dark-field or oblique illumination, many neurons containing neurosecretory material are opalescent in appearance — e.g., the giant dorsal bilateral cells in the third thoracic ganglion of the cockroach.[17]

B. HISTOCHEMICAL METHODS
1. Histofluorescence
a. Acridine Orange (for Neurosecretory Material)

Living nervous tissue can be stained in 0.1 mg/ml acridine orange in saline for 1 min.[18] On observation under UV light with a fluorescence microscope, neurosecretory material appears as red fluorescing granules on a background of pale green cytoplasm and darker green or orange nuclei. The fluorescence bleaches in about 5 min.

b. Falck-Hillarp Method for Biogenic Monoamines

The technique is used for detecting the catecholamines dopamine, noradrenaline, and adrenaline and the indolalkylamine 5-hydroxytryptamine (5-HT, serotonin). It has been used successfully on the brain of *P. americana*.[19,20] The following method is taken from Klemm.[21]

Tissues are normally freeze-dried, but small preparations such as the isolated nerve cord from embryos can be air dried at room temperature or in a vacuum. The frozen or dried specimens are placed in a 1-l glass vessel containing 5 g paraformaldehyde. The water content of the paraformaldehyde should be adjusted beforehand by heating it to 100°C for 1 h and then storing it for 1 week in a dessicator with a relative humidity of 70%. The vessel is sealed and heated to 80°C. Exposure to the resulting formaldehyde vapor for 1 to 1.5 h renders the monoamines fluorescent.

The tissues should be embedded immediately in paraffin wax or plastic in a vacuum.[22] Plastic allows thinner sections, but yields lower total fluorescence intensity and a higher background. Alternatively, tissues can be examined in whole-mounts. Under a fluorescence microscope, catecholamines exhibit a green or blue fluorescence (emission peak at 475 nm) which may appear yellow at high intensities. 5-HT produces yellow fluorescence (520 nm) and fades more rapidly.

c. Glyoxylic Acid Method for Biogenic Monoamines

This method has been developed for visualization of monoamines in whole-mounts of *Rhodnius* ganglia, but is equally applicable to the cockroach.[23] The ganglia are incubated in 5% glyoxylic acid in Ca-free, Mg-rich saline (pH 7) for 1 h at room temperature or for 8 to 12 h at 12°C. Excess liquid is removed from the tissues, which are stored over calcium carbonate for 3 to 5 d in the dark at room temperature. The dried ganglia are covered with a drop of mineral oil, then flash-heated on copper blocks at 85°C for 2 min. This should remove air from the tracheae; if it does not, the heating is repeated, although prolonged heating damages the tissue and increases background fluorescence. 5-HT fluorescence is yellow and fades rapidly; catecholamines fluoresce green.

In both the above techniques, preloading with monoamines or fluorogenic nonmetabolizable monoamine analogues facilitates the visualization of monoaminergic cells. Examples of nonmetabolizable analogues include α-methylnoradrenaline and 6-HT. The ganglia are incubated for 20 to 60 min at room temperature in 0.1 to 0.5 mM concentrations of the compounds.

2. Staining Techniques

a. In Vivo Neutral Red Staining for Biogenic Amines

This technique has been used in *P. americana* by Dymond and Evans[20] and by Adams and co-workers.[24] It has the advantage of allowing further physiological or dye-injection experiments after the amine-containing cells have been localized. Desheathed or intact ganglia are incubated at room temperature in 0.01 to 0.3 mg/ml neutral red in saline or tissue culture medium for 0.5 to 3 h. Alternatively, staining can be carried out at 4°C overnight. A destaining period (incubation in saline or culture medium for 4 to 6 h) reduces nonspecific background staining. Intracellular granules within the cell bodies of amine-containing cells stain red. Whole-mount preparations can be made by fixing in nonalcoholic fixative, followed by rapid dehydration and embedding.

b. Staining of Neurosecretory Cells

Numerous histochemical techniques exist for staining neurosecretory cells (NSC).[25] As an example, we will describe in detail only one of these methods — the paraformaldehyde-fuchsin stain for whole-mounts. This method may not stain all NSC; for a more rigorous study, other staining methods and wax sections should also be used.[25] The tissue is fixed in aqueous Bouin's fixative for 12 to 24 h, washed in lithium salt in 70% alcohol, and transferred to distilled water. A 1:1 mixture of 0.6% potassium permanganate and 0.6% sulfuric acid is used for 15 min to oxidize the material. After rinsing in water, the tissue is bleached in 2.5% sodium metabisulfite until completely white, then washed in distilled water for 5 to 10 min. Next, it is placed in 70% alcohol/paraldehyde-fuchsin (PF) solution (0.125 g PF crystals in 50 ml of 70% alcohol plus 1 ml acetic acid, adjusted with HCl to pH 1.6 to 1.7) for 30 min of staining. Differentiation of the stain in 0.5% HCl in absolute alcohol continues until no more superfluous stain is given off. Finally, the tissue is dehydrated, cleared, and mounted.

C. IMMUNOCYTOCHEMICAL METHODS

Immunocytochemical methods have several advantages over histochemical techniques: they are versatile, yet can be very specific, particularly with the recent advances in monoclonal antibodies whereby unequivocal identification and localization of cell components can be achieved.

1. Fixation

Fixation schedules vary according to the nature of the antigen. In some tissues, antigenicity is only preserved if no fixation is used; in these cases, frozen sections must be employed. Fixatives in general use on insect tissues include (1) 4% paraformaldehyde in buffer or saline for 2 to 12 h at 4 to 20°C,[17,26] with or without the inclusion of 1.4% lysine monohydrochloride and 0.2% sodium metaperiodate;[27] (2) 2.5% glutaraldehyde in 0.05 M phosphate buffer (pH 7.4) for 1 to 2 h (for substances conjugated with glutaraldehyde to a carrier protein);[28] (3) aqueous Bouin's (15:5:1 picric acid-formalin-glacial acetic acid) for 3 to 6 h; (4) 2% formaldehyde and 15% picric acid in 0.1 M phosphate buffer, pH 7.3; and (5) 15 ml picric acid, 5 ml 25% glutaraldehyde, and 0.1 ml glacial acetic acid.[29,30]

2. Tissue Preparation

As mentioned above, in some cases the tissue must be frozen and sections taken with a cryostat microtome. In the case of ganglia which are surrounded by a connective tissue sheath, this should be removed before fixation, using either dissecting instruments or 1 mg/ml collagenase. Other methods of rendering the tissue accessible to the large antibody molecules may be applied after fixation.

In some preparations the whole-mount method can be used. This is particularly suitable for small ganglia or thin sheets of tissue, which are pretreated for 1 to 2 h with a detergent such as 0.3% Triton® X-100 or saponin, made up in the same buffer as will be used later for antibody dilution. Primary antibody incubation times may have to be extended to 1 to 2 d at 4°C. The detergent is included in all later incubation steps in which large molecules are required to penetrate the tissue.

A common method of making the tissue accessible to the antibodies is to take frozen sections of fixed (or unfixed) tissue. The following method is taken from Buchner et al.[31] After fixation the tissue is washed for several hours in 25% sucrose in saline, both to remove the fixative and for cryoprotection. Tissue is mounted in 20% carboxymethylcellulose on a microtome chuck and then frozen in liquid nitrogen. Using a cryostat microtome, 10- to 12-μm sections are cut at −20°C and picked up on cold, gelatinized slides (or slides coated with chrome alum-gelatin). The sections are attached to the slide by briefly melting them and allowing them to air dry at −20°C for 15 min. After removal from the cold chamber, the slides are dried again for 30 min. The sections can be stained immediately or stored in saline at 4°C for several hours. They are incubated overnight at 4°C in the primary antibody.

Similarly, vibratome sections may be taken from pieces of fixed tissue and then incubated in the antisera as described above. Small pieces of tissue can be embedded in 5% agar. This technique has been found particularly useful in combined light and electron microscopy studies using pre-embedding staining, particularly in cases where larger (>1 mm) pieces of nervous tissue are available.

Alternatively, tissue can be fixed, dehydrated, and embedded in paraffin wax, then sectioned at 10 μm. After dewaxing in xylene the sections are rehydrated, then incubated in the antisera. Recently, more hydrophilic embedding media have been developed that eliminate the need for exposure to nonaqueous solvents, which may modify antigenicity. Tissue may be embedded in plastic or resin (e.g., Lowicryl® K4M, LR White, or LR Gold), sectioned at 0.5 to 4 μm, and the plastic sections exposed to antibody. The hydrophilic nature of these resins improves access of the antibodies to the tissue. Some plastics may be removed with solvents to increase access to the antibodies (e.g., Poly/REC, Polysciences, Inc., Warrington, PA).

3. Antibody Staining
a. Peroxidase-Antiperoxidase Method

The following method for peroxidase-antiperoxidase (PAP) staining of whole-mount ganglia is based on that for adult cockroaches and adult and embryonic grasshoppers.[17,32] The method for staining frozen or fixed sections is similar; however, detergent is omitted from the incubation medium and the primary antibody incubation time is reduced to approximately 12 h at 4°C. After overnight fixation at 4°C in 4% paraformaldehyde in 50 mM sodium phosphate buffer, pH 7.3, the tissues are washed, then agitated for 1 h at 4°C in a phosphate-buffered saline (PBS) consisting of 150 mM NaCl and 20 mM sodium phosphate with 0.2 to 0.3% Triton® X-100, pH 7.3 (PBS + T). The tissue is left in PBS + T containing 10% normal goat serum (NGS) or 1% bovine serum albumin (BSA) for 1 h at 4°C to reduce nonspecific staining.

The tissue is incubated in a suitable concentration (approximately 1:500 to 1:5000) of the primary antiserum (raised in an animal other than the goat), diluted in PBS + T with 1% NGS, for 12 to 72 h in the dark at 4°C. After a 1-h wash (six changes) in PBS + T, the tissue is incubated with the secondary unlabeled goat anti-rabbit antiserum, diluted 1:20 in PBS + T with 10% NGS, for 12 h at 4°C (or 1 h at room temperature). It is then washed for 1 h in PBS + T (several changes). Next, the tissue is incubated in a 1:40 dilution of rabbit antiperoxidase or PAP complex in PBS + T with 1% NGS for 12 h at 4°C (or 1 h at room temperature), then washed for 1 h with PBS and brought to room temperature. Finally, it is incubated for 1 h in the dark in 0.25 to 0.5 mg/ml 3,3′-diaminobenzidine (DAB) in PBS, then transferred to the same solution with the addition of 0.005 to 0.01% H_2O_2. The development of the reaction product is stopped by washing in PBS, and the tissues are dehydrated, cleared, and mounted for light microscopy. DAB should be handled with care using a fume hood and gloves. Utensils and work areas exposed to DAB should be washed with sodium hypochlorite (household bleach), which denatures this carcinogen. Old DAB solutions should be denatured prior to disposal.

b. Immunofluorescence

Alternatively, instead of incubating the tissue in unlabeled secondary antibody, a fluorochrome-conjugated secondary antibody is used. Commercial antibodies are labeled with either fluorescein isothiocyanate (FITC), which gives a green fluorescence under UV illumination, or tetramethyl rhodamine isothiocyanate (TRITC), which is red. Specific epifluorescence filter blocks can be obtained for each fluorochrome. Immunofluorescence gives more rapid results than peroxidase techniques, and the use of harmful chromogens is avoided; it is also suitable for double-labeling experiments. However, the method is less sensitive than the PAP technique and the fluorescence gradually fades, although fading is retarded by mounting the specimen in 5% *n*-propyl gallate in glycerol, pH 7.3.

c. ABC method

As an alternative to the PAP method, it is now possible to use a biotinylated secondary antibody followed by incubation with an avidin-biotinylated peroxidase complex (ABC method). The tissue is incubated with the reagents for 1 h at room temperature, and the DAB reaction product is developed as described above. If the tissue contains endogenous biotin, it can be preincubated with excess avidin, followed by excess biotin, before the antisera are applied.

d. Alkaline Phosphatase Labeling

In addition, it is now possible to use alkaline phosphatase as a marker enzyme, either in the form of a biotinylated enzyme-avidin complex or in a modification of the PAP method, using antiphosphatase antisera. This allows the production of red or blue reaction products, which can be distinguished from the brown or black DAB polymer, thus allowing double-labeling experiments to be carried out.[33]

4. Electron Microscopic Immunocytochemistry

A technique for immunocytochemical staining of proctolin in the cockroach has been described by Agricola et al.[30] The ganglia are fixed for 1 to 1.5 h in 4% formaldehyde in 0.1 M phosphate buffer, pH 7.3. For some antigens, 0.1 to 1% glutaraldehyde may be added to the fixative. Tissue can be fixed for electron microscopy in the mixture described by McLean and Nakane:[27] 4% paraformaldehyde, 1.4% lysine monohydrochloride, and 0.2% sodium metaperiodate. After fixation, ganglia are washed in buffer and embedded in 7% agar. Vibratome slices are cut at 100 μm and mounted on slides. The sections are first incubated in the antisera, then in PAP or ABC. Alternatively, the protein A-gold technique may be used.[29] Following incubation, sections are fixed with 1% OsO_4 and processed for electron microscopy.

Small ganglia or other pieces of tissue are sectioned this way less easily. Whole-mount staining techniques can be used for localizing cell-surface antigens in embryonic tissues,[34] but penetration of the antibodies may be a problem at later stages, even after ganglia are desheathed. Care is needed, as permeabilization of the membranes with detergents may have a detrimental effect on ultrastructure. Alternatively, after fixation the tissue can be cryo-protected with 30% sucrose in buffer, then frozen rapidly. The tissue is then allowed to thaw in buffer.

Postembedding staining techniques are effective for staining intracellular antigens. These methods have primarily been used on vertebrate tissues, although a postembedding technique has been described for the cockroach.[29] The use of more hydrophilic embedding media which can be polymerized at low temperatures (e.g., Lowicryl® K4M, LR Gold) has extended the use of this technique to more labile antigens. The main disadvantages of this approach include the need for scrupulous cleanliness when staining the ultrathin sections and the poor ultrastructural preservation caused by the lack of osmication. In the case of a few durable antigens, the tissue may be osmicated and the sections exposed to a strong oxidizing agent, a saturated solution of sodium metaperiodate, before incubation in the antisera.

D. AUTORADIOGRAPHIC METHODS
1. Localization of Functional Activity Using Deoxyglucose

This technique allows the state of physiological activity of nervous tissue to be localized. It relies on the fact that whereas glucose, on being taken up by neurons, enters the glycolytic pathway, deoxyglucose is taken up, but not metabolized. Glucose (and deoxyglucose) uptake is related to the electrical activity of a brain region, but also depends on the surface-to-volume ratio of cell processes in that region. Thus, small-diameter axons will be labeled more strongly than large axons with the same metabolic activity. Autoradiography limits the resolution of the technique to about 2 μm in the light microscope, but the activity of large numbers of neurons can be studied. The temporal resolution of the technique can be as short as 0.75 h for small insects. The method has been described for *Drosophila* by Buchner and Buchner,[35] and these authors have suggested modifications for other species.[36]

The deoxyglucose is introduced into the insect either by feeding or by injection into the hemocoel. In the latter case, 20 μCi/mg are used to give a concentration of 1 to 3 mM in the hemolymph. Higher concentrations may be necessary in the cockroach; prior starvation may increase uptake. The animal is then stimulated (e.g., visual or mechanosensory stimulation) for periods ranging from 0.75 to 8 h. For larger insects such as the cockroach, where the costs of high ^{3}H-deoxyglucose dosage are too great, longer periods of stimulus may be necessary. The ganglia are dissected out of the insect and frozen rapidly, then freeze-dried for several days. The dry tissue is fixed with osmium tetroxide vapor and embedded in Epon® under a slight vacuum. The blocks are sectioned at 2 to 3 μm using dry glass knives. The sections are stuck onto glass slides, which are then clamped against slides covered with dry autoradiography stripping film (see Reference 36 for details). After exposure for a suitable period (e.g., 10 d for 20 μCi), the slides are separated and the autoradiographs developed.

2. Localization of Specific Radiolabeled Probes

Detection of the distribution of specific radiolabeled ligands can be used to map neurotransmitter receptors and ion channels. Figure 2 shows the mapping of putative neuronal nicotinic receptors in the cockroach terminal ganglion.[37] Careful controls are needed in which binding is studied in the presence of a competing ligand. In this way, specific and nonspecific binding can be distinguished.

3. *In Situ* Hybridization Histochemistry

This new and powerful technique uses nucleotide probes (complementary DNA or RNA sequences) to detect specific nucleotide sequences (usually mRNAs) in slide-mounted cells or tissue sections.[38-40] Most probes are radiolabeled for detection by autoradiography, although nonradioactive tags are under development. Quantitative autoradiography allows the measurement of the mRNA levels under various physiological and pharmacological conditions. An example of the methods used is given.[41] Frozen sections of tissue are fixed in cold 95% ethanol for 1 h. Alternatively, wax-embedded tissue may be used. After rehydration (if necessary) the sections are equilibrated in hybridization buffer consisting of 450 mM NaCl, 45 mM sodium citrate (pH 7.2), 50% formamide, 0.02% polyvinyl pyrrolidone, 0.02% BSA, and 0.02% Ficoll®. Each section is incubated for 40 h with 30,000 cpm of (for example) tritium-labeled single-stranded DNA probe in hybridization buffer at 28°C. The sections are then washed with 300 mM NaCl and 30 mM sodium citrate for 4 h to remove nonhybridized probes. The slides are then coated with Kodak® Ilford L4 nuclear emulsion and exposed at 4°C for 12 weeks. For a detailed discussion of the methods for *in situ* hybridization, see Reference 42.

V. NEURONAL MORPHOLOGY

A. GOLGI TECHNIQUES

The Golgi stain allows the shapes of single neurons to be studied. However, this technique differs from those described later in that the experimenter has little or no control over which neurons are stained, although the region of the CNS in which the cells are located can be chosen by allowing only restricted access to staining solutions via holes made in the cuticle. Neurons apparently are impregnated at random, although some types of nerve cells may be refractive to the method. Some neurons may be only partly impregnated. The Golgi technique is not suited to quantitative studies unless the results can be corroborated using some other selective staining method. A variety of Golgi techniques for the insect nervous system have been described in detail by Strausfeld.[43]

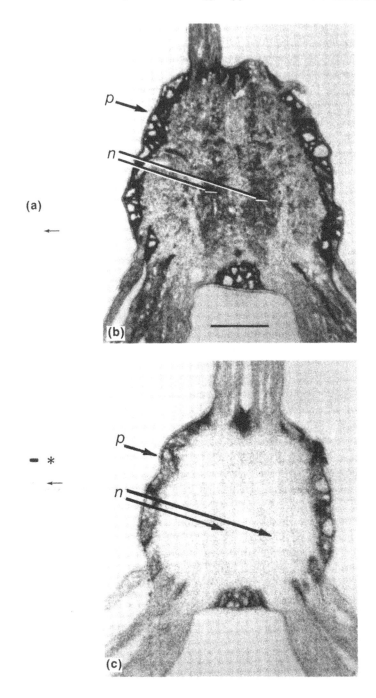

FIGURE 2.   (a) SDS-polyacrylamide gel of 20 μg α-bungarotoxin. Top and bottom of the gel are indicated. The position of the α-bungarotoxin band is marked by an asterisk. (b,c) Distribution of ^{125}I-α-bungarotoxin binding in the terminal (sixth abdominal) ganglion of an adult male cockroach (*Periplaneta americana*). Horizontal frozen sections of the ganglia were prepared and autoradiographs were examined with bright-field illumination. (b) In this section through the central region of the ganglion, the silver grains are primarily located in two central patches in the neuropil (n) and in the periphery (p) of the ganglion. Some of the larger diameter neuronal cell bodies can also be seen in the periphery of the ganglion. (c) Pretreatment of sections with 1 m*M* D-tubocurarine chloride for 15 min removed all but the background binding of ^{125}I-α-bungarotoxin binding from the neuropil (n). Only a portion of the peripheral toxin binding (p) was removed. (Scale bar: 200 μm [b and c].) (From Sattelle, D. B., Harrow, I. D., Hue, B., Pelhate, M., Gepner, J. I., and Hall, L. M., *J. Exp. Biol.*, 107, 473, 1983. With permission.)

1. Buffered Golgi Rapid and Golgi-Colonnier Methods

These methods have proved reliable for a range of insects, including members of the Orthoptera. Fixation is in 2.5% glutaraldehyde in 0.1 M piperazine-N,N'-bis[2-ethanesulfonic acid] (PIPES) buffer at pH 7.0 to 7.4 for 3 to 12 h, followed by washing in several changes of PIPES buffer over 2 h and further washing in several changes of Holmes' buffered chromate, pH 5.4 to 5.6 (2.5% w/v sodium or potassium dichromate in 0.2 M boric acid buffer, adjusted to pH 7.4 with 0.05 M borax), over 2 h. For the Golgi rapid technique, the tissue is washed in Holmes' buffered dichromate-osmium (1 volume 2% osmium tetroxide + 20 volumes buffered chromate) for 12 to 72 h in the dark at room temperature. For the Golgi-Colonnier procedure, buffered dichromate-glutaraldehyde (1 volume 25% glutaraldehyde + 4 volumes buffered dichromate) is used. The tissue is rinsed several times in Holmes' buffered silver nitrate (0.75% w/v silver nitrate in boric acid/borax buffer) until the precipitate no longer appears; it is then left in buffered silver nitrate for 12 to 72 h. The chromation and silver impregnation stages may then be repeated if desired. Dehydration is followed by embedding in Araldite®, Epon®, or Spurr's resin.

2. Golgi for Electron Microscopy

This technique was developed by Ribi[44,45] for insects and works well on the optic lobes of flies and bees. Results on the cockroach CNS are not available. The tissue is fixed for up to 4 h at 4°C in 2% paraformaldehyde and 2.5% glutaraldehyde in Millonig's phosphate buffer (0.164 M monosodium phosphate, adjusted to pH 7.2 to 7.3 with 0.63 M NaOH, plus 0.0024% $MgCl_2$ and 1.3% sucrose). D-Glucose (1%) and $CaCl_2$ (0.009%) may be added to the fixative. Next, the fixed tissue is washed for 3 min in buffer; this is repeated twice. The tissue is then placed in a 1 to 2% OsO_4 solution (containing $CaCl_2$ and D-glucose) for 1 to 2 h at room temperature, followed by three 3-min buffer washes. The next step is to place it in 1% potassium chromate solution for 4 h in the dark at 4°C. The tissue is then washed briefly in distilled water, then in 0.5% silver nitrate solution until no precipitate remains, and, finally, is left in silver nitrate solution in the dark at 4°C for 30 min to 4 h, depending on the size of tissue and the density of impregnation desired.

The tissue is washed repeatedly in distilled water before dehydration in an ethanol series, clearing in propylene oxide, and embedding in Araldite® (see Section VI on electron microscopy). Thick sections (10 to 80 μm) can be cut on a rotary or sledge microtome if the block face is preheated to soften it (using an infrared heating bulb or soldering iron). Examination with light microscopy is followed by reembedding for electron microscopy.

B. DEGENERATION

Experimentally induced Wallerian degeneration is a useful tool for anatomical studies using light and electron microscopy.[46] It has been eclipsed to some extent by other techniques of marking neurons intracellularly, since it has the disadvantage that degenerating neurons cannot be recognized in whole-mounts of ganglia, but it still has a valuable role to play, particularly in studies where it is used along with other marking techniques or in cases where the neurons are inaccessible to other experimental procedures[47] (see also Chapter 11 in this work).

The rationale for using degeneration as a neuroanatomical tool lies in the changes which are produced in neuron processes 1 to 6 d after separation from the cell body. These changes involve first darkening and then shrinkage of the axoplasm and can be detected in sections from tissue prepared for light or electron microscopy following aldehyde-osmium fixation (see Section VI on electron microscopy techniques).

In *P. americana*, enucleated axons do not survive for long periods as they do in other orthopterous insects. However, regeneration of axons from the severed stumps is common after the first postoperative week. Care should be taken to control for possible spontaneous degenerative changes in neurons, along with changes due to aging.

C. INTRACELLULAR MARKERS

It has often been the case that new techniques for marking neurons have been developed first in invertebrate preparations and later are adapted for use in studies on vertebrates. This has been the case with Procion yellow and Lucifer yellow, as well as with cobalt and nickel ions, which were first used as intracellularly injected tracers in large, identified invertebrate neurons that were easily accessible to microelectrode penetration. Horseradish peroxidase, however, was first used in vertebrates as a retrograde axonal tracer and was later used for intracellular injection in vertebrates and invertebrates.

1. Cobalt and Nickel

a. Intracellular Microelectrode Injection

Glass micropipettes are filled with a solution of (1) 1 to 500 mM cobaltous chloride, cobalt nitrate, or cobalt acetate; (2) 3 to 6% hexamine cobalt chloride;[48] or (3) 250 mM nickel chloride. Nickel-filled electrodes are reported to have lower resistances compared to those filled with cobalt, and they are less prone to blockage.[49] These solutions can be combined with KCl or K acetate if lower resistance microelectrodes are necessary, although the presence of potassium ions will reduce the number of metal ions ejected by an iontophoretic current pulse.

These metal ions can be ejected from the microelectrode by means of pressure, but this necessitates a micropipette with a large tip, which may be precluded by the size of the neurons under investigation. A more commonly used technique is that of iontophoretic injection. Positive current should be used; to avoid blockage, the current is applied as 0.5-s pulses at a frequency of 1 to 2 Hz. The amplitude of the current and the injection time vary according to the size of the neurons and the desired density of the final fill. Large cockroach motoneurons impaled in the cell body have been injected with 50-nA pulses for 1 h to produce fills which are easily visible without silver intensification.[50] At the other extreme, embryonic cockroach neurons have been injected with 2-nA pulses for 3 min to give fills which are invisible until after intensification.[4]

It should be borne in mind that both Co^{2+} and Ni^{2+} are Ca^{2+} channel blockers and are toxic to neurons. Even with a good penetration the cell gradually will be poisoned as ions are injected, although many cells are able to sequester some of the ions in mitochondria. Serious damage to the neuron, particularly during impalement, often results in leakage of the metal ions out of the cell. In this regard, the input resistance of the neurons should be considered when selecting the amplitude of the injection current, since the higher the input resistance, the greater the damage sustained by the cell for a given amount of current.

It is most convenient if the amount of current passing out of the electrode is monitored during the injection period, since cobalt- and nickel-filled electrodes have a tendency to become blocked when passing large, positive currents. The current can be monitored using a current pump or a virtual-ground circuit. Blocked electrodes sometimes can be unblocked by temporary reversal of the current, but this strategy rarely is completely successful.

Both cobalt and nickel appear to spread through neurons at a rate of about 3 to 5 mm/h. The preparation may be left at room temperature for approximately 1 to 2 h in saline or tissue culture medium, or at 4°C for several hours, to allow for movement of the metal ions down long axons. Nickel ions are lost after 4 to 6 h.[49]

b. Introduction via Cut Axons

This technique has been described in detail by other authors.[51] It is not as specific as intracellular recording from a neuron followed by injection of dye, but has the advantage of requiring little in the way of apparatus. Basically, the cut ends of selected axons are placed in a reservoir of cobalt (or nickel) chloride solution, which must be isolated to prevent uptake into the rest of the nervous system. The cobalt ions migrate into the axons whether or not current is applied. The method can be applied to either *in vitro* or *in vivo* preparations.

In the first case the ganglia are dissected out of the animal, taking care to leave as long a length of the nerve of interest as possible and not to damage it by pulling. The ganglia are placed in a specially constructed dish containing two chambers connected by a shallow groove. This groove crosses a deeper gutter containing mineral oil, which provides a water-proof seal around the nerve. The ganglia are placed in one chamber, which contains saline or tissue culture medium, and the nerve of interest is draped across the shallow groove connecting the chambers so that its cut end enters the second chamber, containing the cobalt solution. This method also can be applied after sucrose- or oil-gap recording from giant interneurons in the cockroach nerve cord.[52] In that study, the cut ends of the axons were exposed to 2 to 5% cobalt chloride with the application of a 50-nA current for 8 h. This ensured the filling of the fine dendritic branches.

For *in vivo* staining, the nerve of interest is placed in an isolated pool of cobalt chloride solution, either within the body cavity or on the body surface. Such a reservoir is commonly made of purified white petroleum jelly, built up around the nerve using a syringe, and is filled with the cobalt solution. The nerve is then cut. Alternatively, the cut end of the nerve can be drawn into the end of a cobalt chloride-filled suction electrode.[53] Some sensory projections can be filled by cutting the sensory hairs and sealing a drop of cobalt chloride over them with wax[54] or by inserting them into a cobalt chloride-filled micropipette.[55] In cockroaches this is best achieved immediately after the animal has molted, since at this stage sensory hairs retain a lumen which connects to the sensory dendrite.[56]

Choice of cobalt concentration, filling time, and temperature depend on the preparation. The filling solution can be made up in insect Ringer solution or distilled water. For large axons up to 15% cobalt chloride can be used, while for small axons a 1% solution is satisfactory. Blagburn and Beadle[57] achieved satisfactory results filling giant interneurons and cercal sensory axons in the first-instar cockroach using 1% $CoCl_2$ with 0.13 mg/ml BSA (after Strausfeld and Obermayer[58]). The filling time depends on the temperature used and on the size of the neuron. Tyrer et al.[59] suggest that *in vitro* preparations are best left at room temperature for up to 8 h, while *in vivo* preparations are better at low temperatures (4°C) for periods of up to 48 h. In all cases, the preparations should be kept in a humid environment to prevent dessication.

c. Injection into Ganglia

A reservoir of extracellular cobalt ions can be produced inside insect ganglia by insertion of a broken-tipped microelectrode containing cobalt solution into the desired region. Cobalt is allowed to diffuse out of the micropipette for an empirically determined period; this is followed by a further diffusion period after the micropipette has been removed. Some neurons will take up cobalt ions and can be visualized after silver intensification. Bacon and Strausfeld[60] determined that the optimal micropipette tip size for Orthoptera was <1 μm, since larger tips result in an extensive extracellular Co^{2+} pool which later prevents the silver intensification of filled neurons. The micropipette, filled with 5 to 20% cobalt chloride, is held in place using a Plasticine® "micromanipulator" for a period of 30 min, followed by a 30-min diffusion period at room temperature.

d. Preparation for Viewing

The metal ions are precipitated using ammonium or sodium sulfide or rubeanic acid (dithiooxamide).[61] The tissue is first washed in saline to remove any excess of metal ions. Under a fume hood, ammonium or sodium sulfide is added to the saline to give a final concentration of about 1% (approximately one or two drops of concentrated solution usually suffice). After 5 to 10 min, the tissue is washed thoroughly with saline and transferred to fixative. Fresh ammonium sulfide should be used to avoid the formation of soluble polysulfide compounds. Alternatively, the tissue is flooded with a saturated ethanolic solution of rubeanic acid for 5 to 10 min. Rubeanic acid has the advantage of being nontoxic; however, Co- and Ni-rubeanic complexes may be soluble in acidic media or in ethanol and so should not be kept in either one for long periods. Exposure of small ganglia to ethanol may result in rapid shrinkage of the tissue.

In addition, rubeanic acid has the advantage of forming different-colored complexes with different metal ions (Co^{2+}, yellow; Ni^{2+}, blue; 50:50 Co:Ni, red). If sufficient amounts of the metal ions can be introduced into the neurons so as to avoid the need for intensification, then different neurons can be stained different colors using different combinations of metal ions.[62]

The preparation may be treated with one of several different fixatives. Alcoholic Bouin's (1 to 3 h) is perhaps the most effective. Glutaraldehyde-containing fixatives are not suitable if whole-mount intensification is to be used, since they render the ganglionic sheath less permeable to the intensification medium. The preparation is transferred to absolute ethanol and cleared in methyl salicylate, methyl benzoate, or cedarwood oil, then mounted in Canada balsam; alternatively, it can be viewed in the clearing medium.

e. Silver Intensification

Several modifications of silver intensification techniques have been described which allow the enhancement of cobalt (or nickel) deposits in whole ganglia or sections for light and electron microscopy.[59,63,64] The following method of whole-mount intensification for light microscopy is based on the modification of Bacon and Altman[65] and has been used to intensify cobalt- or nickel-filled neurons from orthopterous insects.

After fixation, the tissue is washed in distilled water, then incubated at 48°C (60°C in the original method) in developer stock solution for 1 h. The stock solution consists of 3 g gum arabic (acacia), 10 g sucrose, 0.43 g citric acid, and 0.17 g hydroquinone dissolved in 100 ml distilled water. Commercial powdered gum arabic may be contaminated with metal particles, which will cause immediate precipitation of silver. The tissue is transferred into a solution containing ten parts developer to one part 1% silver nitrate for a further 5 to 15 min at 48°C, until the neuron appears black and the rest of the tissue is yellow or golden brown. The reaction is stopped by washing in warm 30% ethanol, and the tissue is subsequently dehydrated, cleared, and mounted in Canada balsam. The cobalt fills may be photographed, but information from all focal planes of the specimen can best be conveyed in drawings made with a drawing tube or camera lucida (Figure 3).

Recently a new, rapid silver intensification technique has been described which can be carried out at room temperature.[66] This method uses Triton® X-100 as a stabilizer and therefore may not be suitable for electron microscopy. Following precipitation of cobalt sulfide, the tissue is fixed in alcoholic Bouin's for about 30 min for small preparations. Nonspecific silver deposits may be formed if aldehyde fixatives are used. After a 15-min wash in distilled water, the tissue is transferred to 2% sodium tungstate for 30 min, then into the developer solution in a petri dish at room temperature. The developer solution should be made freshly from three stock solutions, mixed in the proportions 8A:1B:1C;* to this is added 1 mg/ml silver nitrate. The intensification can be observed with a dissecting microscope until the cells are visible or the tissue begins to discolor (2 to 10 min). The tissue is then rinsed in distilled water, dehydrated, cleared, and mounted.

f. Electron Microscopy

Several techniques have been developed to intensify cobalt sulfide for electron microscopy. The earliest of these involved intensification of ultrathin sections on the grids or

---

* Solution A consists of 15 ml 1% Triton® X-100, 1.5 g sodium acetate, and 30 ml glacial acetic acid in 355 ml distilled water. Solution B is 5% sodium tungstate. Solution C is 0.25% ascorbic acid in distilled water, made up immediately before use.

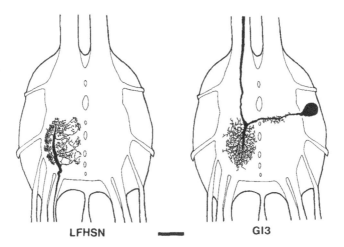

FIGURE 3.   Drawings of identified neurons in the terminal ganglion of the first-instar cockroach. The cells were injected with cobalt chloride via intracellular electrodes, and the cobalt sulfide precipitate was subsequently silver-intensified. Left: the lateral filiform hair sensory neuron (LFHSN); right: giant interneuron 3 (GI3). (Scale bar: 50 μm.)

intensification of 1-μm plastic sections followed by resectioning.[59] Block intensification procedures are less time-consuming and give satisfactory results. We shall describe one of the methods here, but for more details and other variations of the method see Reference 64.

The cobalt is precipitated by placing the tissue for 2 min in four drops of ammonium sulfide in 10 ml 0.05 *M* cacodylate buffer containing 5% sucrose (pH 7.4), followed by washing in buffer for 2 min. The ganglia are fixed in glutaraldehyde in cacodylate buffer (see electron microscopy methods, Section VI). The tissue must be washed thoroughly with 5% sucrose solution to remove cacodylate, which may precipitate silver, and to remove glutaraldehyde, which may increase background staining. As alternative strategies to reduce ultrastructural damage from ammonium sulfide, cobalt precipitation and glutaraldehyde fixation can be combined or sodium sulfite in buffer can be used.

For block intensification, two solutions, A (0.05% hydroquinone, 0.7% citric acid, 2% gum arabic, 1% DMSO, and 5% sucrose in distilled water) and B (0.5% silver nitrate in distilled water), should be made. The tissues are then incubated for 1 h at 20°C in A, followed by 1 h at 37°C in 9.9 volumes of A + 0.1 volume of B, then 1 h at 37°C in 9.75 volumes of A + 0.25 volume of B, then 1.5 h at 37°C in 9.5 volumes of A + 0.5 volume B, and, finally, 1 h at 37°C in 9 volumes of A + 1 volume of B. Increasing the length of the last step to 2 to 3 h will increase the density of the intensification at the expense of increasing the size of the silver grains.

After four 15-min washes in 5% sucrose in the dark at 37°C, the tissue is transferred to buffer for 30 min (or overnight at 4°C) and then to 1% osmium tetroxide. The remaining steps are the same as for conventional electron microscopy. Lightly intensified neurons may not be visible inside osmicated blocks, so for preliminary light microscope reconstruction 1- to 20-μm sections should be taken, then reembedded after examination.

2. Fluorescent Dyes

The use of Procion dyes[67] (in particular, Procion yellow) has largely been superseded by the use of Lucifer yellow CH.[68] The latter is approximately 100 times brighter than Procion yellow and has a molecular weight of 457.3 (compared to 3000).

Lucifer yellow CH is divalent and negatively charged. In contrast with cobalt, the dye has a low toxicity and therefore can be used for longer-term electrophysiological recording.

One advantage of Lucifer yellow is that it can be used to demonstrate dye coupling between cells; this property can often be correlated with the presence of gap junctions and electrotonic coupling. The dye can also be used to kill whole cells or parts by UV illumination of the filled cell.[69] The most common method of filling cells is to inject the dye from an intracellular microelectrode, using negative current, although pressure can be used with large-tipped microelectrodes.

a. Filling from Electrodes

The tip of a fiber-fill micropipette is filled with a 3 to 5% solution of Lucifer yellow CH dissolved in 1% LiCl (precipitation occurs if KCl is used). The barrel of the electrode is filled with 3 M LiCl. The neuron is impaled with the electrode, and the required physiological experiments are carried out. To eject the dye from the electrode, a hyperpolarizing current (either constant or slow pulses) is applied for 3 to 10 min, depending on the size of the cell. The strength of the current may vary from -2 to -20 nA, again depending on the cell size. No diffusion time is necessary; in fact, delays between injection and fixation or viewing should be avoided.

Recently it was discovered that paraformaldehyde-fixed cells can be filled with Lucifer yellow from microelectrodes in the same manner as living cells.[70] This technique is particularly useful for filling very small cells which otherwise might be killed by microelectrode impalement, thus causing leakage of the dye. Since none of the physiological criteria of impalement are present, it is necessary to be able to visualize both cell and electrode tip when inserting the electrode.

b. Filling through Cut Nerves

This involves the same techniques as cobalt backfilling. Special attention should be paid to the tightness of the Vaseline® or grease seal around the Lucifer yellow pool, since the dye can penetrate the neural lamella. Filling takes place within minutes.

c. Fixation and Mounting

The neurons can be examined immediately after filling, but the dye fades rapidly. Permanent preparations are achieved by fixing with buffered aldehydes, 4% paraformaldehyde, or formaldehyde in 0.1 M phosphate buffer, producing less background fluorescence than glutaraldehyde. After fixing, the tissue is washed thoroughly in buffer.

For whole-mounts, the tissue is dehydrated in an alcohol series, cleared, and whole-mounted in methyl benzoate or methyl salicylate. Alternatively, the tissue is taken from the alcohol into acetone, then into Spurr's liquid embedding medium.[71] The latter is most convenient if the tissues are to be sectioned afterward. For sectioning, the Spurr's is polymerized at 70°C for 12 to 16 h. The block will become opaque. After cooling slowly, 10- to 12-μm sections are cut on a sliding microtome, collected on a microscope slide, and mounted in Gurr's Fluoromount. Sections can be counterstained with 1% methylene blue and azure blue.

d. Examination

Whole-mounts or sections are viewed using an epifluorescence microscope equipped with the following combinations of excitation and barrier filters: 450 to 490 nm plus 520 nm; 436 nm plus 490 nm; 350 to 460 nm plus 515 nm. A 270- to 300-nm excitation filter plus a 410- to 580-nm barrier filter will cut out Lucifer yellow fluorescence, leaving only tracheal autofluorescence.[71] Whole-mounts must be mounted on a depression slide or a specially constructed cavity slide. One disadvantage of Lucifer yellow is that bleaching may occur before a detailed drawing of the cell can be made. Serially focused photographs of the filled neuron are taken, allowing serial reconstruction from slides.

e. Lucifer Yellow Antibody

Whole-mounts or frozen sections of the tissue are incubated in antibodies raised against Lucifer yellow-albumin conjugates.[72] The antibodies are then labeled with a secondary antibody tagged with HRP, which is stained with DAB to give a permanent, brown reaction product.

f. Lucifer Yellow for Electron Microscopy

Photoexcitation of Lucifer yellow CH causes the oxidation of DAB, producing a flocculent, brown, electron-dense precipitate.[73] This technique was adapted successfully for use with vertebrate neurons, but to our knowledge has not been attempted in the cockroach. The Lucifer-DAB reaction could prove useful for double-labeling experiments, since it was reported that the precipitate could be distinguished from that produced by HRP. More recently, Sandell and Masland[74] have shown that a wide range of fluorescent molecules (e.g., fluorescein, rhodamine, 5,7-dihydroxytryptamine) can be photoconverted in the same way. The reaction can take place in unfixed or fixed tissue, although a good supply of oxygen to the tissue and strong irradiation are necessary.

The dye Procion Rubin EX-B (Polysciences Ltd.) has been used as an intracellular marker for electron microscopy and can be injected iontophoretically using hyperpolarizing current pulses.[75] To our knowledge this marker has not been used in insects.

g. Lucifer Yellow in Combination with Horseradish Peroxidase

This technique has not, to our knowledge, been tried in insects or vertebrates, but good results have been obtained in the leech nervous system.[76,77] One neuron is injected with 2% HRP (Type VI) in 0.2 M KCl containing 2% fast green FCF (for pressure injection), and the HRP is allowed to move through the cell for several hours. The second neuron is injected with 3% Lucifer yellow CH in 0.1 M LiCl using 5 to 10 nA pulsed negative current for 15 min at 1 Hz. The living tissue is incubated in a saturated solution of DAB in saline (pH 7); after 5 min, a few drops of 3% H_2O_2 are added to the solution, and the reaction is allowed to proceed under microscopic inspection for another 2 to 5 min. Following a rinse in saline and fixation for 1 h with 4% paraformaldehyde in 0.1 M phosphate buffer (pH 7.4), the ganglia are dehydrated in graded ethanol washes and mounted in methyl salicylate under a coverslip sealed with clear fingernail polish. The HRP- and Lucifer yellow-filled cells are examined simultaneously by balancing the intensity of the fluorescent epi-illumination and the transmitted light until both cells are clearly visible.

h. Lucifer Yellow and Other Fluorescent Dyes

Other fluorescent compounds have been used in conjunction with Lucifer yellow, e.g., SITS (4-acetamido-4′-isothiocyanato-stilbene-2,2′-disulfonic acid), which fluoresces blue and can be injected into cells using the same techniques as Lucifer yellow, although higher currents and longer injection times may be necessary.[78] Other such compounds include Texas red (Sulforhodamine 101)[79] and true blue. Strausfeld and Seyan[80] used Texas red in conjunction with Lucifer yellow and the nuclear stain bisbenzimide to stain neurons in the blowfly visual system. These authors reported that attempts to expel Texas red iontophoretically were unsuccessful.

One disadvantage of Lucifer yellow is that, because of its insolubility in K^+-containing solutions, the microelectrode must be filled with LiCl. In some preparations, intracellular Li^+ may be toxic. Also, there is a tendency for Lucifer yellow in the electrode tip to precipitate when exposed to external solutions containing a high concentration of K^+ (e.g., artificial seawater). For these reasons, the fluorescent dye 5,6-carboxyfluorescein recently has become popular as an intracellular marker.[80a] This compound is highly soluble in K^+-containing solutions, but is not as lipid soluble as fluorescein and so stays in the cell. It is strongly fluorescent under UV light (FITC filter blocks can be used).

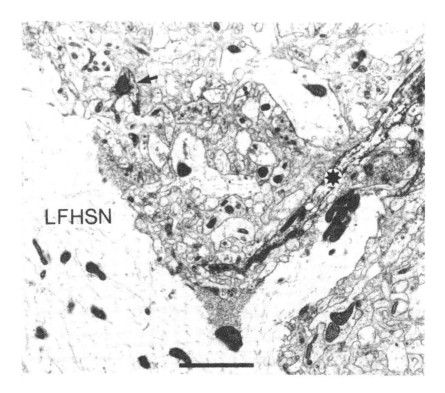

FIGURE 4.   Electron micrograph of the terminal abdominal ganglion neuropil in the first-instar cockroach, sectioned horizontally. The large axon of the lateral filiform hair sensory neuron (LFHSN) is visible on the left of the picture. Giant interneuron 3 has been filled with horseradish peroxidase and an electron-opaque precipitate produced with diaminobenzidine. Dendritic branches of the interneuron (asterisk and arrow) form contacts with the LFHSN axon and receive synaptic input from it. (Scale bar: 2 μm.)

3. Horseradish Peroxidase and Other Enzymes

Intracellular labeling with HRP or other heme proteins such as cytochrome c and myoglobin has the advantage that the reaction product is visible using light and electron microscopy (Figure 4). The disadvantages are that the proteins do not migrate intracellularly as fast, or as far, as cobalt ions or the Lucifer yellow dye. Also, the best method for developing the reaction product involves the use of a probable carcinogen. Methods for introducing heme proteins into insect nerve cells, developing the reaction products, and processing the tissues for light and electron microscopy have been described by Nässel.[81]

a. Intracellular Injection

The tip of a glass micropipette is filled with a solution containing 2 to 10% HRP (type VI is used most frequently) with 0.1 to 0.3 M KCl in 0.2 M Tris or TES buffer, pH 7.4. The microelectrode is inserted into the neuron and, after physiological recording, square positive current pulses are applied, with similar regimes to those described for ejecting cobalt ions. We have found that it may be helpful to pass a few large-amplitude current pulses through the electrode before the cell is impaled. The preparation is left in saline or tissue culture medium for 10 min to 1 h to allow for movement of the protein through the cell. A good supply of oxygen appears to be necessary for transport of the HRP to take place. We rarely obtain good ultrastructural preservation of ganglia if they are left for long periods in saline after they have been desheathed; better results are obtained if the microelectrodes are inserted through the neural lamella which has first been softened by a brief (15 s to 1 min) exposure to 1 mg/ml collagenase or protease.

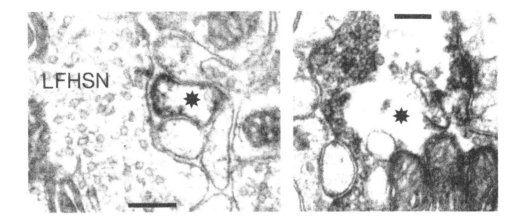

FIGURE 5. Electron micrographs of processes of giant interneuron 3 (GI 3) which has been injected with horseradish peroxidase (asterisks). Left: the axon of the lateral filiform hair sensory neuron (LFHSN) makes synaptic contacts with a GI 3 dendrite. Right: part of a larger branch of GI 3, forming output synapses with processes of unidentified neurons. The HRP-DAB reaction product does not obscure cytoplasmic details in GI 3. (Scale bars: 200 nm.)

The tissue is fixed for 2 h at 4°C in 2.5% glutaraldehyde in 0.1 M cacodylate or phosphate buffer (pH 7.4) containing 0.05 to 0.2 M sucrose. Paraformaldehyde irreversibly reduces the activity of the enzyme, while glutaraldehyde impairs it reversibly. The tissue therefore should be washed overnight in buffer. Alternatively, a 30-min wash in 0.1 M NaBH$_4$ in buffer will remove excess aldehydes if an overnight wash is precluded. The tissue is immersed in 0.1 M Tris buffer (pH 7.4) for 10 min, followed by treatment with 0.5% cobalt chloride (in Tris buffer) for 5 to 10 min to intensify the reaction product. After a 10-min wash in Tris buffer alone, the tissue is washed in cacodylate or phosphate buffer and then transferred to the incubation medium for 20 min to 1 h (until the neuron appears black or brown) at 30 to 37°C in the dark. The incubation medium consists of 10 ml 0.1 M phosphate or cacodylate buffer at pH 7.4, 4 mg ammonium chloride, 20 mg β-D-glucose, 20 to 30 units of glucose oxidase (type V), and 5 mg DAB.[82] After several washes in buffer, the tissue is osmicated and processed for electron microscopy as described in Section VI below.

In filled neurons which appear light brown or brown when examined with a light microscope, the granular DAB reaction product may be scattered throughout the cytoplasm of the filled cell, often preferentially associated with the cell membrane and microtubules, but allowing intracellular organelles to be distinguished (Figure 5). In heavier, black fills, however, the cytoplasmic organelles may be totally obscured by electron-dense deposits which do not penetrate the mitochondria.

VI. ELECTRON MICROSCOPY TECHNIQUES

A. TRANSMISSION ELECTRON MICROSCOPY

For the sake of brevity, techniques for tissue preparation and sectioning are described only briefly. Detailed descriptions of these techniques, as applied to the insect nervous system, have been given by Osborne,[83] and comprehensive reviews of insect neuronal and glial cell ultrastructure are available.[84] The following method has given good results in the embryonic, first-instar, and adult cockroach nervous systems. The tissues are fixed for 2 h at 4°C in 2.5% glutaraldehyde in 0.1 M phosphate or cacodylate buffer (pH 7.4) containing 0.2 M sucrose. Care must be taken with cacodylate since it contains arsenic. All preparative stages should be carried out under a fume hood. After three changes of cacodylate buffer containing sucrose (10 min per wash), the tissue is placed in 1% osmium tetroxide (in

0.1 *M* cacodylate buffer) for 1 to 2 h at 4°C. Following three 10-min washes in cacodylate buffer, the tissue is placed for 1 h in 1% uranyl acetate in distilled water, then washed in distilled water and dehydrated in an alcohol series (30 min each in 50, 70, and 90% ethanol, followed by two 30-min changes of absolute ethanol). The tissue is soaked in two changes of propylene oxide of about 15 min each and transferred to a 1:1 epoxy resin:propylene oxide mixture for 15 min. The tissue is then placed in fresh resin, left at room temperature for 24 h, and subsequently transferred to plastic embedding capsules. The resin is polymerized overnight at 60°C.

After removal from the embedding capsule, the block is mounted in a microtome chuck, trimmed with a razor blade, and sectioned using an ultramicrotome. Sections 1 to 4 μm thick can be used for orientation within the tissue; these are cut using a glass knife, placed on a microscope slide, and stained with toluidine blue. Thin sections (60 to 100 nm in thickness; interference colors, silver to gold) are cut using glass or diamond knives and collected on copper grids. Counterstaining of the sections is done using a saturated aqueous solution of uranyl acetate or a 30% solution in methanol for 2 h or 3 to 7 min, respectively, followed by 3 to 7 min of staining with Reynolds' lead citrate.[85] The sections are washed with 0.001% NaOH and distilled water, and after air drying they are examined with an electron microscope.

B. SCANNING ELECTRON MICROSCOPY

Scanning electron microscopy has limited use in insect neuroanatomical studies, although it is useful for visualizing arrays of cuticular sense organs.[86] Details of the preparative methods will not be given here. The technique has been used to visualize developing axons and their growth cones inside the embryonic grasshopper CNS.[87] This is achieved by preparing the tissue for transmission electron microscopy and embedding it in plastic, sectioning down through the block to the desired area, then etching away the plastic by immersion in a solution of 0.5% sodium methylate in 100% methanol for 2 d followed by a 1-d wash in 100% methanol. The block is critical-point dried and gold coated, as in normal scanning electron microscopy.

C. FREEZE FRACTURE

The freeze-fracture technique is well suited to investigations of the topography of cell membranes and intercellular junctions. Where possible, the results should be compared with those obtained from conventionally thin-sectioned material.

The following method has been used successfully on *Periplaneta* peripheral nerve.[88] Tissue is fixed for 20 to 30 min at room temperature in 2.5% glutaraldehyde in 0.1 *M* phosphate buffer, pH 7.2, containing 6% sucrose. After fixation, the tissue is washed in the same phosphate buffer, then placed in a cryoprotectant solution, 20% glycerol in 0.1 *M* phosphate buffer with 6% sucrose, for 15 to 20 min at room temperature. The tissue is supported with yeast paste and mounted on copper holders which are rapidly immersed in Freon® 22, cooled in liquid nitrogen, and remain there until used.

Fracturing of the tissue is performed in (for example) a Balzers freeze-fracturing device, in a vacuum of 2 μtorr (0.27 kPa) and at a temperature of −100°C. The fractured tissue is shadowed by evaporation from a tungsten-tantalum source followed by backing with carbon. After admission of air, the replicas are strengthened by coating with a drop of 2% collodion in amyl acetate while they are still on the cold table of the fracturing unit. The holders are removed from the Balzers device and the collodion allowed to dry before the replicas and tissue are removed from the holders. Tissue is cleaned from the replicas by immersing them in 2% sodium hypochlorite overnight followed by washing in distilled water. The replicas are then mounted on copper grids and the collodion removed by immersing the grids in undiluted amyl acetate for 30 min. The replicas are examined using a transmission electron microscope.

D. HIGH-VOLTAGE ELECTRON MICROSCOPY

The use of high-voltage electron microscopy (HVEM) can be advantageous in the study of insect nervous systems, particularly in the reconstruction of serial sections of neuropil and in the examination of cuticular sense organs.[89,90] The methods are similar to those used for conventional electron microscopy.[91] Sections for HVEM analysis are taken at 0.2 to 1 μm thickness, depending on the degree of resolution required, and are collected on grids covered with carbonized Formvar® film. Longer exposures to uranyl acetate and lead citrate are necessary to increase the contrast.

REFERENCES

1. **Strausfeld, N. J. and Miller, T. A., Eds.,** *Neuroanatomical Techniques. Insect Nervous System,* Springer-Verlag, New York, 1980.
2. **Strausfeld, N. J.,** *Functional Neuroanatomy,* Springer-Verlag, Berlin, 1983.
3. **Goodman, C. S. and Spitzer, N. C.,** Embryonic development of identified neurones: differentiation from neuroblast to neurone, *Nature (London),* 280, 208, 1979.
4. **Blagburn, J. M., Beadle, D. J., and Sattelle, D. B.,** Development of synapses between identified sensory neurones and giant interneurones in the cockroach, *Periplaneta americana, J. Embryol. Exp. Morphol.,* 86, 227, 1985.
5. **Blagburn, J. M. and Sattelle, D. B.,** Calcium conductance in an identified cholinergic synaptic terminal in the central nervous system of the cockroach, *J. Exp. Biol.,* 129, 347, 1987.
6. **Zawarzin, A. A.,** Histologische Studien über Insekten. 2. Das sensible Nervensystem der Aeschnalarven, *Z. Wiss. Zool. Abt. A,* 100, 245, 1912.
7. **Plotnikova, S. I. and Nevmyvaka, G. A.,** The methylene blue technique: classic and recent applications to the insect nervous system, in *Neuroanatomical Techniques. Insect Nervous System,* Strausfeld, N. J. and Miller, T. A., Eds., Springer-Verlag, New York, 1980, 1.
8. **Pinnock, R. D. and Sattelle, D. B.,** Dissociation and maintenance *in vitro* of neurones from adult cockroach *(Periplaneta americana)* and housefly *(Musca domestica)* neurones, *J. Neurosci. Methods,* 20, 195, 1987.
9. **Altman, J. S. and Bell, E. M.,** A rapid method for the demonstration of nerve cell bodies in invertebrate central nervous systems, *Brain Res.,* 63, 487, 1973.
10. **Altman, J. S.,** Toluidine blue as a rapid stain for nerve cell bodies in intact ganglia, in *Neuroanatomical Techniques. Insect Nervous System,* Strausfeld, N. J. and Miller, T. A., Eds., Springer-Verlag, New York, 1980, 21.
11. **Wigglesworth, V. B.,** The use of osmium in the fixation and staining of tissues, *Proc. R. Soc. Lond. Ser. B,* 147, 185, 1957.
12. **Watson, A. H. D. and Pflüger, H.-J.,** The distribution of GABA-like immunoreactivity in relation to ganglion structure in the abdominal nerve cord of the locust, *Cell Tissue Res.,* 249, 391, 1987.
13. **Blest, A. D.,** Reduced silver impregnations of the Ungewitter type, in *Functional Neuroanatomy,* Strausfeld, N. J., Ed., Springer-Verlag, Berlin, 1980, 119.
14. **Blest, A. D. and Davie, P. S.,** Reduced silver impregnations derived from the Holmes technique, in *Functional Neuroanatomy,* Strausfeld, N. J., Ed., Springer-Verlag, Berlin, 1980, 97.
15. **Gregory, G. E.,** Silver staining of insect central nervous systems by the Bodian protargol method, *Acta Zool. (Stockholm),* 51, 169, 1970.
16. **Gregory, G. E.,** The Bodian protargol technique, in *Neuroanatomical Techniques. Insect Nervous System,* Strausfeld, N. J. and Miller, T. A., Eds., Springer-Verlag, New York, 1980, 75.
17. **Bishop, C. A and O'Shea, M.,** Neuropeptide proctolin (H-Arg-Tyr-Leu-Pro-Thr-OH): immunocytochemical mapping of neurons in the central nervous system of the cockroach, *J. Comp. Neurol.,* 207, 223, 1982.
18. **Beattie, T. M.,** Vital staining of neurosecretory material with Acridine Orange in the insect, *Periplaneta americana, Experientia,* 27, 110, 1971.
19. **Frontali, N.,** Histochemical localization of catecholamines in the brain of normal and drug-treated cockroaches, *J. Insect Physiol.,* 14, 881, 1968.
20. **Dymond, G. R. and Evans, P. D.,** Biogenic amines in the nervous system of the cockroach, *Periplaneta americana:* association of octopamine with mushroom bodies and dorsal unpaired median (DUM) neurones, *Insect Biochem.,* 9, 535, 1979.
21. **Klemm, N.,** Histochemical demonstration of biogenic monoamines (Falck-Hillarp method) in the insect nervous system, in *Neuroanatomical Techniques. Insect Nervous System,* Strausfeld, N. J. and Miller, T. A., Eds., Springer-Verlag, New York, 1980, 51.

22. **Björklund, A., Falck, B., and Owman, C.**, Fluorescence microscopic and microspectrofluorometric techniques for the cellular localization and characterization of biogenic amines, in *The Thyroid and Biogenic Amines*, Rall, J. E. and Kopin, I. J., Eds., Elsevier/North-Holland, Amsterdam, 1972, 318.

23. **Flanagan, T. R. J.**, Monoaminergic innervation in a Hemipteran nervous system: a whole-mount histofluorescence study, in *Functional Neuroanatomy*, Strausfeld, N. J., Eds., Springer-Verlag, Berlin, 1983, 317.

24. **Adams, M. E., Bishop, C. A., and O'Shea, M.**, Strategies for the identification of amine- and peptide-containing neurons, in *Functional Neuroanatomy*, Strausfeld, N.J., Ed., Springer-Verlag, Berlin, 1983, 239.

25. **Panov, A. A.**, Demonstration of neurosecretory cells in the insect central nervous system, in *Neuroanatomical Techniques. Insect Nervous System*, Strausfeld, N. J. and Miller, T. A., Eds., Springer-Verlag, New York, 1980, 25.

26. **Sattelle, D. B., Ho, Y. W., Crawford, G. D., Salvaterra, P. M., and Mason, W. T.**, Immunocytochemical staining of central neurones in *Periplaneta americana* using monoclonal antibodies to choline acetyltransferase, *Tissue Cell*, 18, 51, 1986.

27. **McLean, J. D. and Nakane, P. K.**, Periodate-lysine-paraformaldehyde fixative: a new fixative for immunoelectron microscopy, *J. Histochem. Cytochem.*, 22, 1077, 1974.

28. **Watson, A. H. D. and Burrows, M.**, Immunocytochemical and pharmacological evidence for spiking local interneurons in the locust, *J. Neurosci.*, 7, 1741, 1987.

29. **Eckert, M. and Ude, J.**, Immunocytochemical techniques for the identification of peptidergic neurons, in *Functional Neuroanatomy*, Strausfeld, N. J., Ed., Springer-Verlag, Berlin, 1983, 267.

30. **Agricola, H., Eckert, M., Ude, J., Birkenbeil, H., and Penzlin, H.**, The distribution of a proctolin-like immunoreactive material in the terminal ganglion of the cockroach, *Periplaneta americana* (L.), *Cell Tissue Res.*, 239, 203, 1985.

31. **Buchner, E., Buchner, S., Crawford, G., Mason, W. T., Salvaterra, P. M., and Sattelle, D. B.**, Choline acetyltransferase-like immunoreactivity in the brain of *Drosophila melanogaster*, *Cell Tissue Res.*, 246, 57, 1986.

32. **Keshishian, H. and O'Shea, M.**, The distribution of a peptide neurotransmitter in the postembryonic grasshopper central nervous system, *J. Neurosci.*, 5, 992, 1985.

33. **Mason, D. Y. and Woolston, R.-E.**, Double immunoenzymatic labelling, in *Techniques in Immunocytochemistry*, Bullock, G. R. and Petrusz, P., Eds., Academic Press, London, 1982, 135.

34. **Bastiani, M. J., Harrelson, A. L., Snow, P. M., and Goodman, C. S.**, Expression of fasciclin I and II glycoproteins on subsets of axon pathways during neuronal development in the grasshopper, *Cell*, 48, 745, 1987.

35. **Buchner, E. and Buchner, S.**, Mapping stimulus-induced nervous activity in small brains by ^{3}H-2-deoxy-D-glucose, *Cell Tissue Res.*, 211, 51, 1980.

36. **Buchner, E. and Buchner, S.**, Anatomical localization of functional activity in flies using ^{3}H-2-deoxy-D-glucose, in *Functional Neuroanatomy*, Strausfeld, N. J., Ed., Springer-Verlag, Berlin, 1983, 225.

37. **Sattelle, D. B., Harrow, I. D., Hue, B., Pelhate, M., Gepner, J. I., and Hall, L. M.**, α-Bungarotoxin blocks excitatory synaptic transmission between between cercal sensory neurones and giant interneurone 2 of the cockroach, *Periplaneta americana*, *J. Exp. Biol.*, 107, 473, 1983.

38. **Singer, R. H., Lawrence, J. G., and Villenave, C.**, Optimization of *in-situ* hybridization using isotopic and non-isotopic detection methods, *Biotechniques*, 4, 230, 1986.

39. **Wuenschell, C. W., Fisher, R. S., Kaufman, D. L., and Tobin, A. J.**, *In situ* hybridization to localize mRNA encoding the neurotransmitter biosynthetic enzyme glutamate decarboxylase (GAD) in mouse cerebellum, *Proc. Natl. Acad. Sci. U.S.A.*, 83, 6193, 1986.

40. **Young, W. S., III**, *In-situ* hybridization, histochemistry and the study of the nervous system, *Trends Neurosci.*, November/December, 549, 1986.

41. **Roberts, J. L., Chen, C. C., Dionne, F. T., and Gee, C. E.**, Peptide hormone gene expression in heterogeneous tissues: the pro-opiomelanocortin system, in *Neurotransmitters in Action*, Bousfield, D., Ed., Elsevier, Amsterdam, 1985, 226.

42. **Higgins, G. A. and Wilson, M. C.**, *In situ* hybridization for mapping the neuroanatomical distribution of novel brain mRNAs, in *In Situ Hybridization in Neurobiology: Applications to Neurobiology*, Valentino, K. L., Eberwine, J. H., and Barchas, J. D., Eds., Oxford University Press, New York, 1987, 146.

43. **Strausfeld, N. J.**, The Golgi method: its application to the insect nervous system and the phenomenon of stochastic impregnation, in *Neuroanatomical Techniques. Insect Nervous System*, Strausfeld, N. J. and Miller, T. A., Eds., Springer-Verlag, New York, 1980, 131.

44. **Ribi, W. A.**, A Golgi-electron microscope method for insect nervous tissue, *Stain Technol.*, 51, 13, 1976.

45. **Ribi, W. A.**, Electron microscopy of Golgi-impregnated neurons, in *Functional Neuroanatomy*, Strausfeld, N. J., Ed., Springer-Verlag, Berlin, 1983, 1.

46. Schürmann, F.-W., Experimental anterograde degeneration of nerve fibers: a tool for combined light- and electron-microscopic studies of the insect nervous system, in *Neuroanatomical Techniques. Insect Nervous System*, Strausfeld, N. J. and Miller, T. A., Eds., Springer-Verlag, New York, 1980, 263.

47. Zill, S. N., Underwood, M. A., Rowley, J. C., III, and Moran, D. T., A somatotopic organization of groups of afferents in insect peripheral nerves, *Brain Res.*, 198, 253, 1980.

48. Brogan, R. T. and Pitman, R. M., Axonal regeneration in an identified insect motor neuron, *J. Physiol. (London)*, 319, 34P, 1981.

49. Delcomyn, F., Intracellular staining with nickel chloride, in *Functional Neuroanatomy*, Strausfeld, N. J., Ed., Springer-Verlag, Berlin, 1983, 93.

50. Pitman, R. M., Tweedle, C. D., and Cohen, M. J., Branching of central neurons: intracellular cobalt injection for light and electron microscopy, *Science*, 176, 412, 1972.

51. Altman, J. S. and Tyrer, N. M., Filling selected neurons with cobalt through cut axons, in *Neuroanatomical Techniques. Insect Nervous System*, Strausfeld, N. J. and Miller, T. A., Eds., Springer-Verlag, New York, 1980, 373.

52. Harrow, I. D., Hue, B., Pelhate, M., and Sattelle, D. B., Cockroach giant interneurones stained by cobalt-backfilling of dissected axons, *J. Exp. Biol.*, 84, 341, 1980.

53. Mason, C. A. and Nishioka, R. S., The use of the cobalt chloride-ammonium sulfide precipitation technique for the delineation of invertebrate and vertebrate neurosecretory systems, in *Neurosecretion — The Final Neuroendocrine Pathway: Proceedings*, 6th Int. Symp. Neurosecretion, Knowles, F. and Vollrath, L., Eds., Springer-Verlag, New York, 1974, 48.

54. Tyrer, N. M., Bacon, J. P., and Davies, C. A., Primary sensory projections from the wind-sensitive head hairs of the locust, *Schistocerca gregaria*. I. Distribution in the CNS, *Cell Tissue Res.*, 203, 79, 1979.

55. Bacon, J. P. and Murphey, R. K., Receptive fields of cricket giant interneurones are related to their dendritic structure, *J. Physiol. (London)*, 352, 601, 1984.

56. Bacon, J. P., personal communication, 1987.

57. Blagburn, J. M. and Beadle, D. J., Morphology of identified cercal afferents and giant interneurones in the hatchling cockroach *Periplaneta americana*, *J. Exp. Biol.*, 97, 421, 1982.

58. Strausfeld, N. J. and Obermayer, M., Resolution of intraneuronal and transsynaptic migration of cobalt in the insect visual and central nervous systems, *J. Comp. Physiol.*, 110, 1, 1976.

59. Tyrer, N. M., Shaw, M. K., and Altman, J. S., Intensification of cobalt-filled neurons in sections (light and electron microscopy), in *Neuroanatomical Techniques. Insect Nervous System*, Strausfeld, N. J. and Miller, T. A., Eds., Springer-Verlag, New York, 1980, 429.

60. Bacon, J. P. and Strausfeld, N. J., Nonrandom resolution of neuron arrangements, in *Neuroanatomical Techniques. Insect Nervous System*, Strausfeld, N. J. and Miller, T. A., Eds., Springer-Verlag, New York, 1980, 357.

61. Quicke, D. L. J. and Brace, R. C., Differential staining of cobalt- and nickel-filled neurons using rubeanic acid, *J. Microsc.*, 115, 161, 1979.

62. Hackney, C. M. and Altman, J. S., Rubeanic acid and X-ray microanalysis for demonstrating metal ions in filled neurons, in *Functional Neuroanatomy*, Strausfeld, N. J., Ed., Springer-Verlag, Berlin, 1983, 132.

63. Obermayer, M. and Strausfeld, N. J., Silver-staining cobalt sulfide deposits within neurons of intact ganglia, in *Neuroanatomical Techniques. Insect Nervous System*, Strausfeld, N. J. and Miller, T. A., Eds., Springer-Verlag, New York, 1980, 403.

64. Bassemir, U. K. and Strausfeld, N. J., Block intensification and X-ray microanalysis of cobalt-filled neurons for electron microscopy, in *Functional Neuroanatomy*, Strausfeld, N. J., Ed., Springer-Verlag, Berlin, 1983, 19.

65. Bacon, J. P. and Altman, J. S., A silver intensification method for cobalt-filled neurons in wholemount preparations, *Brain Res.*, 138, 359, 1977.

66. Davis, N. T., Improved methods for cobalt filling and silver intensification of insect motor neurones, *Stain Technol.*, 57, 239, 1982.

67. Stretton, A. O. W. and Kravitz, E. A., Intracellular dye injection: the selection of Procion Yellow and its application in preliminary studies of neuronal geometry in the lobster nervous system, in *Intracellular Staining in Neurobiology*, Kater, S. B. and Nicholson, C., Eds., Springer-Verlag, Berlin, 1973, 21.

68. Stewart, W. W., Functional coupling between cells as revealed by dye-coupling with a highly fluorescent naphthalimide tracer, *Cell*, 14, 741, 1978.

69. Miller, J. P. and Selverston, A. I., Rapid killing of single neurons by irradiation of intracellularly injected dyes, *Science*, 206, 702, 1979.

70. Thomas, J. B., Bastiani, M. J., Bate, C. M., and Goodman, C. S., From grasshopper to *Drosophila*: a common plan for neuronal development, *Nature (London)*, 310, 203, 1984.

71. Strausfeld, N. J., Seyan, H. S., Wohlers, D., and Bacon, J. P., Lucifer Yellow histology, in *Functional Neuroanatomy*, Strausfeld, N. J., Ed., Springer-Verlag, Berlin, 1983, 132.

72. **Taghert, P. H., Bastiani, M., Ho, R. K., and Goodman, C. S.,** Guidance of pioneer growth cones: filopodial contacts and coupling revealed with antibody to Lucifer Yellow, *Dev. Biol.,* 94, 391, 1982.

73. **Maranto, A. R.,** Neuronal mapping: a photooxidation reaction makes Lucifer Yellow useful for electron microscopy, *Science,* 217, 953, 1982.

74. **Sandell, J. H. and Masland, R. H.,** Photocatalysed replacement of fluorescent markers with a diaminobenzidine reaction product, *Soc. Neurosci. Abstr.,* 13, 680, 1987.

75. **Mirolli, M., Cooke, I. M., Talbot, S. R., and Miller, M. W.,** Structure and localization of synaptic complexes in the cardiac ganglion of a portunid crab, *J. Neurocytol.,* 16, 115, 1987.

76. **DeReimer, S. A. and Macagno, E. R.,** Light microscopic analysis of contacts between pairs of identified neurons with combined use of horseradish peroxidase and Lucifer Yellow, *J. Neurosci.,* 1, 650, 1981.

77. **Macagno, E. R., Muller, K. J., Kristan, W. B., Jr., DeReimer, S., Stewart, R., and Granzow, B.,** Mapping of neuronal contacts with intracellular injection of horseradish peroxidase and Lucifer Yellow in combination, *Brain Res.,* 217, 143, 1981.

78. **Kettenmann, H. and Orkand, R. K.,** Intracellular SITS injection dye-uncouples mammalian oligodendrocytes in culture, *Neurosci. Lett.,* 39, 21, 1983.

79. **Pendleton, J. W. and Keshishian, H.,** Sulforhodamine 101 as a fluorescent probe for intracellular dye fills, diffusion backfills, and double labeling with Lucifer Yellow, *Soc. Neurosci. Abstr.,* 11, 440, 1985.

80. **Strausfeld, N. J. and Seyan, H. S.,** Resolution of complex neuronal arrangements in the blowfly visual system using triple fluorescence staining, *Cell Tissue Res.,* 247, 5, 1987.

80a. **Bablanian, G. M. and Page, C. H.,** Characterization of lobster abdominal postural interneurons and extensor and flexor motor neurons using the intracellular dye 5,6-carboxyfluorescein, *Soc. Neurosci. Abstr.,* 13, 680, 1987.

81. **Nässel, D. R.,** Horseradish peroxidase and other heme proteins as neuronal markers, in *Functional Neuroanatomy,* Strausfeld, N. J., Ed., Springer-Verlag, Berlin, 1983, 45.

82. **Watson, A. H. D. and Burrows, M.,** Input and output synapses on identified motor neurons of a locust revealed by the intracellular injection of horseradish peroxidase, *Cell Tissue Res.,* 215, 325, 1981.

83. **Osborne, M. P.,** Electron-microscopic methods for nervous tissues, in *Neuroanatomical Techniques. Insect Nervous System,* Strausfeld, N. J. and Miller, T. A., Eds., Springer-Verlag, New York, 1980, 205.

84. **Lane, N. J.,** Structure of components of the nervous system, in *Comprehensive Insect Biochemistry, Physiology and Pharmacology,* Vol. 5, Kerkut, G. A. and Gilbert, L. I., Eds., Pergamon Press, Oxford, 1985, 1.

85. **Reynolds, E. S.,** The use of lead citrate at high pH as an electron opaque stain in electron microscopy, *J. Cell Biol.,* 17, 208, 1963.

86. **Murphey, R. K., Bacon, J. P., Sakaguchi, D. S., and Johnson, S. E.,** Transplantation of cricket sensory neurons to ectopic locations: arborizations and synaptic connections, *J. Neurosci.,* 3, 659, 1983.

87. **Raper, J. A., Bastiani, M., and Goodman, C. S.,** Pathfinding by neuronal growth cones in grasshopper embryos. II. Selective fasiculation onto specific axonal pathways, *J. Neurosci.,* 3, 31, 1983.

88. **Blanco, R. E.,** personal communication, 1987.

89. **Chi, C. and Carlson, S. D.,** High voltage electron microscopy of the optic neuropile of the housefly, *Musca domestica, Cell Tissue Res.,* 167, 537, 1976.

90. **Moran, D. T. and Rowley, J. C., III,** High voltage and scanning electron microscopy, *J. Ultrastruct. Res.,* 50, 38, 1975.

91. **Chi, C.,** High voltage electron microscopy for insect neuroanatomy, in *Functional Neuroanatomy,* Strausfeld, N. J., Ed., Springer-Verlag, Berlin, 1983, 376.

Section III. Neuron Structure and Function

Chapter 7

COCKROACH AXONS AND CELL BODIES: ELECTRICAL ACTIVITY

Marcel Pelhate, Yves Pichon, and David J. Beadle

TABLE OF CONTENTS

I. INTRODUCTION

The existence of large interneurons in the abdominal nerve cord of *Periplaneta americana* justifies the choice of this insect for electrophysiological investigation. With external bipolar electrodes, the electrical activity originating from these axons is comparatively large (up to 3 mV) and, therefore, easy to record and to discriminate from the background activity. Their size is large enough to allow easy penetration with capillary microelectrodes and recording of full-sized resting and action potentials. Some cell bodies are also quite large and can be studied with the same techniques. Using this preparation, two techniques have been developed which enable very precise and detailed analysis of the biophysical and pharmacological properties of the neuronal membrane: the voltage-clamp technique for studying ionic conductances which underlie the electrical activity of the nerve cell and the patch-clamp technique to analyze activity of single transmembrane proteins, namely, the ionic channels.

In the following sections we shall summarize the results of three categories of experiments done with two neuronal preparations chosen for their ease of use and their accessibility to externally applied molecules: isolated giant axons and cultured neurons. The first section describes the ionic basis of electrical activity in axons and neurons; the second, the effects of channel blockers, channel modulators and insecticides; and the last, the response of the cultured neurons to externally applied neurotransmitters (acetylcholine [ACh], gamma-aminobutyric acid [GABA], and glutamate).

II. IONIC BASIS OF ELECTRICAL ACTIVITY IN AXONS AND CELL BODIES

The ionic basis of electrical activity was fully established on isolated axons of *P. americana* by Pichon,[1-3] using the voltage-clamp technique.

A. RESTING POTENTIAL IN AXONS

When the tip of a microelectrode is pushed through the membrane into the axoplasm of a giant axon of the cockroach, the potential moves abruptly to a new value which is about 70 mV more negative, demonstrating the existence of a resting potential difference between the two faces of the axonal membrane. This value is significantly smaller than E_K, the equilibrium potential for potassium ions, which lies at around -90 mV.[4]

The slope relating the resting potential to the logarithm of the external potassium concentrations is about 42 mV for a tenfold change. For potassium concentrations below 8 mM the curve tends to level off. The slope differs by 17 mV from the 59-mV slope predicted by the Nernst equation for potassium ions:

$$E_K = \frac{RT}{F} \log \frac{[K]o}{[K]i} \tag{1}$$

where R is the gas constant, T is the absolute temperature, F is the Faraday, and [K]o and [K]i are the outside and inside potassium concentrations, respectively. This indicates that the axonal membrane does not behave as a perfect potassium electrode and that ions other than potassium participate in the generation of the resting potential.

B. RESTING POTENTIAL IN NERVE CELL BODIES

The resting potential (Er) recorded in nerve cell bodies is slightly lower than that recorded in axons under similar conditions. In *Periplaneta*[5] it lies at around -60 mV, vs. -46 mV in *Carausius*[6,7] and -50 mV in *Bombyx*.[8]

The relationship between Er and the external potassium concentration is similar to that

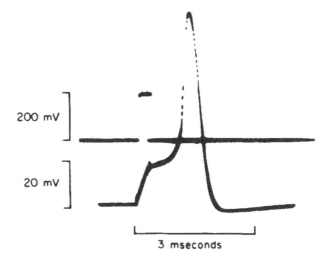

FIGURE 1.   Membrane action potential recorded from an isolated giant axon of *Periplaneta americana* (lower tracing) induced by a short duration depolarizing pulse (upper tracing). (From Pichon, Y. and Boistel, J., *J. Exp. Biol.*, 47, 343, 1967. With permission.)

reported for the axon, the slope of the linear portion of the curve being about 42 mV in excitable nerve cell bodies of the cockroach.[5]

C. IONIC BASIS OF ELECTRICAL ACTIVITY IN AXONS AND NERVE CELL BODIES

The fundamental property of nerve cells is their excitability, i.e., their ability to respond to a depolarization with a graded and/or an all-or-none action potential. In both cases the effect of the stimulus is to modify the ionic permeability of the membrane. These modifications have been studied in isolated axons.

1. Graded Activity and Action Potentials in Axons and Nerve Cell Bodies

When the membrane of a cockroach axon is depolarized above a certain level, a hump appears on top of the initial phase of the catelectrotonic potential. This "local response", which is not actively propagated along the axon, increases with stimulus intensity and, above the critical depolarization level (which lies around −40 mV), gives rise to an all-or-none depolarization, the action potential. The action potential, an example of which is shown in Figure 1, is propagated along the axon. Its amplitude and time course are essentially similar to those recorded in other unmyelinated axons. In a series of experiments on isolated axons of *P. americana,* the mean amplitude of the spike (the large and fast portion of the action potential which follows the stimulus) was found to be approximately 95 mV, the membrane potential changing its polarity (overshooting) by about +35 mV.[9] The maximum rates of rise and fall of the action potential were 1370 and 640 V/s, respectively. The spike was followed by a slight posthyperpolarization (the positive after-potential), followed in turn by an even smaller and slower postdepolarization (the negative after-potential). Both after-potentials were significantly smaller in isolated axons than *in situ.*[9] Such a difference most probably reflects modifications in the driving force for potassium ions and in the geometry (absence of glial cells around the axons in the isolated axonal preparation).

Active electrical responses have also been recorded from nerve cell bodies. In some cases these action potentials arise spontaneously or following depolarization of the cell.[5,7,8,10,13] Most cell bodies in insect ganglia, however, are unexcitable under normal conditions; they can be purely passive or can respond to direct electrical stimulation by graded oscillatory responses. These cells can produce action potentials following various treatments such as axotomy, colchicine, injection of citrate, or superfusion with tetraethylammonium ions.[13,17]

2. Effects of Ions on Action Potentials in Axons and Nerve Cell Bodies

The action potential in cockroach axons is correlated with the presence of sodium ions in the external solution. The slope relating the overshoot of the action potential to the external sodium concentration is not significantly different from the theoretical 59-mV slope for a tenfold change predicted from the Nernst equation for sodium ions:

$$E_{Na} = \frac{RT}{F} \log \frac{[Na]o}{[Na]i} \tag{2}$$

The action potential of nerve cell bodies was found to be sensitive to sodium in a few cases,[5,14] whereas calcium seemed to be responsible for action potential production in all other cases.[7,11,12,15,16]

The ionic channels which carry the inward current are different in axons and cell bodies. Axonal sodium channels were found to be permeable to sodium and lithium ions,[17a] but not to calcium ions. In a careful study of the effects of divalent cations on the maximum rate of rise of the action potential in giant axons of *Periplaneta*, Narahashi[18] found that calcium, magnesium, strontium, and barium ions did not contribute significantly to the action potential, but shifted the sodium inactivation curve (h_∞) toward lower membrane potentials.

Calcium channels from nerve cell bodies are blocked by magnesium ions[16] and are generally much less selective than sodium channels. Another important difference between sodium and calcium channels is the high sensitivity of the former to tetrodotoxin (TTX)[6,9,19] and saxitoxin (STX).[20]

3. Ionic Currents and Conductances in Axons

Voltage-clamp experiments have been carried out on isolated giant axons of *P. americana* using a double oil-gap technique.[21,22] Typical membrane currents generated under those conditions are illustrated in Figure 2. When the membrane potential is held at its resting level (Em = Eh = −60 mV), no current crosses the membrane. When the membrane potential is increased to 0 mV, a transient inward current begins; it reaches a peak value of about −5 mA/cm², then decreases and is followed by an outward current which reaches a plateau of about 6 mA/cm². When the membrane is reduced to its original holding potential value (Eh), the outward current returns to its resting value following an exponential time course.

Such current records can be obtained for various potential levels. The turning on and off of the inward current as well as the turning on of the delayed outward current become faster and faster as the membrane steps are made more positive. The inward current increases first, reaches a maximum for Em = −10 mV, and then decreases. The outward current increases continuously with step size. In all cases the outward tails of current which follow a return to the Eh level are exponential and proportional to the pulse size.

The current densities as well as the current-voltage characteristics are very similar[1,21] to those for the squid axon, described by Hodgkin et al.[23] As in other axons, the peak current is carried by sodium ions.[2,3,9] Since the sodium current is blocked selectively by TTX, it has been possible to separate the two ionic currents and study their kinetics.[3,9] These experiments have shown that the model proposed by Hodgkin and Huxley[24] to describe their experimental findings on the squid axon and to reconstruct most of the reactions of the nerve membrane can be used for the cockroach axon with only minor changes.

The cockroach nerve membrane can be described as being made of a capacitance (Cm) in parallel with three ionic channels: the sodium channel with a variable conductance to

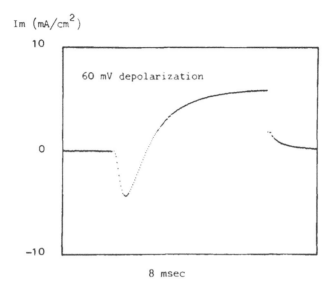

Im (mA/cm^2)

10

60 mV depolarization

0

−10

8 msec

FIGURE 2.   Ionic currents corresponding to a 5-ms depolarizing voltage pulse to 0 mV in a giant axon of *Periplaneta americana* under voltage-clamp conditions. Leak current subtracted. Holding potential (HP) = − 60 mV. (Reprinted with permission from Pichon, Y. and Aschroft, F. M., in *Comprehensive Insect Physiology, Biochemistry and Pharmacology*, Vol. 5, Kerkut, G. A. and Gilbert, L. I., Eds., Copyright 1985, Pergamon Books Ltd.)

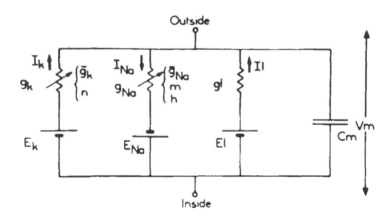

FIGURE 3.   Electrical equivalent circuit of the axonal membrane based on the Hodgkin-Huxley model. (From Pichon, Y, in *Insect Neurobiology*, Treherne, J. E., Ed., Elsevier/North-Holland, Amsterdam, 1974, 73. With permission.)

sodium ions (gNa), which are more concentrated outside than inside (equilibrium potential E_{Na}); the potassium channel with a variable conductance to potassium ions (gK), which are more concentrated inside than outside (equilibrium potential E_K); and a leak channel exhibiting a constant and relatively low conductance (presumably from leaking chloride ions; see Figure 3).

According to the formulation of Hodgkin et al.,[23] the potassium and sodium variable conductances are a function of a maximum conductance ($\bar{g}K$ or $\bar{g}Na$), and their time course and voltage dependence can be described by a set of three time- and voltage-dependent parameters which can vary between 0 and 1. These parameters are potassium activation (n),

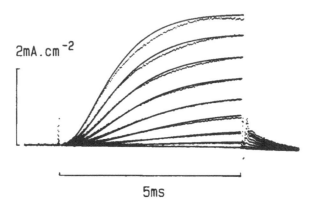

FIGURE 4.   Superimposed tracings of the potassium currents corresponding to 5-ms step depolarizations of the axonal membrane of *Periplaneta americana* under voltage-clamp conditions. The membrane was brought from its holding value of -60 to -50 to $+30$ mV in 10-mV steps. Leak current substracted and Na^+ current blocked with 0.1 μM TTX. (From Pichon, Y., Poussart, D., and Lees, G. V., in *Structure and Function in Excitable Cells,* Chang, D. C., Tasaki, I., Adelman, W. J., and Lenchtag, H. R., Eds., Plenum Press, New York, 1983, 211. With permission.)

sodium activation (m), and sodium inactivation (h). Each corresponds to the probability that a respective charge is located in a given region of the membrane. In cockroach axons, one potassium channel would open when three "n" charges simultaneously move to this region of the membrane. The potassium conductance would then be given by

$$gK = \bar{g}Kn^3 \qquad (3)$$

Figure 4 illustrates the fit between potassium current traces recorded in the presence of 0.1 μM TTX and theoretical curves.[4]

Steady-state values of n, m, and h for different potential values have been calculated.[9] From these data it has been possible to reconstruct, by calculation, the ionic currents corresponding to a step depolarization in a voltage-clamped cockroach axon and to reconstruct the action potential and the underlying conductance changes (Figure 5).

III. PHARMACOLOGICAL PROPERTIES OF COCKROACH AXONS

The majority of investigations on insect axons have used giant axons located in the connectives linking the abdominal ganglia in the central nervous system of *P. americana*.

The electrical properties of these axons and their modification by several drugs and insecticidal compounds were first examined by means of microelectrode recording techniques.[19,25,27] Later, a new recording method, the isolation of a single axon, was introduced by Pichon.[1] The ionic permeability changes underlying the observed electrical properties of these axons were characterized (as described in the previous section) together with the action of some pharmacological agents.[28]

The cockroach axonal model membrane preparation led to the discovery of an important new pharmacological tool in neurophysiology, 4-aminopyridine, a potent specific potassium conductance blocker.[29,32]

In the following section we shall describe the actions of specific sodium or potassium channel blockers, those of sea anemone and scorpion toxins, and the modifications induced by insecticide molecules.

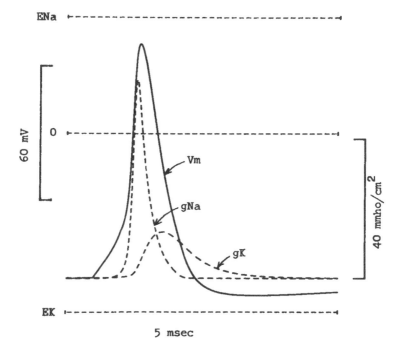

FIGURE 5. Computer reconstruction of a membrane action potential (continuous curve) and underlying conductance changes in a cockroach giant axon (interrupted curves). (Reprinted with permission from Pichon, Y. and Aschroft, F. M., in *Comprehensive Insect Physiology, Biochemistry and Pharmacology*, Vol. 5, Kerkut, G. A. and Gilbert, L. I., Eds., Copyright 1985, Pergamon Books Ltd.)

A. EFFECTS OF SODIUM AND POTASSIUM CHANNEL BLOCKERS
1. Sodium Channel Blockers

TTX, which blocks sodium channels in nerves from a wide range of animals, blocks action potentials in cockroach giant axons.[28,33] Voltage-clamp experiments on isolated axons have shown that TTX, at micromolar concentrations, selectively blocks the sodium current without affecting the delayed potassium current.[33]

STX, a naturally occurring compound obtained from the clam *Saxidomus giganteus,* the mussel *Mytilus californianus,* and from cultures of the dinoflagellate *Gonyaulax catenella,* is also a selective blocker of transient sodium in nerve membranes.[34] The structure of natural STX (nSTX) was established by X-ray crystallography, and the total synthesis of the toxin was reported by Tanino et al.[35] At 0.1 μM, synthetic STX (sSTX) applied externally to the isolated cockroach axon blocks the action potential in <3 min (Figure 6C). sSTX, like TTX, selectively blocks the sodium current of cockroach giant axons. STX is more potent than TTX, and its action is reversed more rapidly than that of TTX.[20] From the concentration dependence of sodium current suppression it is concluded that individual sodium channels are blocked by single molecules of STX which bind reversibly to part of the channel, with a dissociation constant of 3 nM.[20] The relative effectiveness of sSTX, nSTX, and natural TTX (nTTX) on cockroach axons has been determined; apparent K_d values are 3 nM (sSTX), 7 nM (nSTX), and 20 nM (nTTX). In the cockroach axonal preparation either STX or TTX can be used as the sodium channel blocker, although the former is preferred if reversibility is required (see Figure 2).

2. Potassium Channel Blockers

Until now, only a few chemicals and toxins were known to specifically block the voltage-dependent potassium conductance. In cockroach axonal preparations, tetraethylammonium

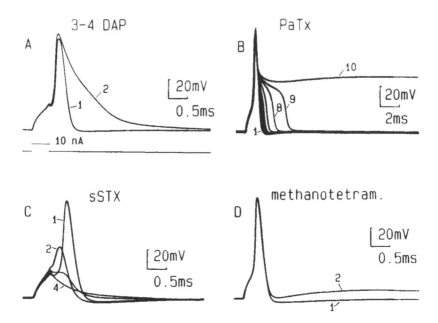

FIGURE 6.   Membrane action potentials of isolated cockroach axons. (A) 1 = control; 2 = 5 min after application of 3,4-diaminopyridine (10 μM). Lower tracing indicates the short duration (0.5-ms) depolarizing current pulse inducing the responses. (B) Superimposed recordings of the membrane action potentials during bath application of *Parasicyonis* toxin (1 μM). An action potential was evoked every 10 s. 1 = control; 8, 9, and 10 = after 3 min 40 s, 3 min 50 s, and 4-min application, respectively. (C) Superimposed recordings of axonal responses during a 3-min application of synthetic STX (0.1 μM), which progressively suppressed the action potential (1 to 4). (D) A control action potential (1) and an action potential followed by a postdepolarization (2) induced by an 8-min application of 2 μM methanotetramethrin.

(TEA$^+$) causes marked prolongation of action potentials and suppresses the delayed potassium current when applied at high concentrations (>0.01 M).[9]

4-Aminopyridine, in micromolar concentrations, slows the repolarization phase of the action potential in the cockroach axon (Figure 6A) and, thus, increases excitability. These effects are the consequence of a specific decrease in the potassium conductance.[29,32] This specific property has been confirmed in similar preparations, and several isomers and derivatives of 4-aminopyridine have been found to be more or less potent and, similarly, more or less selective for the potassium channels.[36] Aminopyridines and their derivatives are widely used today as pharmacological tools. Moreover, particular interactions of potassium channel blockers, i.e., time, voltage, and frequency dependence of the blockade, have revealed novel structure-activity relationships.[37,41] In addition, these molecules have been used to differentiate several populations of potassium channels.[42] The contribution of the cockroach axonal preparation to the discovery of aminopyridine[42] is an illustration of the value of the insect system to neurobiological research.

At the present time, only one natural toxin from the scorpion *Centruroides noxius* has been described as a specific blocker of the potassium channels in the squid axon.[43] Several experiments with similar scorpion toxins[43a] did not reveal any potent natural toxin interacting specifically with potassium channels of the cockroach axon.

B. EFFECTS OF SEA ANEMONE AND SCORPION TOXINS

Among the animal toxins, sea anemone and scorpion toxins have been widely used to study ionic channels. To date, 4 sea anemone toxins and 12 scorpion toxins have been

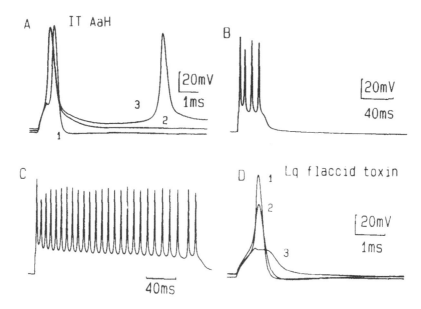

FIGURE 7.   Membrane action potentials of isolated cockroach axons recorded in the presence of scorpion "insect" toxins. (A) 1 = control action potential; 2 and 3 = superimposed recordings of the responses after a 5-min application of 1.3 μ*M* ITAaH (insect toxin from *Androctonus*). (B,C) After 7 and 12 min, respectively. (D) Three superimposed recordings (1 to 3) of the axonal responses during an application of 6 μ*M* insect toxin II, a "flaccid" toxin from *Leiurus* venom.

studied for their membrane effects on the cockroach axon. The sea anemone toxins prolong the action potential considerably (Figure 6B), resulting in long "plateau" potentials,[44,46] through a delay in the turning off of the sodium current. Among the scorpion toxins, α-toxins such as toxin II from *Androctonus australis* (also called Mammal Toxin II [MTII AaH]) and crustacean toxin (CT AaH) cause "plateau" potentials, do not induce repetitive activity, and act in a manner similar to anemone toxins.[46] Another compound much more toxic to insects, called "insect toxin" (IT AaH) by Zlotkin et al.,[47,48] was found in *Androctonus* venom. This material, when applied externally to the isolated cockroach axon, depolarizes by 5 to 10 mV and induces a maintained repetitive activity of "short" action potentials (Figure 7A,B,C).[46] In voltage-clamp procedures IT AaH increases and prolongs the sodium conductance, mainly at negative membrane potential values, through a slowing of the sodium current turnoff.

Two groups of insect-selective toxins have been purified from each of the venoms of *Buthotus judaicus* and *Leiurus quinquestriatus*. The first toxins (Bj IT$_1$ and Lq IT$_1$) induce a fast, excitatory contraction paralysis of fly larvae and have been called "contractive" toxins. The second toxins (Bj IT$_2$ and Lq IT$_2$), causing a slow, depressant, flaccid paralysis of fly larvae, are called "flaccid" toxins.[49,50] The contractive toxins, when applied to the isolated cockroach axon, act like the insect toxin from *Androctonus*, inducing repetitive firing. The underlying phenomenon is also the induction of a slow, voltage-dependent sodium conductance which has its maximum around Em = −30 mV and decreases for larger depolarizations. Conversely, flaccid toxins depolarize the axonal membrane and block the action potential (Figure 7D) by inhibiting sodium conductance.[49,50]

It is not clear from these studies why toxins differing in origin (anemone or scorpion), amino acid sequence, and molecular weight have similar actions, while chemically similar toxins from the same venom show species specificity in insects and modify the sodium channels. Today, scorpion toxins are widely used as pharmacological tools to study populations of ionic channels. In the future, a better understanding of the "discrete" causes of specificity should help in inventing new strategies in pharmacology and insect control and in understanding the development of resistance to neurotoxic agents.

C. EFFECTS OF DDT AND PYRETHROID INSECTICIDES

Several years ago, the action of the organochlorine insecticide dichlorodiphenyltrichloroethane (DDT) on the cockroach axon was described.[9,27] Effects were concentration dependent, ranging from 1 to 10 μM, and had virtually no effect on the resting or action potentials, but postdepolarization increased in amplitude.[9] Larger amounts of DDT (up to 0.5 mM) increased both the amplitude and the duration of the postdepolarization.

The underlying changes in the ionic conductances were complex. A decrease in potassium conductance was noted. The peak current was delayed slightly, and the turning off of the sodium current during the clamp pulse was slowed down. The most obvious change, however, was the occurrence of a long exponential tail of sodium current after the pulse. Analysis of the voltage and time dependency of the characteristic tail of current suggested a close correlation between the development of the tail current and the turning on of the peak sodium conductance.[51] Although all sodium channels open normally (as predicted from the activation and inactivation processes), a small number of them may become insensitive to changes in membrane potential and, thus, remain open longer. The number of open channels was calculated by integrating the sodium conductance against time, and the proportion of modified channels was estimated as 20% of the total population. Assuming an overall exponential decay in conductance for these modified channels, an almost perfect fit to the experimental data was obtained.[51]

Allethrin, a synthetic derivative of pyrethrin, stimulates insect nerves by inducing repetitive discharges[52] and develops large negative after-potentials in cockroach axons.[53] S-Bioallethrin, kadethrin, biotetramethrin, cismethrin, biopermethrin, bioresmethrin, and methanotetramethrin, type I pyrethroids,[56] have been applied to the isolated axon.[54,55] (With regard to methanotetramethrin, see also Reference 58 in Chapter 18.) All induce depolarizing after-potentials of long duration (Figure 6D), slightly depolarize the axonal membrane, and induce repetitive firing in whole cockroach cercal sensory nerves.[56] Only cismethrin develops repetitive activity consisting of slow oscillations of the membrane potential.[54]

Among type II pyrethroids, deltamethrin, a very potent insecticide when applied topically, apparently has much smaller direct effects on the axonal membrane,[54] although the toxicity is well correlated with its ability to depolarize by a block in the open state of a very small population of sodium channels.[55]

Pyrethroids have been found to selectively increase the sodium conductance. The time and voltage dependency of this increase has been analyzed quantitatively for two molecules: S-bioallethrin[51] and methanotetramethrin,[57] both type I pyrethroids. For small concentrations (up to 5 μM), the most obvious effect is the development of low voltage-sensitive sodium conductance which turns off very slowly at the end of the depolarizing voltage pulses, producing large tails of inward current (Figure 8).[55,57] The time and voltage dependencies of this tail have been studied quantitatively during the onset of pyrethroid action. It was found that, during the first minutes, the voltage sensitivity of the tail current followed that of unmodified (fast) sodium conductance quite closely, although the activation curve usually was shifted to the right toward more positive potentials.[55] Later, the voltage sensitivity of the pyrethroid-induced current was found to decrease, as indicated by the decrease in the slope of the activation curve. The extent to which these modifications can account for the observed changes in nerve activity has been evaluated through computer reconstruction of nerve membrane behavior during the early or late phase of poisoning.[57]

The cockroach axon reveals itself as a good model for testing insecticide molecules. Because sodium channels play such a key role in nerve activity, they are and will remain

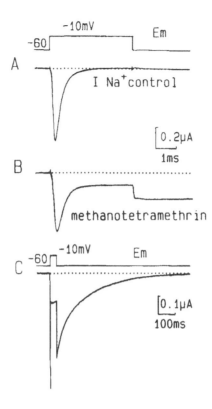

FIGURE 8. Sodium currents in a giant axon from a cockroach. (A) Control sodium current (lower tracing) after suppression of the potassium current in the presence of 3,4-diaminopyridine (0.2 mM). (B) Sodium current recorded after a 10-min application of methanotetramethrin (2 μM). (C) Sodium current during a 40-ms duration pulse to − 10 mV and large "tail" sodium current turning off slowly at the end of the voltage pulse.

one of the best target sites for neurotoxic insecticides. Nevertheless, insecticides are also highly neurotoxic to mammals as well as to insects when applied directly to the nervous system, and it has been demonstrated[57a] that the pyrethroids share the same mechanism of action on voltage-dependent sodium channels in the frog nerve membrane (see Chapter 18 in this work). Much of the selectivity of insecticides would appear to be due to differences in penetration, pharmacokinetics, or metabolism.

IV. PHARMACOLOGICAL PROPERTIES OF COCKROACH NEURONS IN CULTURE

A. EFFECTS OF ACETYLCHOLINE AND RELATED MOLECULES

Cultured cockroach neurons possess α-bungarotoxin binding sites and are depolarized by pressure-applied ACh. The electrical events associated with the effect of ACh and carbamylcholine (CCh) have been analyzed with the patch-clamp technique.[59]

Pressure-injection of micromolar concentrations of the agonists in the vicinity of the neuron induces a depolarization. Under voltage-clamp conditions, in the "whole cell" configuration of the patch-clamp technique, the inward current which underlies this depolarization is associated with a large increase in "noise". The spectrum of these current fluctuations can be fitted with one (CCh, corner frequency = 30 Hz) or two (ACh, corner frequencies = 8 and 150 Hz) lorentzian curves.

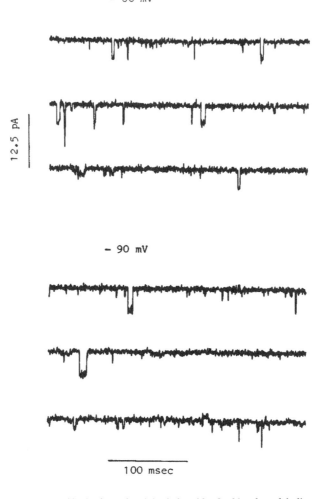

FIGURE 9.   Single channel activity induced by 5 µ*M* carbamylcholine
applied onto the membrane of a cultured cockroach neuron at two holding
potential levels.[59]

Addition of micromolar concentrations of ACh or CCh to the solution contained in the
patch pipette induces two categories of small unitary currents (Figure 9). The reversal
potential of these unitary currents lies between -5 and -15 mV, with a conductance of
ca. 18 to 48 pS. Their mean opening time is clearly longer in CCh than in ACh and, in the
former case, can be fitted by two exponential functions. The time constant of these functions
increases with membrane hyperpolarization.

B. EFFECTS OF GABA AND RELATED MOLECULES

About 60% of cultured neurons respond to externally applied GABA. The response,
which is hyperpolarizing in the majority of the cells, corresponds to an increase in the
chloride permeability of the cell membrane. It is blocked by 10 µ*M* picrotoxin, but is
insensitive to 10 µ*M* bicuculline. In the whole-cell configuration the cells respond to the
external application of GABA or muscimol with a small inward current on which is superimposed a large "noise" (Figure 10). Addition of 2 µ*M* flunitrazepam to the bathing solution
increases the current as well as the current fluctuations (Figure 10D).

The power spectrum of these current fluctuations can be fitted with a single lorentzian

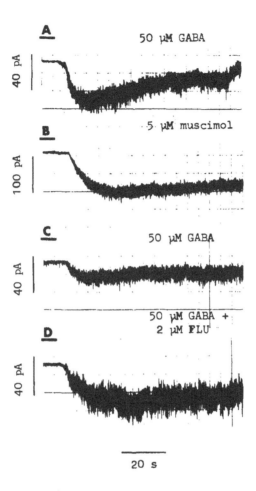

FIGURE 10. Examples of inward currents induced by prolonged application of GABA and muscimol onto insect neurons in the whole-cell configuration of the patch-clamp technique. (From Shimahara, T., Pichon, Y., Lees, G., Beadle, C. A., and Beadle, D. J., *J. Exp. Biol.*, 131, 231, 1987. With permission.)

function with corner frequencies of about 15 Hz (Figure 11). So far it has not been possible to record unitary currents which could be unequivocally related to the effects of GABA or its agonists.[60]

C. EFFECTS OF GLUTAMATE

Iontophoretic application of glutamate in the vicinity of the membrane of cultured cockroach neurons induces, in 30% of the cells tested, a depolarizing or a hyperpolarizing response corresponding to a decrease in membrane conductance. The equilibrium potential of the depolarizing response lies at around 0 mV, whereas that of the hyperpolarizing response is situated between −60 and −80 mV.[61]

L-Glutamate (1 to 10 μM) in the solution contained in the patch pipette in the cell-attached configuration (patch-clamp technique) induces short inward unitary currents. In cells pretreated with 1 μM concanavalin A (Con A) to reduce desensitization, the frequency of these unitary currents is very high and the events very short (mean open time ca. 0.2 ms; Figure 12). Single-event amplitude histograms are very broad, and there are indications that the corresponding channels may exhibit several conductance states.[61]

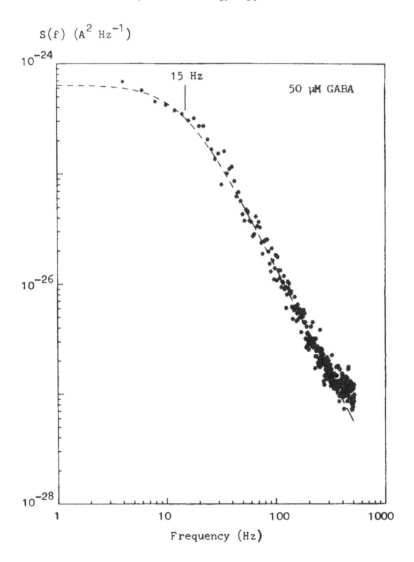

FIGURE 11.   Power spectrum of the noise induced by application of 50 μM GABA onto cultured cockroach neurons: the experimental data points can be fitted with a single lorentzian component with a corner frequency of 15 Hz. (From Shimahara, T., Pichon, Y., Lees, G., Beadle, C. A., and Beadle, D. J., *J. Exp. Biol.*, 131, 231, 1987. With permission.)

V. CONCLUSION

Large axons from four phyla have been used extensively with the voltage clamp technique.

- Molluscs: squid axon[24]
- Arthropods: lobster,[62] crayfish,[63] and cockroach[21]
- Annelids: *Myxicola*[64]
- Vertebrates: amphibians[65] and mammals[66]

As stated by Bertil Hille,[67] "Apparently all axons use the two major channel types first described in the squid giant axon. The simplicity of the excitability mechanism of axons is in accord with the simplicity of their task: to propagate every impulse unconditionally."

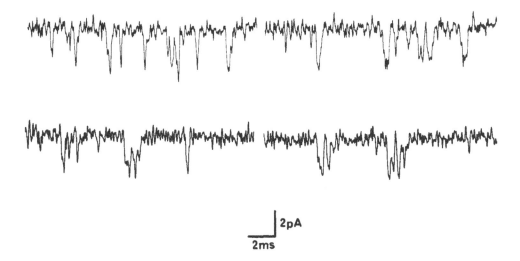

FIGURE 12. Single channel activity evoked by 0.1 m*M* glutamate in the presence of Con A in a cell-attached patch. (From Horseman, B. G., Seymour, C., Bermudez, I., and Beadle, D. J., *Neurosci. Lett.*, 85, 65, 1988. With permission.)

There are many advantages to using cockroaches as laboratory subjects. It takes little effort to rear large numbers, at low cost, in the laboratory; moreover, this is possible throughout the year. A wide range of neuroactive substances, such as aminopyridines, STX, TTX, anemone and scorpion venoms, pyrethroids, and other toxins, have similar effects in both cockroach and vertebrate preparations. The ease of making neural preparations coupled with the facts cited above justifies the use of cockroaches as biological models for vertebrates as well as for study in their own right. It is possible to study both axonal and synaptic responses in the neuron (see Chapter 8 in the present work). The electrophysiological actions of a compound on both voltage- and ligand-gated channels can be studied in such preparations. Newer data obtained with cell cultures of cockroach neurons (see also Chapter 12 in this work) supplement our methods of investigation and show that cockroach central nervous systems are choice animal models for biomedical studies.

REFERENCES

1. **Pichon, Y.,** Application de la technique du voltage imposé à l'étude de la fibre isolée d'insecte, *J. Physiol. (Paris)*, 9, 282, 1967.
2. **Pichon, Y.,** Nature des courants membranaires dans une fibre nerveuse d'insecte: l'axone de *Periplaneta americana*, *C.R. Soc. Biol.*, 162, 2233, 1968.
3. **Pichon, Y.,** Axonal conduction in insects, in *Insect Neurobiology*, Treherne, J. E., Ed., Elsevier/North-Holland, Amsterdam, 1974, 73.
4. **Pichon, Y., Poussart, D., and Lees, G. V.,** Membrane ionic currents, current noise and admittance in isolated cockroach axons, in *Structure and Function in Excitable Cells*, Chang, D. C., Tasaki, I., Adelman, W. J., and Lenchtag, H. R., Eds., Plenum Press, New York, 1983, 211.
5. **Jego, P., Callec, J. J., Pichon, Y., and Boistel, J.,** Etude électrophysiologique de corps cellulaires excitables du VIème ganglion abdominal de *Periplaneta americana;* aspects électriques et ioniques, *C.R. Soc. Biol.*, 164, 893, 1970.
6. **Treherne, J. E. and Maddrell, S. H. P.,** Membrane potentials in the central nervous system of the stick insect, *Carausius morosus, J. Exp. Biol.*, 46, 413, 1967.
7. **Orchard, I. and Finlayson, L. H.,** Electrical properties of identified neurosecretory cells in the stick insect, *Comp. Biochem. Physiol.*, 58, 87, 1977.

8. **Monticelli, G. and Depretts, G.**, Electrophysiological characteristics of *Bombyx mori* L. ventral nerve cord (effects of sodium and potassium on the membrane potential), *Experientia*, 35, 62, 1979.

9. **Pichon, Y.**, Aspects Électriques et Ioniques du Fonctionnement Nerveux chez les Insectes. Cas Particulier de la Chaîne Nerveuse Abdominale d'une Blatte, *Periplaneta americana* L., D.Sc. thesis, University of Rennes, Rennes, France, 1979.

10. **Callec, J. J. and Boistel., J.**, Etude de divers types d'activités électriques enregistrées par microélectrodes capillaires au niveau du dernier ganglion abdominal de *Periplaneta americana* L., *C.R. Soc. Biol.*, 160, 1943, 1966.

11. **Kerkut, G. A., Pitman, R. M., and Walker, R. J.**, Electrical activity in insect nerve cell bodies, *Life Sci.*, 7, 605, 1968.

12. **Orchard, I.**, Calcium-dependent action potentials in a peripheral neurosecretory cell of the stick insect, *J. Comp. Physiol.*, 112, 95, 1976.

13. **Jego, P., Callec, J. J., Pichon, Y., and Boistel, J.**, Etude électrophysiologique de corps cellulaires excitables du vième ganglion abdominal de *Periplaneta americana:* aspects électriques et ioniques, *C.R. Soc. Biol.* 164, 893, 1970.

14. **Goodman, C. and Heitler, W. J.**, Electrical properties of insect neurones with spiking and non-spiking somata: normal, axotomized, and colchicine treated neurones, *J. Exp. Biol.*, 83, 95, 1979.

15. **Pitman, R. M.**, The ionic dependence of action potentials induced by colchicine in an insect motorneurone cell body, *J. Physiol. (London)*, 247, 511, 1975.

16. **Pitman, R. M.**, Calcium-dependent action potentials in the cell body of an insect motorneurone, *J. Physiol. (London)*, 247, 62, 1975.

17. **Pitman, R. M.**, Intracellular citrate and externally applied tetraethylammonium ions produce calcium-dependent action potential in an insect motorneurone cell body, *J. Physiol. (London)*, 291, 327, 1979.

17a. **Pichon, Y. and Pelhate, M.**, unpublished data, 1974.

18. **Narahashi, T.**, Dependence of excitability of cockroach giant axons on external divalent cations, *Comp. Biochem. Physiol.*, 19, 759, 1966.

19. **Narahashi, T.**, The physiology of insect axons, in *The Physiology of the Insect Central Nervous System*, Treherne, J. E. and Beament, J. W. L., Eds., Academic Press, London, 1965, 1.

20. **Sattelle, D. B., Pelhate, M., and Hue, B.**, Pharmacological properties of axonal sodium channels in the cockroach *Periplaneta americana* L. I. Selective block by synthetic saxitoxin, *J. Exp. Biol.*, 83, 41, 1979.

21. **Pichon, Y. and Boistel, J.**, Current-voltage relations in the isolated giant axons of the cockroach under voltage-clamp conditions, *J. Exp. Biol.*, 47, 343, 1967.

22. **Pelhate, M. and Sattelle, D. B.**, Pharmacological properties of insect axons: a review, *J. Insect Physiol.*, 28, 889, 1982.

23. **Hodgkin, A. L., Huxley, A. F., and Katz, B.**, Measurement of current-voltage relations in the membrane of the giant axon of *Loligo*, *J. Physiol. (London)*, 116, 424, 1952.

24. **Hodgkin, A. L. and Huxley, A. F.**, Quantitative description of membrane current and its application to conduction and excitation in nerve, *J. Physiol. (London)*, 117, 500, 1952.

25. **Boistel, J. and Coraboeuf, E.**, Potentiel de membrane et potentials d'action de nerf d'insecte recueillis à l'aide de microélectrodes intracellulaires, *C.R. Acad. Sci.*, 238, 2116, 1954.

26. **Yamasaki, T. and Narahashi, T.**, Studies on the mechanism of action of insecticides. XIV. Intracellular microelectrode recordings of resting and action potentials from the insect axon and the effects of DDT on the action potential, *Botyu-Kagaku*, 22, 305, 1957.

27. **Narahashi, T.**, The properties of insect axons, *Adv. Insect Physiol.*, 1, 175, 1963.

28. **Pichon, Y.**, The pharmacology of the insect nervous system, in *The Physiology of Insecta*, Vol. 4, Rockstein, M., Ed., Academic Press, New York, 1974, 101.

29. **Pelhate, M., Hue, B., and Chanelet, J.**, Effets de la 4-aminopyridine sur le système nerveux d'un insecte: la Blatte (*Periplaneta americana* L.), *C.R. Soc. Biol.*, 166, 1598, 1972.

30. **Pelhate, M., Hue, B., and Chanelet, J.**, Modifications par la 4-aminopyridine des caractéristiques électriques de l'axone géant isolé d'un insecte, la Blatte (*Periplaneta americana* L.), *C.R. Soc. Biol.*, 168, 27, 1974.

31. **Pelhate, M., Hue, B., Pichon, Y., and Chanelet, J.**, Actions de la 4-aminopyrine sur la membrane de l'axone isolé d'insecte, *C.R. Acad. Sci.*, 278, 2807, 1974.

32. **Pelhate, M. and Pichon, Y.**, Selective inhibition of potassium current in the giant axon of the cockroach, *J. Physiol. (London)*, 242, 90, 1974.

33. **Pichon, Y.**, Effets de la tétrodotoxine (TTX) sur les caractéristiques de perméabilité membranaire de la fibre nerveuse isolée d'insecte, *C.R. Acad. Sci.*, 268, 1095, 1969.

34. **Narahashi, T., Haas, H. G., and Therrien, E. F.**, Saxitoxin and tetrodotoxin: comparison of nerve blocking mechanism, *Science*, 137, 1441, 1967.

35. **Tanino, H., Nakata, T., Kaneko, T., and Kishi, Y.**, A stereospecific total synthesis of d,l-saxitoxin, *J. Am. Chem. Soc.*, 99, 2818, 1977.

36. **Pichon, Y., Meves, H., and Pelhate, M.,** Effects of aminopyridines on ionic currents and ionic channel noise in unmyelinated axons, in *Aminopyridines and Similarly Acting Drugs — Effects on Nerves Muscles and Synapses,* Lechat, P., Thesleff, S., and Bowman, W. C., Eds, Pergamon Press, Oxford, 1982, 53.

37. **Yeh, J. Z., Oxford, G. S., Wu, C. H., and Narahashi, T.,** Interactions of aminopyridines with potassium channels of squid axon membranes, *Biophys. J.,* 16, 77, 1976.

38. **Yeh, J. Z., Oxford, G. S., Wu, C. H., and Narahashi, T.,** Dynamics of aminopyridine block of potassium channels in squid axon membrane, *J. Gen. Physiol.,* 68, 519, 1976.

39. **Meves, H. and Pichon, Y.,** The effect of internal and external 4-aminopyridine on the potassium currents in intracellularly perfused squid giant axons, *J. Physiol. (London),* 268, 511, 1977.

40. **Meves, H. and Pichon, Y.,** Modèle d'action de la 4-aminopyridine au niveau des "canaux" potassium de l'axone géant de Calmar (*Loligo forbesi* L.), *C.R. Acad. Sci.,* 284, 1325, 1977.

41. **Kirsch, G. E. and Narahashi, T.,** 3,4-Diaminopyridine. A potent new potassium channel blocker, *Biophys. J.,* 22, 507, 1978.

42. **Dubois, J. M.,** Evidence for the existence of three types of potassium channels in the frog ranvier node membrane, *J. Physiol. (London),* 318, 297, 1981.

42a. **Lechat, P., Thesleff, S., and Bowman, W. C., Eds.,** *Aminopyridines and Similarly Acting Drugs: Effects on Nerves, Muscles and Synapses,* Pergamon Press, Elmsford, New York, 1982.

43. **Carbone, E., Prestipino, G., Spadavecchia, L., Franciolini, F., and Possani, L. D.,** Blocking of the squid axon K^+ channel by noxius toxin: a toxin from the venom of the scorpion *Centruroides noxius,* *Pflügers Arch.,* 408, 423, 1987.

43a. **Pelhate, M.,** unpublished data, 1987.

44. **Pelhate, M., Hue, B., and Sattelle, D. B.,** Pharmacological properties of axonal sodium channels in the cockroach *Periplaneta americana* L. II. Slowing of sodium current turn-off by *Condylactis* toxin, *J. Exp. Biol.,* 83, 49, 1979.

45. **Pelhate, M., Laufer, J., Pichon, Y., and Zlotkin, E.,** Effects of several sea anemone and scorpion toxins on excitability and ionic currents in the giant axon of the cockroach, *J. Physiol. (Paris),* 79, 309, 1984.

46. **Pelhate, M. and Zlotkin, E.,** Actions of insect toxin and other toxins derived from the venom of the scorpion *Androctonus australis* on the isolated giant axons of the cockroach (*Periplaneta americana*), *J. Exp. Biol.,* 97, 67, 1982.

47. **Zlotkin, E., Miranda, F., Kupeyan, C., and Lissitzky, S.,** A new toxic protein in the venom of the scorpion *Androctonus australis* Hector, *Toxicon,* 9, 9, 1971.

48. **Zlotkin, E., Miranda, F., and Lissitzky, S.,** A toxic factor to crustacean in the venom of the scorpion *Androctonus australis* Hector, *Toxicon,* 10, 211, 1972.

49. **Lester, D., Lazarovici, P., Pelhate, M., and Zlotkin, E.,** Purification, characterization and action of two insect toxins from the venom of the scorpion *Buthotus judaicus, Biochim. Biophys. Acta,* 701, 370, 1982.

50. **Zlotkin, E., Kadouri, D., Gordon, D., Pelhate, M., Martin, M. F., and Rochat, H.,** An excitatory and a depressant insect toxin from scorpion venom both affect sodium conductance and possess a common binding site, *Arch. Biochem. Biophys.,* 240, 877, 1985.

51. **Pichon, Y., Guillet, J. Cl., Heilig, U., and Pelhate, M.,** Recent studies on the effects of DDT and pyrethroid insecticides on nervous activity in cockroaches, *Pestic. Sci.,* 16, 627, 1985.

52. **Narahashi, T.,** Effects of the insecticide allethrin on membrane potentials of cockroach giant axons, *J. Cell. Comp. Physiol.,* 59, 61, 1962.

53. **Narahashi, T.,** Nature of the negative after-potential increased by the insecticide allethrin in cockroach giant axons, *J. Cell. Comp. Physiol.,* 59, 67, 1962.

54. **Laufer, J., Roche, M., Pelhate, M., Elliott, M., Janes, N. F., and Sattelle, D. B.,** Pyrethroid insecticides: actions on insect axons, *J. Insect Physiol.,* 30, 341, 1984.

55. **Laufer, J., Pelhate, M., and Sattelle, D. B.,** Actions of pyrethroid insecticides on insect axonal sodium channels, *Pestic. Sci.,* 16, 651, 1985.

56. **Gammon, D. W., Brown, M. A., and Casida, J. E.,** Two classes of pyrethroid action in the cockroach, *Pestic. Biochem. Physiol.,* 15, 181, 1981.

57. **Pichon, Y., Pelhate, M., and Heilig, U.,** Sodium channels: primary targets for insecticide action?, *Am. Chem. Soc. Proc.,* 356, 212, 1987.

57a. **Vijverberg, H. P. M. and de Weille, J. R.,** The interaction of pyrethroids with voltage-dependent Na channels, *Neurotoxicology,* 6, 23, 1985.

58. **Pichon, Y. and Aschroft, F. M.,** Nerve and muscle: electrical activity, in *Comprehensive Insect Physiology, Biochemistry and Pharmacology,* Vol. 5, Kerkut, G. A. and Gilbert, L. I., Eds, Pergamon Press, Oxford, 1985, 85.

59. **Beadle, C. A., Beadle, D. J., Pichon, Y., and Shimahara, T.,** Patch-clamp and noise analysis studies of cholinergic properties of cultured cockroach neurones, *J. Physiol. (London),* 371, 145P, 1985.

60. **Shimahara, T., Pichon, Y., Lees, G., Beadle, C. A., and Beadle, D. J.,** Gamma-aminobutyric acid receptors on cultured cockroach brain neurones, *J. Exp. Biol.,* 131, 231, 1987.

61. **Horseman, B. G., Seymour, C., Bermudez, I., and Beadle, D. J.,** The effects of L-glutamate on cultured insect neurones, *Neurosci. Lett.,* 85, 65, 1988.

62. **Julian, F. J., Moore, J. W., and Goldman, D. E.,** Membrane potentials of the lobster giant axon obtained by use of the sucrose-gap technique, *J. Gen. Physiol.,* 45, 1195, 1962.

63. **Shrager, P.,** Ionic conductance changes in voltage clamped crayfish axons at low pH, *J. Gen. Physiol.,* 64, 666, 1974.

64. **Goldman, L. and Schauf, C. L.,** Quantitative description of sodium and potassium currents and computed action potentials in *Myxicola* giant axons, *J. Gen. Physiol.,* 61, 361, 1973.

65. **Dodge, F. A. and Frankenhaeuser, B.,** Membrane currents in isolated frog nerve fibre under voltage clamp conditions, *J. Physiol. (London),* 143, 76, 1958.

66. **Chiu, S. Y., Ritchie, J. M., Rogart, R. B., and Stagg, D.,** A quantitative description of membrane currents in rabbit myelinated nerve, *J. Physiol. (London),* 292, 149, 1979.

67. **Hille, B.,** *Ionic Channels of Excitable Membranes,* Sinauer Associates, Sunderland, MA, 1984.

Chapter 8

ELECTROPHYSIOLOGY AND PHARMACOLOGY OF SYNAPTIC TRANSMISSION IN THE CENTRAL NERVOUS SYSTEM OF THE COCKROACH

Bernard Hue and Jean-Jacques Callec

TABLE OF CONTENTS

I. INTRODUCTION

In experimental neuropharmacology, studies are based on the use of various nerve preparations. The objective of such research is the development of more refined methods of recording that minimize the use of complicated nervous structures. Obviously, giant cells offer a breadth of possibility for investigation at the single-cell level, but they present specific anatomy and physiology. In this respect, Arthropoda, the largest phylum in the animal kingdom, including Insecta, are well known for their contribution to electrophysiology.

The attractiveness of the giant fiber's relatively accessible structure and suitable size has focused research on the study of synaptic transmission in the terminal abdominal ganglion (A6) of the American cockroach, *Periplaneta americana*. This preparation has provided us with the earliest and the most numerous studies on insect synaptic function and is particularly useful for studies including the effects of diverse neuroactive agents on the synaptic membrane (e.g., neurotransmitter receptors and ion channels).

In this chapter we present evidence that this simple insect nervous preparation is suitable for neuropharmacological applications to biomedical research.

II. GENERAL PHYSIOLOGY AND PHARMACOLOGY

The central nervous system (CNS) of insects, confined to the nerve cord, consists of a pair of ganglia fused in the midline and linked together by paired interganglionic connectives. Each part is surrounded by the nerve sheath comprising an outer layer of connective tissue, the neural lamella. In ganglia, below the neural lamella, the perineurium is formed from a single inner layer of glial cells. The cell bodies of unipolar moto- and interneurons form more or less well-defined groups in the outer glial layer. Synaptic areas lie within the neuropil, which is usually restricted to the core of the ganglion.[1] The physiological importance of the nerve sheath has been demonstrated, especially in the cockroach nerve cord, where it is regarded as a specialized type of blood-brain barrier which restricts the penetration of extraneously applied substances within the ganglion.[2-4]

A. ANATOMY OF GIANT INTERNEURONS

Detailed neuroanatomical studies performed many years ago revealed the structural organization of the intraganglionic processes of giant interneurons (GI) of *P. americana*. Most of these investigations were related to the cercal afferent-GI system of the A6 ganglion. Cobalt chloride impregnation of fibers has been done either at the pre- or postsynaptic level, resulting in staining both cercal afferents and GI. Filling of GI was achieved with microelectrodes inserted into the GI cell body[5] or by retrograde migration via an isolated giant axon.[6,7] The morphologies of the three largest GI (referring to the diameter of the axon) of the ventral group were grossly similar. The cell body of each GI lies contralaterally to its axon, on the lateral edge of the ganglion. A neurite projects from each cell body and crosses to the opposite side of the ganglion. Numerous fine processes branch from each neurite in the contralateral half of the ganglion and near the midline (Figure 1). Each GI has many primary dendrites from which fine branches are seen to ramify through the neuropil constituting the dendritic tree. In contrast, the morphology of GI of the dorsal group is less homogeneous.[5,7] The simultaneous staining of afferent fibers issued from cercal nerves X and XI (nomenclature of Roeder et al.[8]) and ventral GI 1, 2, and 3 provided evidence that terminals of presynaptic fibers were distributed in the ipsilateral part of the A6 ganglion containing the dendritic tree. An overlap between the projections of cercal nerves X and XI and the dendritic area of GI was observed (Figure 1).

The localization of central synapses in this ganglion resulted not only from neuroanatomical studies, but especially from electrophysiological investigations developed by Callec.[9-11]

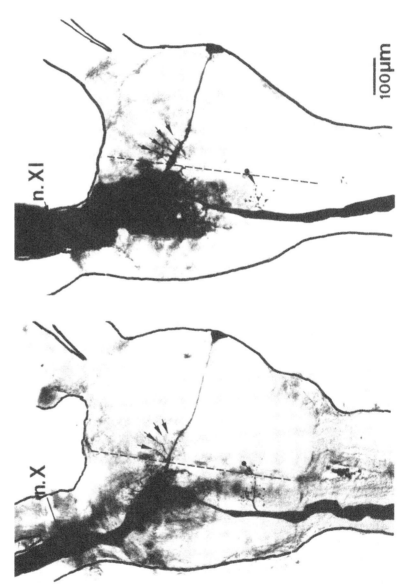

FIGURE 1.   Photomicrograph of cobalt sulfide staining of caudal nerves (n. X and n. XI) and giant interneuron 2 in the A6 ganglion of

The main excitatory and inhibitory pathways of the cercal afferent-GI synaptic system have been identified accurately, promoting this preparation as a suitable model for detailed pharmacological investigations.[12]

B. ELECTROPHYSIOLOGICAL METHODS

Investigation of the electrophysiology of synaptic transmisssion in the CNS of the cockroach was pioneered by Roeder et al.[13] Electrophysiology studies of GI in the A6 ganglion started with the use of microelectrodes[14] and later advanced with the implementation of a more suitable technique, the single-fiber oil-gap method.[9]

1. Single-Fiber Oil-Gap Method

This technique necessitates the microdissection of a small length of a giant axon between the sixth and fifth abdominal ganglia of the nerve cord.

a. Principle

The principle of this method is based on the measurement of potential between two points via external electrodes separated by a high-resistance medium. The potential measured (V) by these electrodes will deviate from the true membrane potential (V_m) by a short-circuiting factor, such that

$$V = \frac{R_e}{R_e + R_i} V_m$$

where R_e = longitudinal resistance of the external medium and R_i = longitudinal resistance of the internal medium.

For this method (Figure 2), paraffin oil is employed to increase Re; the short-circuiting factor, $\dfrac{R_e}{R_e + R_i}$, is then close to unity.

Moreover, V sampled by the electrodes fades if the axon is dissected too far from synaptic sites. The space constant of GI is 1300 μm.[15] Recording of synaptic events in such favorable conditions exhibits high amplitude and is devoid of noise due to the low resistance of the recording electrode (<1 MΩ).

b. Dissection

The experiments normally were performed on adult male cockroaches. A portion of the nerve cord, including the A6 ganglion, the cercal nerves, and the cerci, was removed from the insect and placed on a glass slide in a drop of saline. If necessary, the nerve sheath was removed from the dorsal part of the A6 ganglion. One of the connectives linking the fifth and sixth ganglia was desheathed, and one of the GI of the ventral group was isolated under a binocular microscope. It appeared that selection during dissection of a particular GI was reasonably accurate. All other axons in the connective were cut, and the preparation was transferred to a Perspex experimental chamber. Finally, the intact connective was severed, leaving only one isolated axon linking the fifth and sixth abdominal ganglia bathing in paraffin oil.

A simplification of the oil-gap technique may be performed by recording all or a part of the electrical activity on the connective linking the fifth and sixth ganglia. It is referred to as the mannitol-gap technique since the resistance of the external medium is elevated with sucrose.[16] It is technically less demanding and readily enables the construction of dose-response curves.

In most of the experiments reported here, the A6 ganglion was permanently superfused by normal saline containing 208 mM NaCl, 3.1 mM KCl, and 5.4 mM CaCl$_2$. The pH value

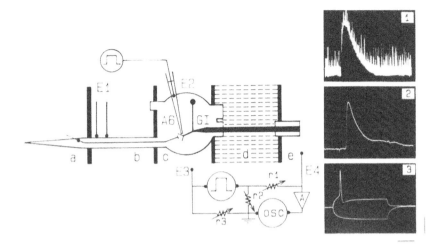

FIGURE 2. Schematic representation of the experimental recording apparatus used in the single-fiber oil-gap method.[93] Left: the experimental chamber and recording circuit; a = cercus compartment, b = cercal nerve compartment with electrical stimulating electrodes (E_1), c = A6 ganglion compartment irrigated with saline (E_3, reference electrode), d = gap compartment filled with paraffin oil in which the isolated giant axon is floating, e = recording electrode (E_4) compartment; iontophoretic electrode (E_2) is impaled within the neuropil of the A6 ganglion. A wheatstone bridge included in the recording circuit allows application of depolarizing or hyperpolarizing currents through the membrane; r1, r2, and r3 are variable resistances used to balance the wheatstone bridge. A = amplifier, osc = oscilloscope. Right: samples of recordings obtained with the single-fiber oil-gap method; 1 = ACh iontophoretic-induced potential superimposed by numerous unitary EPSPs, 2 = composite EPSP, 3 = action potential and hyperpolarizing potential produced by passing DC currents through the GI membrane. Horizontal scale: 1, 2 s; 2, 10 ms; 3, 10 ms. Vertical scale: 1, 2 mV; 2, 2 mV; 3, 40 mV.

was adjusted to 7.2 with either sodium bicarbonate buffer or HEPES. In some recent experiments 26 mM sucrose was added to the saline.[7]

2. Experimental Procedure

As we focus on pharmacological aspects, some details concerning the experimental procedure must be specified.

Because access to molecules within the neuropil most often is restricted by the nerve sheath, most of the pharmacological tests have been done on a desheathed A6 ganglion. Several modes of application of drugs have been used. First, the drugs can be solubilized in saline and applied externally by superfusion. Second, a large-tipped glass micropipette may be used for local topical application. Furthermore, iontophoretic injection has been employed for polar compounds such as acetylcholine (ACh). This method, developed by Callec[9] and later refined,[17] was particularly useful in demonstrating the putative postsynaptic effect of drugs.

C. ELECTROPHYSIOLOGY OF CERCAL AFFERENT-GIANT INTERNEURON SYNAPSES

The pharmacological investigations on synaptic mechanisms mentioned in this chapter will be devoted to bioelectrical activity and especially to postsynaptic potentials.

The electrophysiology of synapses has been studied particularly for GI 1, 2, and 3, which receive sensory information from afferent fibers reaching the A6 ganglion through cercal nerves X and XI. Cercal nerve X contains two groups of fibers; one emerges from a

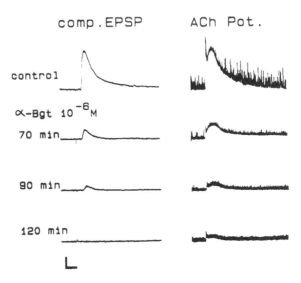

FIGURE 3.   Effects of 1 μ*M* α-bungarotoxin on composite (comp.) EPSP and ACh iontophoretic-induced potential (ACh Pot.) in the A6 ganglion of the cockroach. A parallel decrease in amplitude of both phenomena is observed. Vertical scale: 1 mV; horizontal scale: 10-ms Comp. EPSP and 0.5-s ACh Pot.

chordotonal organ located at the base of each cercus. Electrical stimulation delivered on cercal nerve X triggers an inhibitory postsynaptic potential (IPSP) in GI. The other group is formed with axons derived from mechanoreceptors located on the ipsilateral half of the supra-anal plate and a few campaniform sensilla on the lateral extended portion of the same plate. The activation of this latter group of fibers evokes an excitatory postsynaptic potential (EPSP) in GI.[18]

Cercal nerve XI contains sensory fibers which originate from mechanoreceptors of the cerci and which project mainly in the ipsilateral half of the A6 ganglion. The mechanical or electrical activation of a mechanoreceptor or the electrical stimulation of presynaptic fibers give rise, respectively, to a monosynaptic unitary or composite EPSP in GI. The synaptic delay was estimated to be approximately 0.68 ms;[9] nevertheless, delayed polysynaptic depolarizations have been observed in response to increased stimulation.

The ionic mechanisms underlying the IPSP and EPSP were investigated by classical methods, i.e., estimation of the reversal potential under current-clamp conditions and modification of ionic composition of the saline. It appeared that the IPSP is mainly driven by chloride currents,[7,9] whereas the EPSP, whose reversal potential cannot be evaluated easily, is generated by sodium and potassium currents.[9]

D. PHARMACOLOGY OF RECEPTORS

1. Cholinergic Receptors

Extensive data have considerably strengthened the conclusion that ACh is an excitatory transmitter in the cockroach CNS.[4,9,19-24] It appeared that ACh effects were similar in part to those induced by nicotine and muscarine, suggesting the hypothesis of the existence of at least three distinct putative cholinergic receptors: nicotinic, muscarinic, and mixed. In cockroach GI, this pharmacological relationship has been characterized using radioligand binding, electrophysiology, and histoautoradiography.

As in other nervous tissues, the nicotinic probe α-bungarotoxin (α-Bgt) was used extensively in the characterization of the nicotinic synaptic component of insects (Figure 3). Recently reviewed[25] data concerning the biochemistry of insects indicate that the ^{125}I-α-Bgt binding component from extracts prepared from *Drosophila* heads closely resembles a nicotinic ACh receptor. Properties similar to those described for *Drosophila* have been reported in other insects, particularly in the abdominal nerve cord of *P. americana*. Table 1, derived from data of Lummis and Sattelle,[26] summarizes the pharmacology of α-Bgt binding sites of various animal tissue extracts.

Quinuclidinyl benzilate (QNB), a muscarinic ligand,[34] has been used in studies on CNS extracts of the locust[32] and of *P. americana*.[26] A saturable specific component of ³H-QNB binding is present in these insects. Quite recently, a specific binding of *N*-methyl-³H-scopolamine was found in the cockroach nerve cord.[35] Pharmacological properties of this component have led us to the conclusion that there is a muscarinic receptor in the insect CNS. Furthermore, a mixed "nicotinic-muscarinic" putative receptor has been characterized from the nerve cord of the cockroach *Gromphadorhina portentosa*.[36]

The American cockroach CNS preparation is of particular interest because it readily enables direct comparison between ligand binding data and electrophysiological experiments on GI. It was shown that α-Bgt (10 nM), when bath-applied to the A6 ganglion, fully and irreversibly blocked unitary as well as composite EPSPs without any modification of the resting potential and membrane resistance.[37,38] Of particular significance was the finding that the active concentration of α-Bgt used in physiological studies was close to its K_d determined in binding experiments.[39] In contrast, QNB was found ineffective on synaptic transmission, even at concentrations as high as 1 μM.

Studies performed on neuronal tissue of head and thorax of *Drosophila*[40] had shown that the ¹²⁵I-α-Bgt binding is localized in synaptic areas of the neuropil. Parallel studies carried out in the cockroach revealed important binding on either half of the A6 ganglion,[38] corresponding to the bilateral localization of synaptic areas between cercal nerve endings and GI dendritic trees.

The good agreement between biochemical, electrophysiological, and autoradiographic results enhances the conclusion that the ¹²⁵I-α-Bgt binding component represents nicotinic receptors involved in postsynaptic functions in cercal afferent-GI synapses. There is no electrophysiological evidence for other (muscarinic) receptors.

2. GABA Receptors

The presence of γ-aminobutyric acid (GABA)[41,42] and its synthesizing enzyme, glutamic acid decarboxylase (GAD),[43] in relatively large amounts in the nerve cord of the cockroach, coupled with successful electrophysiological tests,[9,10] suggested to us that GABA was an inhibitory neurotransmitter in the insect CNS.

In order to characterize the pharmacological properties of the GABA receptor, biochemical investigations have been developed recently. They revealed specific properties quite different from those described in the vertebrate CNS. For example, it was shown that the separate classes of GABA receptors (i.e., GABA$_A$ and GABA$_B$) were not found in the insect CNS (Table 2). Furthermore, the recent investigation of binding of the benzodiazepine ligand ³H-flunitrazepam and that of ³⁵S-*t*-butylbicyclophosphorothionate, a ligand which binds to the barbiturate/picrotoxin receptor site of the vertebrate GABA receptor complex, confirmed that insect receptors possess pharmacological properties distinct from mammalian GABA receptor subtypes.[35]

Electrophysiological data have revealed that GABA receptors are involved in inhibitory pathways between cercal nerve fibers and GI. When GABA was either bath-applied (3 to 10 mM) or iontophoretically injected into the A6 ganglion, a decrease in membrane resistance and hyperpolarization of the postsynaptic membrane of GI occurred. These effects were reversibly antagonized with picrotoxin.[9,10] Bath-applied high concentrations of GABA were required to induce an effect. It was later determined that this was due to an active uptake mechanism localized principally in glial cells.[45] The ionic mechanisms underlying effects

TABLE 1
Pharmacological Profiles of Putative Nicotinic Cholinergic Receptors Determined by Inhibition of Radiolabeled α-Bungarotoxin Binding to Tissue Extracts

Ligand	K_i(M) (binding)							V_{50}(M) (electrophysiology)
	Chick optic lobe[27]	Rat diaphragm[28]	Aplysia californica (CNS ganglia)[29]	Limulus polyphemus (brain)[30]	Drosophila melanogaster (heads)[31]	Locusta migratoria (head ganglia)[32]	Periplaneta americana (nerve cords)[26]	Periplaneta americana (D_f)[33]
α-Bungarotoxin	—	—	—	—	500 pM	700 pM	850 pM	64 nM
Nicotine	120 nM	—	—	7 µM	800 nM	40 nM	3.5 µM	2.3 µM[a]
d-Tubocurarine	250 nM	200 nM	2 µM	2 µM	2 µM	30 µM	850 nM	80 µM
Gallamine	3.8 µM	—	2 µM	—	50 µM	—	14 µM	1.5 mM
Acetylcholine	37 µM	470 nM	200 µM	—	20 µM	400 µM	18 µM[b]	11 µM[a]
Atropine	85 µM	—	30 µM	300 µM	50 µM	100 µM	20 µM	100 µM
Decamethonium	27 µM	2.1 µM	200 µM	20 µM	—	800 µM	130 µM	2.8 mM
Hexamethonium	250 µM	120 µM	300 µM	—	—	800 µM	140 µM	800 µM
Succinylcholine	—	—	—	—	—	—	280 µM	2.8 mM
Pancuronium	—	110 nM	—	—	—	—	210 µM	150 µM
Mecamylamine	—	—	—	>1 mM	—	1 mM	25 µM	—
Quinuclidinyl benzilate	—	—	—	—	—	—	140 µM	160 µM

Note: V_{50} was the concentration that produced 50% inhibition of the depolarizing response to iontophoretic application of acetylcholine onto the cell body membrane of the fast coxal depressor motoneuron (D_f).

a Concentration that produced 50% maximum depolarization of D_f.[26]

b IC_{50} used in estimation of K_i was determined from preincubation experiments.

TABLE 2
IC$_{50}$ Values (Concentrations which Inhibit 50% of Specific
^{3}H-GABA Binding) for Various Ligands in the Cockroach
CNS Preparation Compared with Values for a Vertebrate
(Rat Brain) Preparation

		IC$_{50}$(M)	
		Vertebrate CNS	
Ligand	Insect CNS	GABA$_A$	GABA$_B$
GABA	130 ± 50 nM	120 ± 15 nM	80 ± 10 nM
Muscimol	710 ± 50 nM	15 ± 3 nM	5.4 ± 1.1 μM
Thiomuscimol	30 ± 3 μM	*120 ± 10 nM	n.t.
THIP	390 ± 60 μM	*130 ± 5 nM	i.a.
3-APS	390 ± 20 μM	110 ± 15 nM	10 ± 1 μM
Isoguvacine	450 ± 90 μM	190 ± 90 nM	i.a.
Bicuculline methiodide	i.a.	27 ± 3.6 μM	i.a.
Baclofen	i.a.	i.a.	130 ± 50 nM
P4S	i.a.	*34 ± 2 nM	i.a.
Nipecotic acid	i.a.	i.a.	i.a.
Picrotoxin	i.a.	i.a.	i.a.
β-Alanine	i.a.	n.t.	n.t.

Note: Insect IC$_{50}$ values (mean ± SEM) were determined with data from three or
more experiments with each analogue using at least five different concentrations. Vertebrate data are taken from Reference 95, with the exception of data
marked "*", which are from Reference 44. For vertebrate data: i.a. (inactive),
IC$_{50}$ > 100 μM; n.t., not tested. For insect data: i.a., IC$_{50}$ > 1 μM. ^{3}H-
GABA concentrations: insect CNS = 30 nM; vertebrate CNS — GABA$_A$ =
5 nM, GABA$_B$ = 10 nM. THIP = 4,5,6,7-tetrahydroisoxazolo[5,4-c]pyridin-
3-ol; P4S = piperidine-4-sulfonic acid; 3-APS = 3-aminopropanesulfonic
acid.

From Lummis, S. C. R. and Sattelle, D. B., *Neurosci. Lett.*, 9, 287, 1986. With
permission.

of GABA and IPSP were mainly chloride dependent[7] (Figure 4). Autoradiographic analysis
of ^{3}H-GABA binding sites in Na$^+$-free saline revealed a neuropilar localization of GABA
receptor sites, which is in agreement with that of synaptic areas of the A6 ganglion.[26]
Pharmacological tests also showed that bicuculline methiodide, commonly used as a GABA-
receptor ligand in vertebrates, did not antagonize the IPSP or GABA effects.[7] Finally, the
GABA receptor of the insect CNS seems to be formed from a complex molecule linked to
the chloride ion channel. The GABA-sensitive site should be associated with benzodiazepine
and barbiturate receptor sites linked to the chloride ionophore. Although this configuration
was close to that described in vertebrates, the insect GABA receptor complex differs slightly
in pharmacological properties. Furthermore, it has been suggested that GABA modulates
ACh release at the cercal afferent-GI synapse via presynaptic inhibition.[46,47]

3. Other Receptors

The widespread distribution of the biogenic amine octopamine in the insect CNS does
not necessarily account for a putative role in central neurotransmission, even if it is now
well established that octopamine regulates neuromuscular transmission in the locust.[48] In
the A6 ganglion of the cockroach, neutral red staining has been particularly successful in
demonstrating the presence of octopamine in large dorsal unpaired median (DUM) cells.[49]

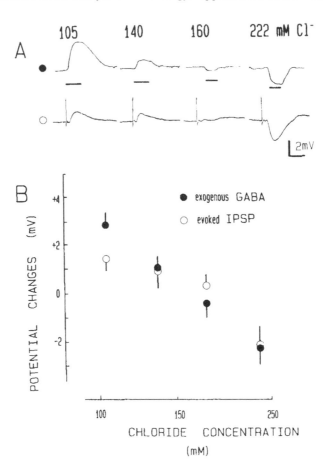

FIGURE 4.   Effect of changing external chloride concentration on the
evoked IPSP and the bath-applied GABA potential in a GI in the A6
ganglion of the cockroach *Periplaneta americana*. Decreasing chloride
ion concentration reverses the two potentials at about the same value.
Each point on the semilogarithmic plot is the mean ± SE of three to
six experiments. Temperature, 20°C; GABA concentration, 20 m*M*.
Horizontal scale: exogenous GABA, 2 min; evoked IPSP, 10 ms.

However, the lack of electrophysiological data in this field does not permit definitive conclusions on a putative role for octopamine in the regulation of cercal afferent-GI synaptic
transmission.

III. SYNAPTIC ACTION OF NEUROACTIVE AGENTS

In this third section of this chapter we shall mainly be concerned with events at the
cercal afferent-GI synapse of the cockroach, but will also briefly consider other pharmacological interactions which influence the insect CNS.

A. 4-AMINOPYRIDINE

It is now well known that some aminopyridines (APs), such as 4-aminopyridine (4-AP),
are selective blockers of the potassium channel at the axonal membrane. This was first
demonstrated at the GI axonal membrane of the cockroach by Pelhate et al.[50] Moreover,
APs were known to facilitate synaptic transmission at central synapses[51] as well as at

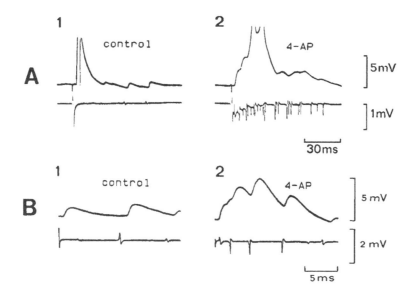

FIGURE 5.   Effects of 50 μM 4-AP on the composite EPSP (A) and unitary EPSP (B) recorded in GI no. 2. Note in A2 and B2 that EPSPs are enhanced and reinforced by action potentials spontaneously evoked in presynaptic afferents (cercal nerve XI) after 10 min of action. For each panel, the upper trace represents the postsynaptic event and the lower trace the presynaptic control derived from ipsilateral cercal nerve XI by external electrodes. (From Hue, B., D.Sc. thesis, University of Angers, Angers, France, 1983. With permission.)

peripheral synapses.[52] At the cercal afferent-GI synapse of the cockroach, APs (1 to 10 μM) — particularly 4-AP, the most potent isomer — increased the postsynaptic background activity, i.e., the frequency and amplitude of unitary EPSPs and IPSPs. At the presynaptic level, 4-AP triggered action potentials in cercal nerve XI. These action potentials appeared singly or repetitively, especially after subthreshold electrical stimulation of cercal nerve XI (Figure 5). Monosynaptic correlation with unitary EPSPs was observed. The synaptic depressant effect of Mg^{2+} was antagonized by 4-AP. All of these events were observed without changes in the resting potential, the postsynaptic membrane resistance, the sensitivity of postsynaptic cholinoreceptors, or the equilibrium potentials of ions involved in postsynaptic events. It was suggested that APs facilitate central synaptic transmission by causing an increase in transmitter release.[51] Higher concentrations of AP (100 μM to 10 mM) always induced depolarization of postsynaptic membranes, even after blockade of muscarinic and nicotinic receptors by atropine and *d*-tubocurarine, respectively.[53] 2- and 3-AP were approximately ten times less potent than 4-AP and produced a more reversible depolarization. These postsynaptic depolarizing effects of APs were not induced by activation of postsynaptic cholinoreceptors.

B. AMANTADINE AND PIRIBEDIL

We shall report here the synaptic actions of two agents (amantadine and piribedil) whose properties in the pharmacology of the vertebrate CNS have resulted in therapies for Parkinson's disease. Both molecules have been tested in synaptic transmission of *P. americana* in order to elucidate a putative role for cholinergic receptors.

The well-known antiviral agent amantadine[54] has also been used as an anti-Parkinson's drug.[55,56] The latter use was correlated with its pharmacological properties in the field of catecholamine metabolism, uptake, synthesis, or release.[57] However, binding experiments on membrane preparation from the electric organ of *Torpedo ocellata* showed that amantadine

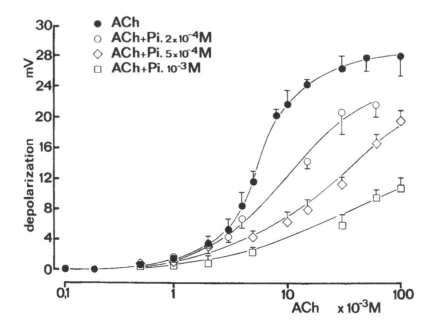

FIGURE 6.   Semilogarithmic dose-response curves of the depolarizing action of ACh before and after pretreatment with three doses of piribedil (Pi.).[61] Each point is the mean ± SE of ten experiments.

interacted with the ionophore of the ACh-receptor complex,[58] and pharmacological investigations performed on the heads of flies demonstrated that binding of ^{3}H-H$_{12}$-histrionicotoxin was inhibited by amantadine.[59]

On the cercal afferent-GI synapses of the cockroach, amantadine (20 μM to 1 mM) reduced the amplitude of both unitary and composite EPSPs, resulting in a blockade of synaptic transmission in a dose-dependent manner. A Hill coefficient of 0.94 calculated from dose-response curves revealed no cooperativity (i.e., an interaction of order greater than one) in the binding of amantadine to its site of action. Moreover, a shift to the right in the ACh dose-response curve was obtained after treatment of the A6 ganglion with amantadine (50 μM). No protection against α-Bgt (50 μM) blocking action was observed, indicating separate sites of binding for each molecule. Finally, tests on ACh extrasynaptic receptors in the cell body of the fast coxal depressor motoneuron (D$_f$) of the cockroach led to the conclusion that interaction of amantadine with the ACh-receptor complex was strongly dependent on membrane potential. An action of amantadine at the open ACh receptor-ion channel complex has been suggested.[60]

A parallel approach was carried out to investigate the putative interaction of piribedil (1-[2-pyrimidyl]4-piperonyl piperazine) with the ACh receptor of the cockroach CNS.[61] This molecule, known to possess a dopaminergic agonist action,[62] also has been promoted as an anti-Parkinson's drug.[63-65] However, a single agonist dopaminergic action of piribedil could not yield the observed pharmacological effects induced by this molecule.[66] The cockroach CNS has been employed to demonstrate the synaptic effect of this drug. Piribedil (10 μM to 1 mM), which did not alter nervous conduction in pre- and postsynaptic fibers, depressed unitary and composite EPSPs in a dose-dependent relationship. No fluctuations in resting membrane potential or postsynaptic membrane resistance were noted. The piribedil dose-dependent shift to the right of the ACh semilogarithmic dose-response curve revealed a noncompetitive interaction of piribedil with postsynaptic ACh receptors (Figure 6).

C. TAURINE

Taurine has been proposed as a putative neurotransmitter in the vertebrate CNS, where

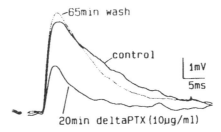

FIGURE 7.   Effect of δ-philanthotoxin on the amplitude of subthreshold composite EPSP.[74] At a concentration of 10 μg/ml, the toxin causes a slowly reversible blockage of 50%.

pharmacological and electrophysiological investigations have shown that this amino acid mimics the action of GABA. Therefore, taurine has been considered a partial agonist of GABA. In insects, where GABA is also a possible inhibitory transmitter, taurine has been tested using mannitol- and oil-gap methods on the cercal afferent-GI synapse of the cockroach, where it depresses synaptic transmission.[67-69] On the postsynaptic membrane, taurine has hyperpolarizing effects brought about by an increase in chloride conductance.

On the presynaptic cercal afferents, taurine produces a depolarization antagonized by picrotoxin. Furthermore, inhibition of the taurine synaptic blockade by theophylline suggests competition between these molecules in the excitation-secretion coupling.

D. INVERTEBRATE AND VERTEBRATE TOXINS

The effects of toxins and venoms extracted from various animal species on synaptic transmission have been tested in the cockroach. We shall report two examples in order to illustrate both groups.

Evidence suggests that most of the solitary wasps (Hymenoptera Aculeata) sting their prey (i.e., insects, spiders) into or in the direction of the CNS.[70,71] The question then arises whether or not the venom injected into the ganglia can interfere with central synaptic transmission and/or axonal conduction. Morphological changes induced by the venom of *Philanthus triangulum* have been detected in the CNS of worker honeybees.[72]

The electrophysiological demonstration of central synaptic action of the venom and of the most active toxin, δ-philanthotoxin (δ-PTX), was performed on the cockroach A6 ganglion.[73,74] The venom of *P. triangulum*, at a concentration lower than the wasp might inject into the ganglion, induced a slight depolarization of the postsynaptic membrane of a GI test preparation. This was accompanied by a decrease in membrane resistance and evoked a composite EPSP. We now have increasing evidence that the main site of action of δ-PTX is at the central synaptic level. It appears that δ-PTX (20 μg/ml) does not affect the excitability of the cockroach giant axon significantly, but causes a slowly reversible block of synaptic transmission from cercal nerve XI to GI without any change in the resting membrane potential (Figure 7). Furthermore, investigations on the putative interaction with the ACh-receptor ionophore complex of GI have suggested that δ-PTX could interfere with the ion channel linked to the receptor protein. The ion channel blocking action of δ-PTX might be facilitated by an appropriate agonist, causing massive opening of channels in the postsynaptic membrane. In fact, it has been established that venoms of Hymenoptera often contain considerable amounts of agonists such as ACh, histamine, serotonin, and kinins.[70,71] When measured under δ-PTX treatment, the decay time of iontophoretic-induced ACh potential was not modified. This suggests that, as was first thought,[75] *Philanthus* toxin does not inhibit acetylcholinesterase in the insect CNS.

Small peptides extracted from the venom of the scoliid wasp *Megascolia flavifrons* have been studied recently using cockroach GI.[76] As a result of this study, five different active fractions have been recognized. Three fractions contained identified substances: histamine, Thr[6] bradykinin, and Thr[6] bradykinin-Lys-Ala (megascoliakinin). Two fractions were active, but have not been identified chemically; both of them, as well as bradykinin, blocked the cercal afferent-GI synapses.

A novel probe, κ-bungarotoxin (κ-Bgt), isolated from the venom of *Bungarus multicinctus* has been used on the cockroach CNS recently.[77] While α-toxins bind to pharmacologically nicotinic sites in autonomic ganglia and the CNS, this binding usually fails to inhibit the action of ACh at neuronal nicotinic receptors. For example, α-Bgt fails to block nicotinic transmission in rat and chick sympathetic ganglia and in the chick ciliary ganglion.[78] Nevertheless, there are CNS nicotinic receptors that are sensitive to this particular toxin, notably the nicotinic receptors of the fish and amphibian tectum[79] and those of certain invertebrates, including insects.[38] κ-Bgt blocked nicotinic synaptic transmission without affecting electrical excitability in the chick ciliary ganglion.[80] κ-Bgt, therefore, appeared to be a specific probe for postsynaptic neuronal nicotinic receptors in a wide variety of neuronal preparations.

In the cockroach CNS, the toxin (κ-Bgt) was applied by superfusion to the A6 ganglion. It was noted that κ-Bgt (0.1 to 1 μ*M*) reduced and then blocked EPSPs and the iontophoretically induced ACh potential. The mean time required to achieve a full block was 180 min for 0.1 μ*M* κ-Bgt. The fact that the ACh potential was fully blocked indicated an unequivocally postsynaptic action on nicotinic GI postsynaptic receptors. These results clearly demonstrate, for the first time, that insect neuronal nicotinic receptors are recognized by this novel probe. It is now conceivable that, unlike the situation in vertebrates, both κ- and α-Bgt bind to the same receptor molecule in invertebrates. In this context, there is evidence that insect receptors may have a different subunit composition than their vertebrate counterparts.[81,82] These latest results provide evidence for evolutionary changes in the recognition site of neuronal receptor protein.

E. INSECTICIDES

Most insecticides currently in use act on nervous tissue. Consequently, central synaptic effects involved in their neurotoxicity have been of considerable interest. In this section we address the neurophysiological effects of selected insecticides on cockroach GI synapses.

1. Isothiocyanate Compounds

The synthetic insecticide 2-isothiocyanotoethyltrimethylammonium iodide was characterized as a potent inhibitor of choline acetyltransferase purified from fly heads.[83] A direct putative postsynaptic effect on nicotinic receptors of the cockroach had been investigated in parallel experiments using biochemistry[84] and electrophysiology.[85] It was demonstrated that 2-isothiocyanotoethyltrimethylammonium iodide induced half-maximal blockage of ^{125}I-α-Bgt binding to insect extracts at concentrations similar to those required to produce half-maximal blockage of the composite EPSP recorded at the cercal afferent-GI synapses. The conclusion drawn was that the synaptic effects of this molecule could be explained by its binding to the α-Bgt-sensitive ACh receptors, as it inhibits toxin binding at physiologically relevant concentrations.

2. Nicotine

Still in use, nicotine is one of the oldest known insecticides. It is well known that nicotine acts at the nicotinic receptor level in the central and peripheral nervous systems. Although direct interaction between nicotine and ACh receptors by biochemical methods had been demonstrated,[86,87] the electrophysiological effects remained unknown. The electrophysiological proof was provided by experiments carried out on cercal afferent-GI synapses,[84] in which the concentration of nicotine needed to inhibit 50% of the composite EPSP was estimated to be 200 n*M*.

3. Nereistoxin

Nereistoxin (NTX) was initially isolated from a marine worm, *Lumbriconereis heteropoda*;[88] subsequently, synthetic NTX was prepared.[89] NTX is a potent neurotoxin in insects. At low concentrations (20 n*M* to 1 μ*M*), NTX produced partial blockage of cholinergic synaptic transmission in the A6 ganglion of the cockroach. Only partial recovery was observed. Further analysis performed in voltage-clamp experiments on the cockroach D_f indicated a potent voltage-dependent blocking action of NTX. Thus, it was concluded that NTX acted preferentially at the ionic channel linked to the ACh receptor.[90]

4. Pyrethroids

The main effect of pyrethroids on axonal membranes is to stabilize the sodium channel in the open state[91] (see also Chapter 18 of this work). Although this action may principally explain pyrethroid neurotoxicity, other properties were observed, especially at the cholinergic synaptic level. For example, using biochemical methods, Abbassy et al.[92] presented evidence of interaction of pyrethroids with the nicotinic ACh receptor complex of *Torpedo*. The effects of two pyrethroids (deltamethrin and tralomethrin) on the ACh receptor of GI have been investigated recently.[93] Both compounds, at 1 μ*M* concentrations, produced identical effects on synaptic transmission. Thus, in <10 to 20 min there was first observed a drastic increase in the amplitude of composite EPSP eliciting action potentials in the GI. In most cases, this was accompanied by a weak depolarization which eventually reached a plateau (up to 5 mV). This early facilitating phase was rapidly followed by a decrease of the EPSP leading to an irreversible blockade of the synapses. Furthermore, these pyrethroids induced a parallel decrease of the iontophoretic ACh-induced potential (Figure 8). At the presynaptic level, it was noted that deltamethrin strongly depolarized the cercal nerve endings.[94] Thus, two synaptic sites were suggested to account for the observed action of these molecules. A presynaptic site would be associated with a transient increase in synaptic efficiency and a long-term decrease in ACh release. The slowly developed synaptic interaction with postsynaptic ACh receptor could be ascribed to an irreversible block of the synapse.

IV. CONCLUSION

The cercal afferent-GI synapse of the cockroach, monitored with the latest electrophysiological techniques, provides a particularly suitable neurophysiological preparation for the routine study of the effects of numerous neuroactive agents, such as putative neurotransmitters, fractions of venoms and toxins, insecticides, and drugs. Cockroaches are readily available and, furthermore, do not present any ethical problems.

Finally, even though during the last 15 years enormous progress has been made in the area of GI anatomy, synaptic connections, and pharmacology of neurotransmitters, we need to devote more energy to increasing our knowledge of the physiology of the insect CNS.

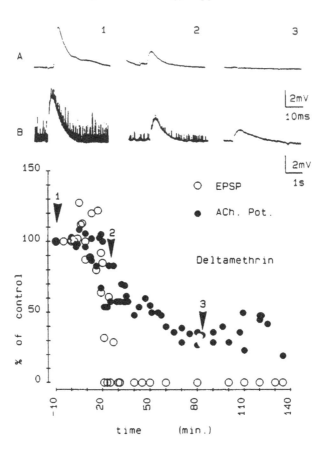

FIGURE 8.   Time course of action of deltamethrin (1 μM) on the composite EPSP and the depolarizing potential resulting from ionto-phoretic application of ACh (ACh. Pot.).[93] Normalized values (referred to control) were plotted against time. Control value was that obtained 10 min before the beginning of pyrethroid application. Test solution was applied during a space-time of 90 min from t = 0. Recording of EPSP (A) and ACh. Pot. (B) occurred at three periods (arrows) of the experiment. Open circles: EPSP amplitude; filled circles: ACh. Pot. amplitude.

REFERENCES

1. **Wigglesworth, V. B.**, The nutrition of the central nervous system of the cockroach *Periplaneta americana*. The mobilization of reserves, *J. Exp. Biol.*, 37, 500, 1960.
2. **Smith, D. S. and Treherne, J. E.**, Electron microscopical localization of acetylcholinesterase activity in the central nervous system of an insect *(Periplaneta americana)*, *J. Cell Biol.*, 26, 445, 1965.
3. **O'Brien, R. D.**, *Insecticides, Action and Metabolism*, Academic Press, New York, 1967.
4. **Pitman, R. M.**, Transmitter substances in insects: a review, *Comp. Gen. Pharmacol.*, 2, 347, 1971.
5. **Daley, D. L., Vardi, N., Appignani, B., and Camhi, J. M.**, Morphology of the giant interneurons and cercal nerve projections of the American cockroach, *J. Comp. Neurol.*, 196, 41, 1981.
6. **Harrow, I. D., Hue, B., Pelhate, M., and Sattelle, D. B.**, Cockroach giant interneurones stained by cobalt-backfilling of dissected axons, *J. Exp. Biol.*, 84, 341, 1980.
7. **Hue, B.**, Electrophysiologie et Pharmacologie de la Transmission Synaptique dans le Système Nerveux Central de la Blatte, *Periplaneta americana*, D.Sc. thesis, University of Angers, Angers, France, 1983.
8. **Roeder, K. D., Tozian, L., and Weiant, E. A.**, Endogenous nerve activity and behaviour in the mantis and cockroach, *J. Insect Physiol.*, 15, 45, 1960.

9. **Callec, J. J.**, Etude de la Transmission Synaptique dans le Système Nerveux Central d'un Insecte *(Periplaneta americana)*. Aspects Électrophysiologiques et Pharmacologiques, D.Sc. thesis, University of Rennes, Rennes, France, 1972.

10. **Callec, J. J.**, Synaptic transmission in the central nervous system of insects, in *Insect Neurobiology*, Treherne, J. E., Ed., Elsevier, New York, 1974, 119.

11. **Callec, J. J.**, Synaptic transmission in the central nervous system, in *Comprehensive Insect Physiology, Biochemistry and Pharmacology*, Vol. 5, Kerkut, G. A. and Gilbert, L. I., Eds., Pergamon Press, Oxford, 1985, 139.

12. **Callec, J. J. and Callec, M.**, Les poisons du système nerveux, *Pour la Science*, 92, 90, 1985.

13. **Roeder, K. D., Kennedy, N. K., and Samson, E. A.**, Synaptic conduction to giant fibers of the cockroach and the action of anticholinesterases, *J. Neurophysiol.*, 10, 1, 1947.

14. **Callec, J. J. and Boistel, J.**, Etude de divers types d'activités électriques enregistrées par microélectrodes capillaires au niveau du dernier ganglion abdominal de la Blatte, *Periplaneta americana* L., *C.R. Soc. Biol.*, 160, 1943, 1966.

15. **Boistel, J.**, Quelques caractéristiques électriques de la membrane de la fibre nerveuse au repos *(Periplaneta americana)*, *C.R. Soc. Biol.*, 153, 1009, 1959.

16. **Callec, J. J. and Sattelle, D. B.**, A simple technique for monitoring the synaptic actions of pharmacological agents, *J. Exp. Biol.*, 59, 725, 1973.

17. **Callec, J. J., David, J. A., and Sattelle, D. B.**, Ionophoretic application of acetylcholine onto the dendrites of an identified giant interneurone (G.I. 1) in the cockroach *Periplaneta americana*, *J. Insect Physiol.*, 28, 1003, 1982.

18. **Bernard, J., Gobin, B., and Callec, J. J.**, A chordotonal organ inhibiting giant interneurones in the sixth abdominal ganglion of the cockroach, *J. Comp. Physiol.*, 153, 377, 1983.

19. **Farley, R. D. and Milburn, N. S.**, Structure and function of the giant fibre system in the cockroach, *Periplaneta americana*, *J. Insect Physiol.*, 15, 457, 1969.

20. **Colhoun, E. H.**, Distribution of choline acetylase in insect conductive tissue, *Nature (London)*, 182, 1378, 1958.

21. **Colhoun, E. H.**, Acetylcholine in *Periplaneta americana*. I. Acetylcholine levels in nervous tissue, *J. Insect Physiol.*, 2, 108, 1958.

22. **Colhoun, E. H.**, Acetylcholine in *Periplaneta americana*. II. Acetylcholine and nervous activity, *J. Insect Physiol.*, 2, 117, 1958.

23. **Sattelle, D. B., McClay, A. S., Dowson, R. J., and Callec, J. J.**, The pharmacology of an insect ganglion: actions of carbamylcholine and acetylcholine, *J. Exp. Biol.*, 64, 13, 1976.

24. **Dagan, D. and Sarne, Y.**, Evidence for cholinergic nature of cockroach giant fibers. Use of specific degeneration, *J. Comp. Physiol.*, 126, 157, 1978.

25. **Sattelle, D. B.**, Acetylcholine receptors, in *Comprehensive Insect Physiology, Biochemistry and Pharmacology*, Vol. 11, Kerkut, G. A. and Gilbert, L. I., Eds., Pergamon Press, Oxford, 1985, 395.

26. **Lummis, S. C. R. and Sattelle, D. B.**, Binding of N-[propionyl-^{3}H]-propionylated α-bungarotoxin and L-[benzilic-4,4'-^{3}H]-quinuclidinyl benzilate to CNS extracts of the cockroach *Periplaneta americana*, *Comp. Biochem. Physiol.*, 80C, 75, 1985.

27. **Wang, G. K., Molinaro, S., and Schmidt, J.**, Ligand responses of α-bungarotoxin binding sites from skeletal muscle and optic lobe of the chick, *J. Biol. Chem.*, 253, 8507, 1978.

28. **Colquhoun, D. and Rang, H. P.**, Effects of inhibitors on the binding of iodinated α-bungarotoxin to acetylcholine receptors in rat muscle, *Mol. Pharmacol.*, 12, 519, 1976.

29. **Shain, W., Greene, L. A., Carpenter, D. O., Sykowski, A. J., and Vogel, Z.**, *Aplysia* acetylcholine receptors: blockade by a binding of α-bungarotoxin, *Brain Res.*, 72, 225, 1974.

30. **Thomas, W. E., Brady, R. N., and Townsel, J. G.**, A characterization of α-bungarotoxin binding in the brain of the horseshoe crab *Limulus polyphemus*, *Arch. Biochem. Biophys.*, 187, 53, 1978.

31. **Dudai, Y.**, Properties of an α-bungarotoxin-binding cholinergic nicotinic receptor from *Drosophila melanogaster*, *Biochim. Biophys. Acta*, 539, 505, 1978.

32. **Breer, H.**, Properties of putative nicotinic and muscarinic cholinergic receptors in the central nervous system of *Locusta migratoria*, *Neurochem. Int.*, 3, 43, 1981.

33. **David, J. A. and Sattelle, D. B.**, Actions of cholinergic pharmacological agents on the cell body membrane of the fast coxal depressor motoneurone of the cockroach *(Periplaneta americana)*, *J. Exp. Biol.*, 108, 119, 1984.

34. **Yamamura, H. I. and Snyder, S. H.**, Muscarinic cholinergic binding in rat brain, *Proc. Natl. Acad. Sci. U.S.A.*, 71, 1725, 1985.

35. **Lummis, S. C. R. and Sattelle, D. B.**, Binding sites for [^{3}H]GABA, [^{3}H]flunitrazepam and [^{35}H]STBPS in insect CNS, *Neurosci. Lett.*, 9, 287, 1986.

36. **Aziz, S. A. and Eldefrawi, M. E.**, Cholinergic receptors of the central nervous system of insects, *Pestic. Biochem. Physiol.*, 3, 168, 1973.

37. **Sattelle, D. B., David, J. A., Harrow, I. D., and Hue, B.,** Actions of α-bungarotoxin on identified central neurones, in *Receptors for Neurotransmitters, Hormones and Pheromones in Insects,* Sattelle, D. B., Hall, L. M., and Hildebrand, J. G., Eds., Elsevier/North-Holland, Amsterdam, 1980, 125.

38. **Sattelle, D. B., Harrow, I. D., Hue, B., Pelhate, M., Gepner, J. I., and Hall, L. M.,** α-Bungarotoxin blocks excitatory synaptic transmission between cercal sensory neurones and giant interneurone 2 of the cockroach *Periplaneta americana, J. Exp. Biol.,* 107, 473, 1983.

39. **Sattelle, D. B.,** Acetylcholine receptors of insects, *Adv. Insect Physiol.,* 15, 215, 1980.

40. **Schmidt-Nielsen, B. K., Gepner, J. I., Teng, N. N. H., and Hall, L. M.,** Characterization of an α-bungarotoxin binding component from *Drosophila melanogaster, J. Neurochem.,* 29, 1013, 1977.

41. **Ray, J. W.,** The free amino acid pool of the cockroach *(Periplaneta americana)* central nervous system and the effect of insecticides, *J. Insect Physiol.,* 10, 587, 1964.

42. **Ray, J. W.,** The free amino acid pool of cockroach *(Periplaneta americana)* central nervous system, in *The Physiology of the Insect Central Nervous System,* Treherne, J. E. and Beament, J. W. L., Eds., Academic Press, London, 1965, 31.

43. **Oliver, G. W. O., Taberner, P. V., Rick, J. T., and Kerkut, G. A.,** Changes in GABA level, GAD and ChE activity in CNS of an insect during learning, *Comp. Biochem. Physiol.,* 38B, 529, 1971.

44. **Krosgaard-Larsen, P. and Falch, E.,** GABA agonists: development and interactions with the GABA receptor complex, *Mol. Cell. Biochem.,* 31, 105, 1981.

45. **Hue, B., Gabriel, A., and Le Patezour, A.,** Autoradiographic localization of [³H]-GABA accumulation in the sixth abdominal ganglion of the cockroach, *Periplaneta americana* L., *J. Insect Physiol.,* 28, 753, 1982.

46. **Hue, B. and Callec, J. J.,** Presynaptic inhibition in the cercal afferent-giant interneurone synapses of the cockroach, *Periplaneta americana* L., *J. Insect Physiol.,* 29, 741, 1983.

47. **Blagburn, J. M. and Sattelle, D. B.,** Presynaptic depolarization mediates presynaptic inhibition at a synapse between an identified mechanosensory neurone and giant interneurone in the first instar cockroach, *Periplaneta americana, J. Exp. Biol.,* 127, 135, 1987.

48. **Evans, P. D.,** Octopamine, in *Comprehensive Insect Physiology, Biochemistry and Pharmacology,* Vol. 11, Kerkut, G. A. and Gilbert, L. E., Eds., Pergamon Press, Oxford, 1985, 499.

49. **Dymond, G. R. and Evans, P. D.,** Biogenic amines in the nervous system of the cockroach, *Periplaneta americana:* association of octopamine with mushroom bodies and dorsal unpaired median (DUM) neurones, *Insect Biochem.,* 9, 535, 1979.

50. **Pelhate, M., Hue, B., Pichon, Y., and Chanelet, J.,** Action de la 4-aminopyridine sur la membrane de l'axone isolé d'insecte, *C.R. Acad. Sci.,* 278, 2807, 1974.

51. **Hue, B., Pelhate, M., Callec, J. J., and Chanelet, J.,** Synaptic transmission in the sixth ganglion of the cockroach: action of 4-aminopyridine, *J. Exp. Biol.,* 65, 517, 1976.

52. **Molgo, J., Lemeignan, M., and Lechat, P.,** Effects of 4-aminopyridine at the frog neuromuscular junction, *J. Pharmacol. Exp. Ther.,* 203, 653, 1977.

53. **Hue, B., Pelhate, M., Callec J. J., and Chanelet, J.,** Postsynaptic effects of 4-aminopyridine in the sixth abdominal ganglion of the cockroach, *Eur. J. Pharmacol.,* 49, 327, 1978.

54. **Jackson, G. G., Muldoon, R. L., and Akers, L. W.,** Serological evidence for prevention of influenza infection in volunteers by an anti-influenza drug, adamantanamine hydrochloride, *Antimicrob. Agents Chemother.,* 3, 703, 1963.

55. **Schwab, R. S., England, A. C., Poskancer, D. C., and Young, R. R.,** Amantadine in the treatment of Parkinson's disease, *J. Am. Med. Assoc.,* 208, 1168, 1969.

56. **Grelak, R. P., Clark, R., Stump, J. M., and Vernier, V. G.,** Amantadinedopamine interaction: possible mode of action in Parkinsonism, *Science,* 169, 203, 1970.

57. **Brown, F. and Redfern, P. H.,** Studies on the mechanism of action of amantadine, *Br. J. Pharmacol.,* 58, 561, 1976.

58. **Warnick, J. E., Maleque, M. A., Bakry, N., Eldefrawi, A. T., and Albuquerque, E. X.,** Structure activity relationships of amantadine. I. Interaction of the *N*-alkyl analogues with the ionic channels of the nicotinic acetylcholine receptor and electrically excitable membrane, *Mol. Pharmacol.,* 22, 82, 1982.

59. **Eldefrawi, A. T., Shaker, N., and Eldefrawi, M. E.,** Binding of acetylcholine receptor/channel probes to housefly head membranes, *Neuropharmacol. Insects,* 88, 137, 1982.

60. **Artola, A., Callec, J. J., Hue, B., David, J. A., and Sattelle, D. B.,** Actions of amantadine at synaptic and extrasynaptic cholinergic receptors in the central nervous system of the cockroach, *Periplaneta americana, J. Insect Physiol.,* 30, 185, 1984.

61. **Hue, B., Pelhate, M., and Chanelet, J.,** Effets du piribédil sur la transmission synaptique dans le système nerveux central d'un insecte, la Blatte (*Periplaneta americana* L.), *J. Pharmacol.,* 4, 455, 1981.

62. **Corrodi, H., Farnebo, L. O., Fuxe, K., Hamberger, B., and Understedt, U.,** ET 495 and brain catecholamine mechanisms: evidence for stimulation of dopamine receptors, *Eur. J. Pharmacol.,* 20, 195, 1972.

63. **Truelle, J. L., Chanelet, J., Bastard, J., Six, P., and Emile, J.,** Enregistrement polygraphique du tremblement et des phénomènes d'hypertonie dans la maladie de Parkinson. Application à l'action de différentes drogues, *Nouv. Presse Med.,* 6, 2987, 1977.

64. **Chase, T. N., Woods, A. C., and Glaubiger, G. A.,** Parkinson disease treated with a suspected dopamine receptor agonist, *Arch. Neurol.,* 30, 383, 1974.

65. **Corsini, G. U., Del Zompo, M., Spissu, A., Mangoni, A., and Gessa, G. L.,** Parkinsonism by haloperidol and piribedil, *Psychopharmacology,* 59, 139, 1978.

66. **Chanelet, J. and Lonchampt, P.,** Action de l'acétylcholine sur la moelle épinière du chat, *J. Physiol. (Paris),* 63, 186A, 1971.

67. **Hue, B., Pelhate, M., and Chanelet, J.,** Sensitivity of postsynaptic neurons of the insect central nervous system to externally applied taurine, in *Taurine and Neurological Disorders,* Barbeau, A. and Huxtable, R. J., Eds., Raven Press, New York, 1978, 225.

68. **Hue, B., Pelhate, M., and Chanelet, J.,** Pre- and postsynaptic effects of taurine and GABA in the cockroach central nervous system, *J. Can. Sci. Neurol.,* 6, 243, 1979.

69. **Hue, B., Pelhate, M., and Callec, J. J.,** GABA and taurine sensitivity of cockroach giant interneurone synapses: indirect evidence for the existence of a GABA uptake mechanism, *J. Insect Physiol.,* 27, 357, 1981.

70. **Piek, T.,** Insect venoms and toxins, in *Comprehensive Insect Physiology, Biochemistry and Pharmacology,* Vol. 11, Kerkut, G. A. and Gilbert, L. I., Eds., Pergamon Press, Oxford, 1985, 595.

71. **Piek, T.,** *Venoms of the Hymenoptera,* Academic Press, New York, 1986.

72. **Rathmayer, W.,** Das Paralysierungsproblem beim Bienenwolf *Philanthus triangulum* F. (Hym. Sphec.), *Z. Vgl. Physiol.,* 45, 413, 1962.

73. **Piek, T., May, T. E., and Spanjer, W.,** Paralysis of locomotion in insects by the venom of the digger wasp *Philanthus triangulum* F., in *Insect Neurobiology and Pesticide Action,* Society for Chemical Industry, London, 1980, 219.

74. **Piek, T., Mantel, P., and Van Ginkel, C. J. W.,** Megascoliakinin, a bradykinin-like compound in the venom of *Megascolia flavifrons* (Hymenoptora: Scoliidae), *Comp. Biochem. Physiol.,* 78C, 473, 1984.

75. **Piek, T., Spanjer, W., Veldsema-Currie, R. D., Van Groen, T., De Haan, N., and Mantel, P.,** Effect of the venom of the digger wasp, *Philanthus triangulum* F., on the sixth abdominal ganglion of the cockroach, *Comp. Biochem. Physiol.,* 71C, 159, 1982.

76. **Piek, T., Hue, B., Mony, L., Nakajima, T., Pelhate, M., and Yasuhara, T.,** Block of synaptic transmission in insect CNS by toxins from the venom of the wasp *Megascolia flavifrons* (Fab.), *Comp. Biochem. Physiol.,* 87C, 287, 1987.

77. **Chiappinelli, V. A., Hue, B., Mony, L., and Sattelle, D. B.,** κ-Bungarotoxin blocks nicotinic transmission at an identified invertebrate central synapse, *J. Exp. Biol.,* 141, 61, 1989.

78. **Chiappinelli, V. A.,** Actions of snake venom toxins of neuronal nicotinic receptors and other neuronal receptors, *Pharmacol. Ther.,* 30, 1, 1986.

79. **Freeman, J. A., Schmidt, J. T., and Oswald, R. E.,** Effect of bungarotoxin on retinotectal synaptic transmission in the goldfish and the toad, *Neuroscience,* 5, 929, 1980.

80. **Dryer, S. E. and Chiappinelli, V. A.,** Kappa-bungarotoxin: an intracellular study demonstrating blockade of neuronal nicotinic receptors by a snake neurotoxin, *Brain Res.,* 289, 317, 1983.

81. **Sattelle, D. B. and Breer, H.,** Purification by affinity-chromatography of a nicotinic acetylcholine receptor from the CNS of the cockroach *Periplaneta americana, Comp. Biochem. Physiol.,* 82C, 349, 1985.

82. **Breer, H., Kleene, R., and Behnke, D.,** Isolation of a putative nicotinic acetylcholine receptor from the central nervous system of *Locusta migratoria, Neurosci. Lett.,* 46, 323, 1984.

83. **Baillie, A. C., Corbett, J. R., Dowsett, J. R., Sattelle, D. B., and Callec, J. J.,** Inhibitors of choline acetyltransferase as potential insecticides, *Pestic. Sci.,* 6, 645, 1975.

84. **Gepner, J. I., Hall, L. M., and Sattelle, D. B.,** Insect acetylcholine receptors as a site of insecticide action, *Nature (London),* 276, 188, 1978.

85. **Sattelle, D. B. and Callec, J. J.,** Actions of isothiocyanates on the central nervous system of *Periplaneta americana, Pestic. Sci.,* 8, 735, 1977.

86. **Eldefrawi, M. E., Abbassy, M. A., and Eldefrawi, A. T.,** Effects of environmental toxicants on nicotinic acetylcholine receptors: action of pyrethroids, *Cellular and Molecular Neurotoxicology,* Narahashi, T., Ed., Raven Press, New York, 1984, 177.

87. **Baillie, A. C. and Wright, K.,** Biochemical pharmacology, in *Comprehensive Insect Physiology, Biochemistry and Pharmacology,* Vol. 11, Kerkut G. A. and Gilbert, L. I., Eds., Pergamon Press, Oxford, 1985, 323.

88. **Nitta, S.,** Uber Nereistoxin, einen giftigen Bestandteil von *Lumbriconereis heteropoda* Marenz (Eunicidae), *Yakagaku Zasshi,* 54, 648, 1934.

89. **Okaichi, T. and Hashimoto, Y.,** The structure of nereistoxin, *Agric. Biol. Chem.,* 26, 224, 1962.

90. **Sattelle, D. B., Harrow, I. D., David, J. A., Pelhate, M., Callec, J. J., Gepner, J. I., and Hall, L. M.,** Nereistoxin: actions on a CNS acetylcholine receptor/ion channel in the cockroach *Periplaneta americana, J. Exp. Biol.,* 118, 37, 1985.

91. **Lund, A. E. and Narahashi, T.,** Kinetics of sodium channel modification as the basis for the variation in the nerve membrane effects of pyrethroids and DDT analogs, *Pestic. Biochem. Physiol.*, 20, 203, 1983.

92. **Abbassy, M. A., Eldefrawi, M. E., and Eldefrawi, A. T.,** Allethrin interactions with the nicotinic acetylcholine receptor channel, *Life Sci.*, 31, 1547, 1983.

93. **Hue, B. and Mony, L.,** Actions of deltamethrin and tralomethrin on cholinergic synaptic transmission in the central nervous system of the cockroach *(Periplaneta americana)*, *Comp. Biochem. Physiol.*, 86, 349, 1987.

94. **Hue, B. and Mony, L.,** Effets pré et postsynaptiques de la deltamethrin dans le système nerveux central de la Blatte, *Periplaneta americana*, *C.R. Groupe Fr. Pestic.*, 16, 35, 1987.

95. **Bowery, N. J., Hill, D. R., and Hudson, A. L.,** Characteristics of $GABA_B$ receptor binding sites on rat whole brain synaptic membranes, *Br. J. Pharmacol.*, 78, 191, 1983.

Chapter 9

NEUROMUSCULAR TRANSMISSION IN THE COCKROACH

Hiroshi Washio

TABLE OF CONTENTS

I. INTRODUCTION

Nerve cells are characterized by two specialized properties which distinguish them from all other cells. The first is the propagation of impulses along the axons of nerve cells, depending upon the ionic fluxes across the neuronal membrane. The second is the transmission of nerve impulses to other nerve and muscle cells. Therefore, to understand the neural control of insect behavior, research on synaptic and neuromuscular transmission as well as on the ion channels mediating neural ion fluxes may be considered essential, one of the central themes of insect neurobiology.

In this chapter on neuromuscular transmission in the cockroach, the reader will be referred to recent results obtained from cockroach preparations as well as from locust preparations. Many refined studies on neuromuscular transmission have been made in locust muscles. Recently, the patch-clamp technique[1,2] has been used to examine electrical changes associated with the opening and closing of a single ion channel activated by L-glutamate in locust muscle fibers.[3,4] These works undoubtedly have contributed to the great progress in the understanding of insect neuromuscular transmission.

With our work on insect preparations we recognize two research strategies: the comparative and the general. For example, neuromuscular systems of insects as well as crustaceans have an advantage for work on inhibitory mechanisms because of the existence of peripheral inhibitory innervation. The substantial evidence that γ-aminobutyric acid (GABA) is an inhibitory neurotransmitter was obtained in studies on arthropod neuromuscular preparations. In this context, you are referred to Hoyle's description[5] of the relationship between comparative and general studies: "Comparative and general studies should always proceed together, being continually and reciprocally related."

The muscle and muscular activity, including neuromuscular transmission of cockroaches, have been reviewed previously by Delcomyn.[6] An excellent account of the pharmacology of neurotransmission and neuromodulation of insect muscle has been offered by Piek.[7] A more detailed treatment of the major problems of neurotransmitters in insect muscles has been provided by Usherwood.[8]

II. TRANSMISSION AT EXCITATORY NEUROMUSCULAR JUNCTIONS

Much work on excitatory neuromuscular junctions has been done with locusts.[9-12] Although not yet confirmed, it is generally agreed that L-glutamate is the transmitter at excitatory neuromuscular junctions in insect skeletal muscles.[8,13] In the cockroach *Periplaneta americana*, bath application of L-glutamate was found to cause muscle contraction.[14] The threshold concentration was 340 nM, and higher concentrations brought about greater and more sustained contractions.

Iontophoretic application of L-glutamate by Beránek and Miller[9] and by Usherwood and Machili[10] was made to record depolarizations of the membrane from locust leg muscles. Usherwood and Machili[10] also worked on the metathoracic retractor unguis muscle of the giant South American cockroach, *Blaberus giganteus,* and the results were identical depolarizations. Transient depolarizations (glutamate potentials) to brief application of L-glutamate were recorded only at the synaptic sites. Beránek and Miller[9] found that the equilibrium potential for L-glutamate coincided with the equilibrium potential for miniature excitatory postsynaptic potentials (mEPSPs) and lay between -10 and -25 mV (Figure 1). The pharmacological properties of excitatory neuromuscular junctions in the locust and cockroach have been studied by Usherwood and Machili.[10] L-Glutamate was the most active excitatory substance tested, while D-glutamate was inactive. Among the other amino acids tested, L-aspartate, N-methyl-D,L-aspartate, L-asparagine, and L-glutamine were about 1000 times less

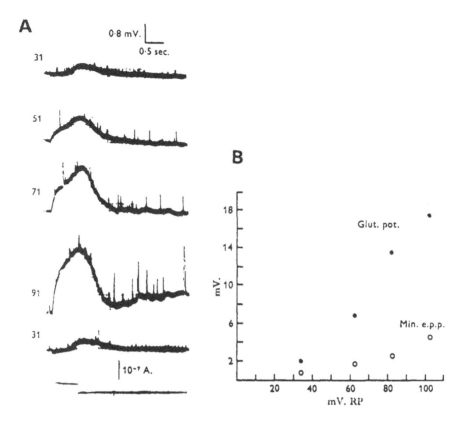

FIGURE 1. (A) Glutamate potentials and miniature EJPs recorded simultaneously by an intracellular electrode, while the resting membrane potential is displaced to various values which are indicated over the corresponding recording in mV. (B) Diagrammatic representation of the experiment from A. Ordinate, amplitude of glutamate potentials (Glut. Pot.), and miniature EJPs (Min. e.p.p.) in mV; abscissa, corresponding resting potentials. The minus signs of membrane potential are omitted. (From Beránek, R. and Miller, P. L., *J. Exp. Biol.*, 49, 83, 1968. With permission.)

potent. Four acidic amino acids (glycine, alanine, aspartate, and L-glutamate) were identified in the perfusate from isolated retractor unguis muscles, but only L-glutamate increased in concentration during nerve stimulation.[10]

At the larval neuromuscular junction of the fruit fly *Drosophila melanogaster*, Jan and Jan[15] also found that iontophoretically applied L-glutamate produced muscle depolarization when L-glutamate was applied close to the synaptic junctions. Other putative transmitter substances and chemicals were tested and found ineffective in mimicking the neurally evoked excitatory junction potential (EJP); they were D-glutamate, L-glutamine, L-aspartate, L-asparagine, glycine, octopamine, serotonin, GABA, acetylcholine (ACh), and substance P. The reversal potentials of the neurally evoked EJP and L-glutamate were virtually identical (-1 and -0.2 mV, respectively). Thus, it was concluded that L-glutamate is the excitatory transmitter at *Drosophila* larval neuromuscular junctions.[15]

The ionic basis underlying the neurally evoked EPSP and the depolarization evoked by L-glutamate at locust extensor tibiae muscles was later studied by Anwyl[11,16] using the voltage-clamp technique. He showed that the reversal potentials of the excitatory postsynaptic current (EPSC) and the glutamate current were $+3$ and $+4$ mV, respectively, and that the neural stimulation and L-glutamate caused an increase in permeability to Na and K, with an approximately equal permeability increase to both of these ions, but not to Cl. This conclusion was derived from the finding that the reversal potential of the synaptic current was approximately midway between the equilibrium potentials for Na and K when the concentration of Na, K, and Cl ions in the perfusate was altered. Additionally, a very small permeability increase to Ca was found when the external concentration of Na was lowered.

Analysis of the membrane noise which resulted from the activation of glutamate receptors at the locust nerve-muscle junction was made by Anderson et al.[17,18] Their work showed that glutamate channels had a conductance of about 125 pS and a mean open time of about 2 ms at a temperature of 21°C. The duration of a channel opening was found to decrease with hyperpolarization of the membrane, which was in marked contrast to the end-plate channels opened by ACh.[17,18] Direct measurement of the currents passing through glutamate-activated single channels in the extrajunctional membrane of locust muscle was first made by Patlak et al.[3] and Cull-Candy et al.[4] using the patch-clamp technique[1,2] (Figure 2). The conductance of the single extrajunctional channels was between 130 and 150 pS, which was similar to the conductance of junctional channels obtained in the membrane noise analysis.[18] The study on the kinetics of the channel's opening and closing revealed that the open times of the channels are distributed roughly exponentially with a mean time constant of about 2 ms, which was again similar to the mean open time measured by noise analysis and by the decay of miniature excitatory junction currents (mEJCs).[16,17]

The presence of glutamate receptors on the nonsynaptic (extrajunctional) membrane was shown in the extensor tibiae muscle of the locust *Schistocerca gregaria*.[19,20] Iontophoretic application of L-glutamate to the extrajunctional membrane was found to produce a biphasic response — depolarization (D-response) followed by hyperpolarization (H-response). Cull-Candy[19,20] showed that glutamate and ibotenate, the glutamate analogue, are able to distinguish two subpopulations of extrajunctional glutamate receptors; glutamate produces a two-component response, but ibotenate produces only an H-response. Study of the equilibrium potential (ca. −60 mV) for the H-response indicates that an increase in chloride conductance is responsible. An increase in the permeability to sodium and potassium is responsible for the extrajunctional D-response. So far, there have been no reports regarding the presence of extrajunctional glutamate receptors in the cockroach muscle fiber.

The long-term application of transmitter molecules to postsynaptic receptors has been known to reduce the amplitude of junctional responses. This process has been called desensitization by Katz and Thesleff,[21] who worked on the frog motor end plate. Desensitization of glutamate receptor populations at excitatory neuromuscular junctions in the extensor tibiae muscle of *S. gregaria* has been studied by Daoud and Usherwood.[22] Their work using a double-barreled iontophoretic electrode indicates that the time course of onset of and recovery from desensitization is very rapid and that the rate of onset of desensitization increases with the concentration of L-glutamate. Later, however, Anis et al.[23] showed that the kinetics of recovery from desensitization are agonist dependent, which is different from the desensitization of ACh receptors on vertebrate muscle.[24] Application of L-glutamate (1 m*M*) has been found to completely abolish the extrajunctional H- and D-receptor responses on locust leg muscle fibers.[19,20] On the other hand, concanavalin A (Con A), the lectin from the seed of the jack bean (*Canavalin ensiformis*), has been shown to specifically inhibit the desensitization of junctional and extrajunctional D-receptors, but not of extrajunctional H-receptors.[25,26] The effect of Con A has been found to persist up to 7 h after washing the muscle with Con A-free saline. It is conceivable that Con A may bind directly to junctional and D-receptor molecules, thereby inhibiting desensitization transitions in these molecules.[25,26]

It is generally accepted that ACh is not involved in neuromuscular transmission in insects. Therefore, it seems reasonable to conclude that curare (*d*-tubocurarine) does not affect the insect neuromuscular junctions. In fact, 1 m*M* curare failed to alter the indirect response of leg muscles in *P. americana*.[27] Also, curare had no effect on the spontaneous miniature end-plate potentials (mEPPs) in the locust and cockroach muscle fibers, even at concentrations as high as 10 m*M*.[28] However, Yamamoto and Washio[29] recently reported that curare

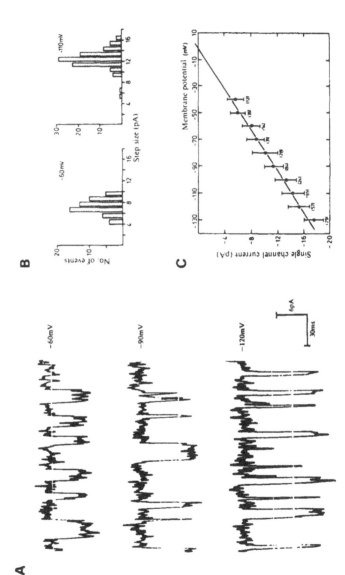

FIGURE 2. (A) The currents through single glutamate-activated channels in the metathoracic extensor tibiae muscle of a 6-d denervated *L. migratoria*. The muscle cell was voltage clamped, the membrane potentials being held at the value shown next to each trace. A patch electrode was filled with physiological saline containing 100 μ*M* L-glutamate. All three records were from the same patch. The temperature was 21°C. The cell had been treated previously for 15 min with 1 μ*M* Con A in saline, then washed with normal saline for more than 30 min. (B) Histograms of measured step sizes of the *single-channel currents at two membrane potentials, −60 and −110 mV. (C) The mean of the measured step sizes and the standard deviation plotted as a function of the holding potential. Numbers in brackets are the number of individual events analyzed at each potential. The solid*

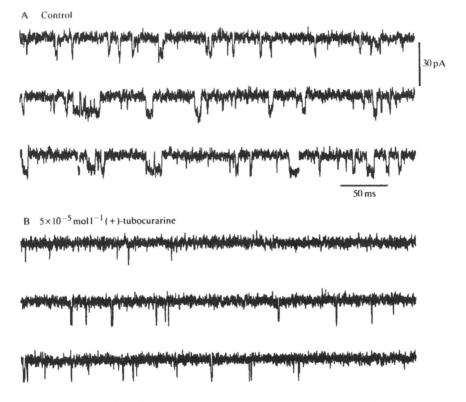

FIGURE 3.   Single-channel recordings from an adult locust extensor tibiae muscle fiber showing channels gated by 100 μM sodium L-glutamate in the absence (A) and in the presence (B) of (+)-tubocurarine (TC). A and B are parts of a continuous recording from separate recording sites on a muscle fiber obtained using megohm seals, with the muscle fiber voltage clamped at −100 mV. The data were filtered at 3 kHz. Comparison of B with A shows the absence of long open events when TC is present in the patch electrode along with sodium L-glutamate. (From Kerry, C. J., Ramsey, R. L., Sansom, M. S. P., Usherwood, P. N. R., and Washio, H., *J. Exp. Biol.*, 127, 121, 1987. With permission.)

blocked transmission at the neuromuscular junction of ventral longitudinal muscle fibers of the larval mealworm, *Tenebrio molitor*, by antagonizing the transmitter at the postsynaptic sites in a voltage-dependent manner. The noncompetitive antagonism of curare against glutamate-induced depolarization has been taken as evidence that the drug blocks ionic channels mediated by the transmitter rather than acting on the receptor itself. Also, Cull-Candy and Miledi[30] have observed that curare and gallamine, both cholinergic antagonists, have an increasingly pronounced effect on the excitatory junction current (EJC) and the glutamate-induced current at hyperpolarized membrane potentials in the extensor tibiae muscle fiber of the *S. gregaria* metathoracic leg. Thus, the result suggests that cholinergic antagonists may be able to block amino acid-activated channels by acting after receptor activation. More recently, the effect of curare on the extrajunctional ionic channel gated by glutamate D-receptors[18] in voltage-clamped locust muscle fibers has been examined using the patch-clamp technique,[30] as shown in Figure 3. (+)-Tubocurarine (5 to 500 μM) caused a concentration-dependent decrease in the mean open time, as well as in the probability of the channel being in the open state, for channels opened by 100 μM glutamate. The results support the view that the cholinergic antagonist curare blocks glutamate receptor channels on locust muscle membrane when they are open, in a manner similar to the action of local anesthetics.[31]

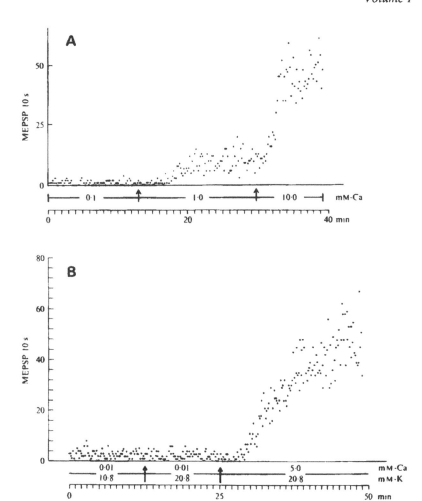

FIGURE 4.   Time course of the effect of the external calcium concentration (m*M*) on the mEPSP frequency in high-potassium (20.8 m*M*) saline (A) and effect of external calcium and potassium concentration on mEPSP frequency (B) in the coxal depressor muscle of *P. americana*. Arrows indicate the changeover points. (From Washio, H. M. and Inouye, S. T., *J. Exp. Biol.*, 75, 101, 1978. With permission.)

III. RELEASE OF TRANSMITTER FROM EXCITATORY MOTONEURON TERMINALS

The release of transmitter from nerve terminals on the insect muscle is quantal in nature. The excitatory mEPPs in insect muscle fibers were first recorded intracellularly from the extensor tibiae in *B. giganteus* by Usherwood.[32] The frequency of the spontaneous discharges is significantly increased by calcium and decreased by magnesium.[32] The interaction of calcium with potassium was studied in *P. americana*.[33] The effect of raising the potassium concentration in saline containing 10 μ*M* $CaCl_2$ was negligible. When calcium was added to the saline, the acceleration of the frequency was rapid and substantial. Thus, the increase in the rate of spontaneous transmitter release upon depolarization is highly dependent on the extracellular calcium concentration in the cockroach muscle (Figure 4), as it is in vertebrate neuromuscular junctions.[34-38]

Calcium ions are essential for the release of transmitter quanta at the neuromuscular junction,[39] and this action of calcium is antagonized competitively by magnesium.[40-42] The effect of magnesium as a competitive antagonist of the spontaneous release of transmitter was studied in cockroach muscle.[43] In this work, in the absence of magnesium the slope of double logarithmic plots of miniature EPSP (mEPSP) frequency vs. calcium concentration varied from 0.5 to 2.4 over ten preparations. Although the slope was much less than the value obtained in the *Drosophila* larval neuromuscular junction,[44] in the presence of 2 mM Mg^{2+} the slope remained the same, but the frequency of mEPSPs was depressed. The results are taken as evidence that Ca^{2+} and Mg^{2+} compete for a common site.[40,41] The equilibrium dissociation constant for Mg^{2+} was obtained from the concentration of extracellular Ca ions that produces a frequency of mEPSPs in the presence of 2 mM Mg^{2+} identical to that in the absence of Mg.[43] The value for Mg^{2+} was 4.0 mM. In comparison, the value for Co^{2+} ranged from 0.40 to 0.65 mM, implying that Co^{2+} is much more potent than Mg^{2+} in suppressing spontaneous transmitter release (Figure 5).

On the other hand, at relatively low extracellular calcium concentrations the depressed mEPSP frequency produced by Co^{2+} was followed by an increase in frequency.[43] The delayed accelerating action of Co^{2+} may result from the entry of Co^{2+} into the nerve terminal via voltage-sensitive channels and subsequent activation of the release process.[45-47]

In frog neuromuscular junctions the spontaneous release of transmitter was shown to occur at random intervals, indicating no interaction between consecutive potentials. By counting the number of mEPPs per unit time interval, Gage and Hubbard[49] showed that spontaneous transmitter release followed a Poisson distribution under a variety of experimental conditions at the mammalian neuromuscular junction. Recently, a more sophisticated statistical analysis[50] has been applied to the mode of transmitter release at vertebrate[51,52] and invertebrate[53] neuromuscular junctions and at the mammalian ganglionic synapse.[54]

A great deal of statistical analysis has been done on the release of transmitter at excitatory neuromuscular junctions in the insect muscle.[55-59] In locust neuromuscular junctions[55,56,58] it has been shown that spontaneous transmitter release is rarely a random process, bursts of mEPSPs being recorded frequently. Hodgkiss and Usherwood[58] analyzed spontaneous and evoked transmitter release from terminals of excitatory motor axons on locust muscle fibers using intra- and extracellular recording combined with the Ca-electrode technique.[60-62] This enabled them to activate the release of transmitter from single nerve terminals on locust muscle fibers innervated multiterminally. Intracellular recording of spontaneous mEPSPs at active spots on these muscle fibers showed nonrandom occurrence with frequent bursts of mEPSPs. Later, Washio and Inouye[59] analyzed the time course of mEPSPs detected at the cockroach neuromuscular junctions, using focal extracellular recording by the method of Cox and Lewis.[50] The time intervals between mEPSPs showed that the variance-to-mean curve for the extracellular data sets lay significantly above the line predicted for a Poisson process (Figure 6B). In contrast with the curves for the extracellular data, those for the intracellular data sets in which the frequency of the mEPSPs did not change with time were found to lie around the line for a Poisson prediction (Figure 6A). In comparison with the intensity function, the departure from the random process appeared to be mainly due to a large positive correlation of intervals.[59] Shuffling procedures, which randomize the order of intervals, made the series almost random. This was clearly shown in variance-to-mean curves (Figure 6C) for shuffled series which lay around the straight line for a Poisson prediction. These results suggest that observed transmitter release may be the pooled output of a finite number of transmitter release sites,[48,50,51,63] each of which releases quanta at insect neuromuscular junctions in a manner more clustered in nature than Poisson.

IV. TRANSMISSION AT INHIBITORY NEUROMUSCULAR JUNCTIONS

Evidence that GABA is an inhibitory transmitter at the metathoracic extensor tibiae muscles of the locust *S. gregaria* and the grasshopper *Romalea microptera* was presented

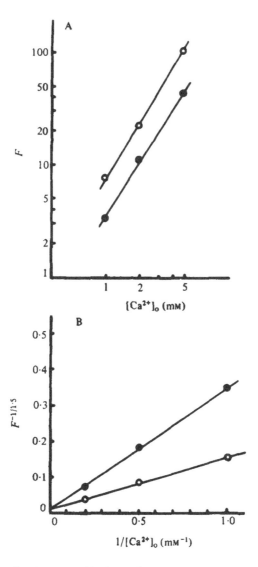

FIGURE 5. The competitive interaction of calcium and cobalt in spontaneous transmitter release in elevated-K^+ (20.8 mM) saline in the coxal depressor muscle of *P. americana*. (A) Double logarithmic plots of mEPSP frequency per 10 S (F, in ordinate) against the external calcium concentration in the absence (O) and presence (●) of 0.5 mM Co ions. (B) Modified Lineweaver-Burk plot of data shown in A, in the absence (O) and presence (●) of 0.5 mM Co ions. (From Washio, H., *J. Exp. Biol.*, 98, 353, 1982. With permission.)

by Usherwood and Grundfest.[64] They showed that the postsynaptic effect of stimulating the inhibitory neuron resulted from an increased permeability to chloride ions and that picrotoxin blocked both the inhibitory postsynaptic potential (IPSP) and the effect of GABA, but did not affect the EPSP.

The iontophoretic application of GABA to locust coxal adductor muscle fibers revealed that the sensitive sites for GABA corresponded to the inhibitory synapses and that the reversal potential for GABA was close to the resting potential, which was similar to the reversal potential for the IPSP.[65] Thus, the similarity of action of GABA to that of the inhibitory transmitter has been shown in the locust muscle.

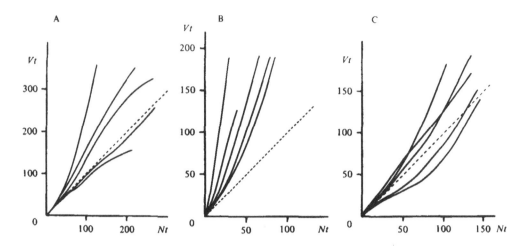

FIGURE 6. The variance-to-mean curves calculated from the data recorded intracellularly from five cells. (A) Recorded extracellularly at five active spots before (B) and after (C), applying a shuffling procedure. The dashed lines show the relationship expected for a Poisson process. The abscissa is the mean number, Nt, of the events occurring in any duration of time t. The ordinate is the variance, Vt, of the event in that duration. (From Washio, H. M. and Inouye, S. T., *J. Exp. Biol.*, 87, 195, 1980. With permission.)

In *P. americana*, Kerkut et al.[14] showed that bath application of GABA at the threshold concentration (5 μg/ml) inhibited the leg's response to electrical stimulation by the thoracic ganglion. Iontophoretic application of GABA was found to decrease both neurally evoked contractions and contraction caused by bath application of glutamate in leg muscles of *P. americana*. The amplitude and frequency of the mEPSPs were also decreased by iontophoretic application of GABA.[66]

Iontophoretic application of GABA to the extensor tibiae muscle of *S. gregaria* was performed by Cull-Candy and Miledi[67] using a voltage-clamp technique. In their work, the presence of both junctional and extrajunctional GABA receptors was determined. The equilibrium potential for GABA action was close to − 60 mV, similar to the chloride equilibrium potential for these muscle fibers (Figure 7C,D). Interestingly, extrajunctional GABA receptors have even been demonstrated on a fiber that lacks inhibitory innervation.[67] The pharmacological property of the extrajunctional GABA receptors is identical to the junctional GABA receptors, as picrotoxin and Cl-deficient medium abolish the extrajunctional GABA responses.[67]

Analysis of the membrane current noise resulting from the activation of GABA receptors showed that for junctional and extrajunctional GABA receptors the mean channel lifetime was 3 to 4 ms at − 80 mV, longer than that of glutamate receptors.[67] The single-channel conductance was about 22 pS at − 80 mV and 21°C, much smaller than that of the excitatory glutamate-activated channels in locust muscle fibers. Inhibitory synaptic currents in voltage-clamped locust muscle fibers were also recorded in the extensor tibiae muscle where the excitatory synapses were desensitized with glutamate to abolish EJCs (Figure 7A,B). The time constant of decay of the nerve-impulse-evoked inhibitory junction currents (IJCs) was found to be 7.7 ms at − 80 mV, roughly three times longer than that of the EJC at − 80 mV.[68] The relationship between amplitude and membrane potential was nonlinear at hyperpolarization (Figure 7B). The nonlinear relationship was also observed between the amplitude of GABA-induced membrane current and the membrane potential (Figure 7D). Interestingly, a direct recording of miniature IJCs (mIJCs) was made in some fibers. The time constant of decay of mIJCs, measured from spectral analysis of these currents, was 6.6 ms at − 75 mV, roughly equivalent to that of the IJCs.

In *P. americana* muscles, Smyth et al.[69] recorded the miniature IPSPs (mIPSPs) from

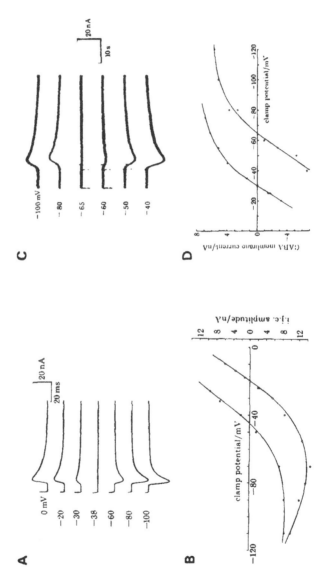

FIGURE 7. (A) Inhibitory junctional currents (IJCs) at various holding potentials in the presence of 1 mM glutamate in the extensor tibiae muscle of *S. gregaria* at 22°C. An upward direction of the current is outward. Decay time (and rise time) is decreased upon hyperpolarization. (B) Dependence of IJC amplitude on membrane potential in two fibers. In one of the fibers (open circle) the equilibrium potential has been shifted from its normal level to −20 mV by holding the cell at less negative potentials. (C) GABA-induced membrane currents over a range of clamp-holding potentials. An upward direction of the current is inward. The equilibrium potential for GABA action occurs close to −65 mV. (D) Relation between amplitude of GABA-induced membrane currents and membrane potential in two muscle fibers. Peak conductance change induced by GABA becomes progressively smaller as the membrane potential is hyperpolarized. (A and B from Cull-Candy, S. G., *Proc. R. Soc. London Ser. B*, 221, 375, 1984;

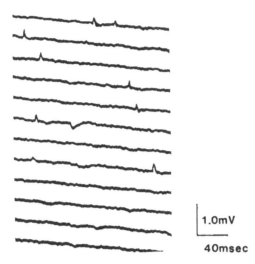

FIGURE 8.   Miniature excitatory and inhibitory junction potentials recorded from the posterior coxal levator (182 D) muscle of *P. americana*.[71a] Membrane potential, −42 mV.

metathoracic femoral muscles. The results suggest that the release of the inhibitory transmitter is quantal in nature. They reported the occurrence of small spontaneous hyperpolarizations and a decrease in amplitude in the saline containing low Ca and high Mg concentrations. Since the inhibitory equilibrium potential is generally close to the resting potential, it is difficult to record the mIPSP from the insect muscle fiber innervated by the inhibitory motoneuron. Later, Atwood et al.[70] showed the recording location and the electrical response in the metathoracic extensor tibiae muscle of *P. americana*. They recorded the hyperpolarizing IPSP from the group of fibers at the distal end of the muscle which were innervated by fast, slow, and inhibitory axons. However, they could not record the mIPSP in this muscle fiber.

Recently, Washio[71] recorded mIPSPs from posterior coxal levator (182C,D)* and coxal depressor (177d,e) muscles in the metathoracic legs of *P. americana* using K-citrate electrodes (Figure 8). Frequent depolarizing and occasional hyperpolarizing potential changes were recorded at the level of a resting potential (−45 mV). The hyperpolarizing spontaneous potentials had a relatively low frequency (<1.0 per second) and a slow time course. When the muscle was bathed in 100 μM glutamate saline the depolarizing spontaneous potentials gradually disappeared. On the other hand, hyperpolarizing spontaneous potentials were abolished in Cl-free saline and a saline containing 1 μM picrotoxin. Increasing extracellular K and/or Ca concentrations increased the frequency of the hyperpolarizing spontaneous potentials. The effect of high K concentration may be due to an increase in intracellular Ca concentration of the inhibitory nerve terminal, similar to the excitatory nerve terminal described above. Injection of hyperpolarizing DC current through a recording K-citrate electrode using a conventional bridge circuit converted the hyperpolarizing potential to a depolarizing potential. The reversal potential of the miniature potentials was about 10 mV more negative than the resting potential when recorded with a K-citrate electrode. All results indicate that the hyperpolarizing spontaneous potentials observed in these cockroach muscle fibers are due to the spontaneous release of transmitter from inhibitory nerve terminals and that the transmitter substance significantly increases the Cl permeability of the postsynaptic membrane.

It is of interest to note that GABA acts presynaptically on the excitatory nerve terminals

* Notation of Carbonell.[72]

of the phallic neuromuscular system of *P. americana.*[73] Application of GABA causes the reduction of the amplitude of EPSPs and the absence of change in the amplitude histogram of mEPPs, and it suggests that the phallic muscles are innervated by inhibitory axons which act presynaptically to reduce the release of the excitatory transmitter.[73]

V. DEGENERATION AND REGENERATION OF NEUROMUSCULAR TRANSMISSION

Systematic studies on the effect of denervation on the neuromuscular transmission of the extensor tibiae muscle of the locust *S. gregaria* have been done by Usherwood.[74,75] In his experiments the effects of independently cutting the fast and slow axons which innervate the extensor tibiae were studied. The results suggest that the motor nerve innervating the locust muscle exerts a "trophic" influence on this muscle. Impulse transmission failed within about 2 weeks after nerve section at 20°C. It is important to note that the site of the breakdown was in the region of the nerve-muscle junction, since the isolated peripheral portions of the severed motor nerves were still able to conduct impulses and the muscle was electrically excitable after transmission had failed.[75] In the metathoracic extensor tibiae muscle of the locust, the fibers were permanently silent electrically after the failure of transmission. Thus, a resumption of miniature activity during the later postoperative days was not observed. However, in the metathoracic coxal muscles of *P. americana*, locomotor activity resumed and the miniature EJPs (mEJPs) and evoked electrical response reappeared within about 40 d after the leg nerves were injured.[76]

Cockroach nerves and muscles are more readily induced to regenerate than those of the locust. *P. americana* was found to possess a great ability to regenerate nerves, even in the adult animal[77] (see also Chapter 11 in this work). The time course of degeneration and regeneration was studied carefully in the coxal depressor muscle of *P. americana.*[78] This time course depended largely on the temperature. Spontaneous subthreshold activity ceased in about 2 d at 26°C (16 d at 12°C) in the coxal depressor muscle. Also, the onset of the failure was partly dependent on the length of the transected distal stump, since the normal stage (the time during which neuromuscular transmission in the denervated muscle remains normal after nerve section) was shorter in the mesothoracic than in the metathoracic coxal muscle at the same temperature.[78] In the coxal depressor muscle of the metathoracic leg the earliest reappearance of mEPSPs was observed 13 d after the nerve crush at 26°C, and the resumption of the miniature potentials generally was accompanied by a response to nerve stimulation (Figure 9). The result was in contrast to the finding on the frog end plate, at which no response to nerve stimulation was observed while mEPPs were resumed.[79] The Schwann cell, a type of glial cell, established close contact with the muscle fiber at this stage, suggesting that the spontaneous release of ACh from the Schwann cell caused the mEPPs.[80,81] On the other hand, in the denervated coxal muscle of *P. americana*, three cellular layers — the glial cells (which surround the nerve terminal on one side and separate it from the hemolymph), the nerve terminal, and the muscle fiber[82] — reappeared. However, the nerve terminals were fewer in number per section of the regenerating preparation, and the glial cells seemed to take up much more space in the synaptic cleft within 18 d of nerve crush (Figure 10).[83] At this stage the mEPSPs were observed accompanied by a response to nerve stimulation. Thus, the resumed spontaneous release of transmitter in the cockroach muscle may be derived from the regenerative nerve terminals, not from the glial cells which once engulfed the nerve terminal.

Insect muscle fibers are innervated multiterminally, and some of them are polyneural. The effect of denervation on the dual nervous control at the peripheral level was examined using red fibers of the metathoracic retractor unguis (RU) muscle of *S. gregaria.*[85] These fibers are supplied by both fast excitatory and inhibitory motor axons. For some parts of

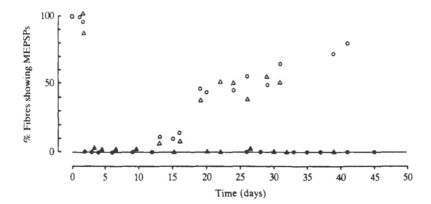

FIGURE 9.   Time course of changes in neuromuscular transmission after axons were crushed at 26°C. Abscissa, time in days after operation; ordinate, percentage of fibers impaled which showed mEPSPs from one muscle. At least 25 muscle fibers were tested. Data are from the metathoracic (circles) and mesothoracic (triangles) coxal muscles of *P. americana*. Open symbols show fibers with transmission functioning; filled symbols indicate fibers in which no response to nerve stimulation was obtained. (From Washio, H., *J. Exp. Biol.*, 120, 143, 1986. With permission.)

the study the RU muscle was denervated by sectioning metathoracic nerve 5, while for others it was done by sectioning excitatory and inhibitory axons selectively, keeping operated animals at 30°C.[85] Excitatory motor nerve terminal degeneration was complete 2 to 3 d after nerve section. Interestingly, however, IPSPs were recorded from the denervated red fibers of RU muscles up to 18 d after nerve section. These results suggested that the inhibitory inputs to the RU muscle were intact despite the disappearance of the excitatory innervation. The difference between disappearance time courses of the excitatory and inhibitory innervation was probably caused by more rapid degeneration of the large excitatory axons compared with the small inhibitory ones.[84,85] In general, the inhibitory axons are much smaller in diameter than the excitatory axons in insect muscle. However, in coxal depressor muscles 177d and e of *P. americana*, which are innervated by one slow motoneuron and three common inhibitory motoneurons,[86] both excitatory and inhibitory transmission failed almost simultaneously at 27°C within about 24 h after chronic denervation of both excitatory and inhibitory axons by severing nerve 5.[71a] mEJPs and miniature IJPs (mIJPs) disappeared at almost the same time. The cross-sectional figure of nerve 5r1 in *P. americana*, shown by Pearson and Iles,[86] indicates that the largest is a fast axon, the second is a slow one, and the three smaller ones are all inhibitory. If the degeneration of the motoneuron is dependent on its diameter, the degeneration of the inhibitory axons innervating the coxal muscle should be delayed in comparison with the excitatory axon. Further studies are needed to evaluate the effect of denervation of the inhibitory and excitatory axons in the coxal depressor muscle in relation to the size of axons which innervate the muscle.

In the chronically denervated muscles of the locust, sensitivity of the muscle membrane to L-glutamate was prominently enhanced and the receptor sites of L-glutamate spread along the fibers.[12,87] Usherwood[87] has shown that a high density of glutamate receptors is found at the extrajunctional membrane, but there is no increase in the glutamate sensitivity of the synaptic membrane in denervated muscles. The changes in glutamate sensitivity were first observed about 5 d after nerve section and were most noticeable 32 d after nerve section at 30°C.[87] Later, Cull-Candy[12] worked in detail on the glutamate sensitivity of the extrajunctional membrane of denervated locust muscles and found that after denervation the entire extrajunctional sensitivity to L-glutamate was increased about 100 times above control levels. Distribution of extrajunctional receptor sensitivity following denervation revealed that the increased sensitivity resulted from an increase in D-receptors, not in H-receptors.

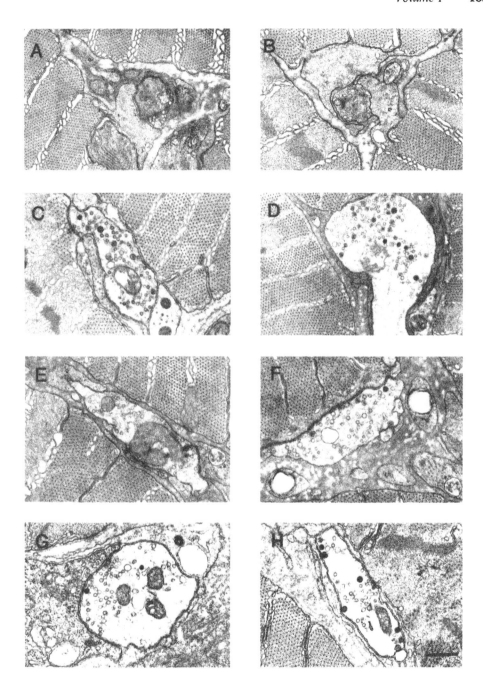

FIGURE 10.   Cross section of degenerating and regenerating nerve terminal following nerve crush in the coxal depressor muscle of *P. americana*; (A,B) 4 d, (C,D) 18 d, (E,F) 24 d, (G,H) 36 d after nerve crush. The glial cell is in close contact with the muscle in A and B. A three-cellular arrangement is clearly seen in C and D. Also, many large dense-core vesicles are visible in C and D. Accumulation of synaptic vesicles at the presynaptic membrane is seen in G and H. The bar in H indicates 0.5 μm for all parts of this figure. (Reprinted with permission from *Comp. Biochem. Physiol.*, Vol. 86A, Washio, H. and Nihonmatsu, I., Structural changes in the denervated neuromuscular junction in coxal muscles of the cockroach *Periplaneta americana*, ©1987, Pergamon Journals Ltd.)

Recently, the effect of denervation on the mechanical responses to various concentrations of L-glutamate in the RU muscle of *P. americana* has been examined and compared with contractions induced by high-potassium saline.[88] By 16 to 20 d after denervation, the tension ratios of the maximum glutamate-to-potassium contractions were significantly higher than that of the innervated muscles. Thus, it has been suggested that the sensitivity of the muscle membrane to L-glutamate in cockroach muscle fibers is increased in the denervated muscle in accordance with the results obtained from previous work in the locust muscle fibers.[12,87]

VI. CONCLUDING SUMMARY

L-Glutamate and GABA fulfill most of the criteria for the identification of excitatory and inhibitory transmitters, respectively, in both the insect and the crustacean neuromuscular junction. Also, L-glutamate and GABA are regarded as the major excitatory and inhibitory transmitters, respectively, in many regions of the mammalian brain. It has now been about 30 years since L-glutamate was found to depolarize and excite individual neurons in the mammalian central nervous system (CNS).[89] At present, glutamate receptors have been classified into at least three types of receptors based on agonist specificity in the mammalian CNS. These are *N*-methyl-D-aspartate (NMDA), kainate, and quisqualate. However, there has been no clear evidence to show the involvement of these glutamate receptors in insect muscles.

On the other hand, GABA receptors on vertebrate neurons have been characterized as being of two distinct types, GABA$_A$ and GABA$_B$.[90,91] GABA$_A$ receptors are linked to changes in chloride conductance and are antagonized by bicuculline, whereas GABA$_B$ receptors are inactive to bicuculline.[91] GABA$_B$ is sensitive to baclofen as an agonist, but GABA$_A$ is not.[91] Unfortunately, there has been little work to date on GABA receptors on insect muscle. Recently, GABA receptors on the locust muscle have been reported to lack sensitivity to bicuculline, unlike the vertebrate GABA$_A$ receptor,[92] suggesting that there are pharmacological differences in GABA receptors on insect muscle fibers and neurons, since bicuculline reduces GABA-induced chloride currents in an identified motoneuron of *P. americana*.[93] Thus, there is evidence for considerable heterogeneity of GABA receptors in both vertebrates and invertebrates. However, the recent findings[94] on the interaction between the GABA receptor-chloride channel complex in the CNS of mammals and the sites of action of pyrethroid insecticides have contributed to characterizing the properties of the GABA receptor site in insects.[95] A detailed pharmacological study on the insect GABA receptor-ionophore complex in the peripheral and central nervous systems may provide possibilities for the design of novel insecticides[93] because we have been confronted with the fact that the study of insect neuromuscular junctions has been impeded by the lack of an antagonist which can be used to block excitatory junctional transmission by a specific postsynaptic action.

ACKNOWLEDGMENTS

I thank Drs. F. Delcomyn and I. Huber for giving me the opportunity to write this chapter. Thanks are also due to Mrs. Matsumura for her preparation of the manuscript.

REFERENCES

1. **Neher, E. and Sakmann, B.,** Single-channel currents recorded from membrane of denervated frog muscle fibers, *Nature (London)*, 260, 799, 1976.
2. **Neher, D., Sackmann, B., and Steinback, J. H.,** The extracellular patch clamp: a method for resolving currents through individual open channels in biological membrane, *Pflugers Arch.*, 375, 219, 1978.

3. **Patlak, J. B., Gration, K. A. F., and Usherwood, P. N. R.,** Single glutamate-activated channels in locust muscle, *Nature (London),* 278, 643, 1979.
4. **Cull-Candy, S. G., Miledi, R., and Parker, I.,** Single glutamate-activated channels recorded from locust muscle fibres with perfused patch-clamp electrodes, *J. Physiol. (London),* 321, 195, 1980.
5. **Hoyle, G.,** *Muscles and Their Neural Control,* John Wiley & Sons, New York, 1983.
6. **Delcomyn, F.,** Muscle and muscular activity, in *The American Cockroach,* Bell, W. J. and Adiyodi, K. G., Eds., Chapman and Hall, London, 1981, chap. 11.
7. **Piek, T.,** Neurotransmission and neuromodulation of skeletal muscles, in *Comprehensive Insect Physiology, Biochemistry and Pharmacology,* Vol. 11, Kerkut, G. A. and Gilbert, L. I., Eds., Pergamon Press, Oxford, 1985, 55.
8. **Usherwood, P. N. R.,** Amino acids as neurotransmitters, in *Advances in Comparative Physiology and Biochemistry,* Vol. 7, Lowenstein, O., Ed., Academic Press, New York, 1978, 227.
9. **Beránek, R. and Miller, P. L.,** The action of iontophoretically applied glutamate on insect muscle fibres, *J. Exp. Biol.,* 49, 83, 1968.
10. **Usherwood, P. N. R. and Machili, P.,** Pharmacological properties of excitatory neuromuscular synapses in the locust, *J. Exp. Biol.,* 49, 341, 1968.
11. **Anwyl, R.,** Permeability of the post-synaptic membrane of an excitatory glutamate synapse to sodium and potassium, *J. Physiol. (London),* 273, 367, 1977.
12. **Cull-Candy, S. G.,** Glutamate sensitivity and distribution of receptors along normal and denervated locust muscle fibres, *J. Physiol. (London),* 276, 165, 1978.
13. **Usherwood, P. N. R. and Cull-Candy, S. G.,** Pharmacology of somatic nerve-muscle synapses, in *Insect Muscle,* Usherwood, P. N. R., Ed., Academic Press, London, 1975, chap. 4.
14. **Kerkut, G. A., Shapira, A., and Walker, R. J.,** The effect of acetylcholine, glutamic acid and GABA on the contractions of the perfused cockroach leg, *Comp. Biochem. Physiol.,* 16, 37, 1965.
15. **Jan, L. Y. and Jan, Y. N.,** L-Glutamate as an excitatory transmitter at the *Drosophila* larval neuromuscular junction, *J. Physiol. (London),* 262, 215, 1976.
16. **Anwyl, R.,** The effect of foreign cations, pH and pharmacological agents on the ionic permeability of a excitatory glutamate synapse, *J. Physiol. (London),* 273, 389, 1977.
17. **Anderson, C. R., Cull-Candy, S. G., and Miledi, R.,** Glutamate and quisqualate noise in voltage-clamped locust muscle fibres, *Nature (London),* 261, 151, 1976.
18. **Anderson, C. R., Cull-Candy, S. G., and Miledi, R.,** Glutamate current noise: postsynaptic channel kinetics investigated under voltage clamp, *J. Physiol. (London),* 282, 219, 1978.
19. **Cull-Candy, S. G. and Usherwood, P. N. R.,** Two populations of L-glutamate receptors on locust muscle fibres, *Nature (London),* 246, 62, 1973.
20. **Cull-Candy, S. G.,** Two types of extrajunctional L-glutamate receptors in locust muscle fibres, *J. Physiol. (London),* 255, 449, 1976.
21. **Katz, B. and Thesleff, S.,** A study of ''desensitization'' produced by acetylcholine at the motor end plate, *J. Physiol. (London),* 138, 63, 1957.
22. **Daoud, M. A. R. and Usherwood, P. N. R.,** Desensitization and potentiation during glutamate application to locust skeletal muscle, *Comp. Biochem. Physiol.,* 59C, 105, 1978.
23. **Anis, N. A., Clark, R. B., Gration, K. A. F., and Usherwood, P. N.R.,** Influence of agonists on desensitization of glutamate receptors on locust muscle, *J. Physiol. (London),* 312, 345, 1981.
24. **Rang, H. P. and Ritter, J. M.,** On the mechanism of desensitization at cholinergic synapses, *Mol. Pharmacol.,* 6, 357, 1970.
25. **Mather, D. A. and Usherwood, P. N. R.,** Concanavalin A blocks desensitization of glutamate receptors on insect muscle fibres, *Nature (London),* 259, 409, 1976.
26. **Mather, D. A. and Usherwood, P. N. R.,** Effect of concanavalin A on junctional and extrajunctional L-glutamate receptors on locust skeletal muscle fibres, *Comp. Biochem. Physiol.,* 59C, 151, 1978.
27. **Roeder, K. D. and Weiant, E. A.,** The electrical and mechanical events of neuromuscular transmission in the cockroach, *Periplaneta americana* (L.), *J. Exp. Biol.,* 27, 1, 1950.
28. **Usherwood, P. N. R.,** Spontaneous miniature potentials from insect skeletal muscle fibres, *J. Physiol. (London),* 169, 149, 1963.
29. **Yamamoto, D. and Washio, H.,** Curare has a voltage-dependent blocking action on the glutamate synapse, *Nature (London),* 281, 372, 1979.
30. **Cull-Candy, S. G. and Miledi, R.,** Blocks of glutamate-activated synaptic channels by curare and gallamine, *Proc. R. Soc. London Ser. B,* 218, 111, 1983.
31. **Kerry, C. J., Ramsey, R. L., Sansom, M. S. P., Usherwood, P. N. R., and Washio, H.,** Single-channel studies of the action of (+)-tubocurarine on locust muscle glutamate receptor, *J. Exp. Biol.,* 127, 121, 1987.
32. **Usherwood, P. N. R.,** Spontaneous miniature potentials from insect muscle fibres, *Nature (London),* 191, 814, 1961.

33. **Washio, H. M. and Inouye, S. T.**, The effect of calcium and magnesium on the spontaneous release of transmitter at insect motor nerve terminals, *J. Exp. Biol.*, 75, 101, 1978.

34. **del Castillo, J. and Katz, B.**, Quantal components of the end-plate potential, *J. Physiol. (London)*, 124, 560, 1954.

35. **del Castillo, J. and Katz, B.** Statistical factors involved in neuromuscular facilitation and depression, *J. Physiol. (London)*, 124, 574, 1954.

36. **del Castillo, J. and Katz, B.**, Changes in end-plate activity produced by presynaptic polarization, *J. Physiol. (London)*, 124, 586, 1954.

37. **Hubbard, J. I., Jones, S. F., and Landau, E. M.**, On the mechanism by which calcium and magnesium affect the spontaneous release of transmitter from mammalian motor nerve terminals, *J. Physiol. (London)*, 194, 355, 1968.

38. **Cooke, J. D., Okamoto, K., and Quastel, D. M. J.**, The role of calcium in depolarization-secretion coupling at the motor nerve terminal, *J. Physiol. (London)*, 228, 459, 1973.

39. **Katz, B.**, *The Release of Neural Transmitter Substances*, Charles C Thomas, Springfield, IL, 1969.

40. **Jenkinson, D. H.**, The nature of the antagonism between calcium and magnesium ions at the neuromuscular junction, *J. Physiol. (London)*, 138, 434, 1957.

41. **Dodge, F. A. and Rahamimoff, R.**, Co-operative action of calcium ions in transmitter release at the neuromuscular junction, *J. Physiol. (London)*, 193, 419, 1967.

42. **Hubbard, J. I., Jones, S. F., and Landau, E. M.**, On the mechanism by which calcium and magnesium affect the release of transmitter by nerve impulses, *J. Physiol. (London)*, 196, 75, 1968.

43. **Washio, H.**, A dual effect of cobalt ions on the spontaneous release of transmitter at insect motor nerve terminals, *J. Exp. Biol.*, 98, 353, 1982.

44. **Jan, L. Y. and Jan, Y. N.**, Properties of the larval neuromuscular junction in *Drosophila melanogaster*, *J. Physiol. (London)*, 262, 189, 1976.

45. **Kita, H. and Van der Kloot, W.**, Action of Co and Ni at the frog neuromuscular junction, *Nature New Biol.*, 245, 52, 1973.

46. **Alnaes, E. and Rahamimoff, R.**, On the role of mitochondria in transmitter release from motor nerve terminals, *J. Physiol. (London)*, 248, 285, 1975.

47. **Kita, H. and Van der Kloot, W.**, Effect of the ionophore X-537A on acetylcholine release at the frog neuromuscular junction, *J. Physiol. (London)*, 259, 177, 1976.

48. **Fatt, P. and Katz, B.**, Spontaneous subthreshold activity at motor nerve endings, *J. Physiol. (London)*, 117, 109, 1952.

49. **Gage, P. W. and Hubbard, J. I.**, Evidence for Poisson distribution of miniature end-plate potentials and some implications, *Nature (London)*, 208, 395, 1965.

50. **Cox, D. R. and Lewis, P. A. W.**, *The Statistical Analysis of Series of Events*, Methuen, London, 1966.

51. **Hubbard, J. I. and Jones, S. F.**, Spontaneous quantal transmitter release: a statistical analysis and some implications, *J. Physiol. (London)*, 232, 1, 1973.

52. **Cohen, I., Kita, H., and Van der Kloot, W.**, The stochastic properties of spontaneous quantal release of transmitter at the frog neuromuscular junction, *J. Physiol. (London)*, 236, 341, 1974.

53. **Cohen, I., Kita, H., and Van der Kloot, W.**, Stochastic properties of spontaneous transmitter release at the crayfish neuromuscular junction, *J. Physiol. (London)*, 236, 363, 1974.

54. **Bornstein, J. C.**, Spontaneous multiquantal release at synapse in guinea-pig hypogastric ganglia: evidence that release can occur in bursts, *J. Physiol. (London)*, 282, 375, 1978.

55. **Usherwood, P. N. R.**, Transmitter release from insect excitatory motor nerve terminals, *J. Physiol. (London)*, 227, 527, 1972.

56. **Rees, D.**, The spontaneous release of transmitter from insect nerve terminals as predicted by the negative binomial theorem, *J. Physiol. (London)*, 236, 129, 1974.

57. **Washio, H. and Inouye, S. T.**, The mode of spontaneous transmitter release at the insect neuromuscular junction, *Can. J. Physiol. Pharmacol.*, 53, 679, 1975.

58. **Hodgkiss, J. P. and Usherwood, P. N. R.**, Transmitter release from normal and degenerating locust motor nerve terminals, *J. Physiol. (London)*, 285, 113, 1978.

59. **Washio, H. M. and Inouye, S. T.**, The statistical analysis of spontaneous transmitter release at individual junctions on cockroach muscle, *J. Exp. Biol.*, 87, 195, 1980.

60. **Katz, B. and Miledi, R.**, Propagation of electric activity in motor nerve terminals, *Proc. R. Soc. London Ser. B*, 161, 453, 1965.

61. **Katz, B. and Miledi, R.**, The measurement of synaptic delay and the time course of acetylcholine release at the neuromuscular junction, *Proc. R. Soc. London Ser. B*, 161, 483, 1965.

62. **Katz, B. and Miledi, R.**, The effect of calcium on acetylcholine release from motor nerve terminals, *Proc. R. Soc. London Ser. B*, 161, 496, 1965.

63. **Vere-Jones, D.**, Simple stochastic models for the release of quanta of transmitter from a nerve terminal, *Aust. J. Stat.*, 8, 53, 1966.

64. **Usherwood, P. N. R. and Grundfest, H.,** Peripheral inhibition in skeletal muscle of insects, *J. Neurophysiol.*, 28, 497, 1965.

65. **Usherwood, P. N. R.,** Action of iontophoretically applied gamma-aminobutyric acid on locust muscle fibres, *Comp. Biochem. Physiol.*, 44A, 663, 1973.

66. **Kerkut, G. A. and Walker, R. J.,** The effect of iontophoretic injection of L-glutamic acid and γ-amino-N-butyric acid on the miniature end-plate potentials and contractures of the coxal muscles of the cockroach *Periplaneta americana* L., *Comp. Biochem. Physiol.*, 20, 999, 1967.

67. **Cull-Candy, S. G. and Miledi, R.,** Junctional and extrajunctional membrane channels activated by GABA in locust muscle fibres, *Proc. R. Soc. London Ser. B*, 211, 527, 1981.

68. **Cull-Candy, S. G.,** Inhibitory synaptic currents in voltage-clamped locust muscle fibres desensitized to their excitatory transmitter, *Proc. R. Soc. London Ser. B*, 221, 375, 1984.

69. **Smyth, T., Jr., Greer, M. H., and Griffiths, J. G.,** Insect neuromuscular synapses, *Am. Zool.*, 13, 315, 1974.

70. **Atwood, H. L., Smyth, T., Jr., and Johnston, H. S.,** Neuromuscular synapsis in the cockroach extensor tibiae muscle, *J. Insect Physiol.*, 15, 529, 1969.

71. **Washio, H.,** Inhibitory miniature potentials in insect leg muscle fibers, *J. Physiol. Soc. Jpn.*, 49, 378, 1987.

71a. **Washio, H.,** unpublished data, 1987.

72. **Carbonell, C. S.,** The thoracic muscles of the cockroach, *Periplaneta americana*, *Smithson. Misc. Collect.*, 107, 1, 1947.

73. **Parnas, I. and Grossman, Y.,** Presynaptic inhibition in the phallic neuromuscular system of the cockroach *Periplaneta americana*, *J. Comp. Physiol.*, 82, 23, 1973.

74. **Usherwood, P. N. R.,** Response of insect muscle to denervation. I. Resting potential changes, *J. Insect Physiol.*, 9, 247, 1963.

75. **Usherwood, P. N. R.,** Response of insect muscle to denervation. II. Changes in neuromuscular transmission, *J. Insect Physiol.*, 9, 811, 1963.

76. **Jacklet, J. W. and Cohen, M. J.,** Nerve regeneration: correlation of electrical, histological and behavioral events, *Science*, 156, 1640, 1967.

77. **Bodenstein, D.,** Studies on nerve regeneration in *Periplaneta americana*, *J. Exp. Zool.*, 136, 89, 1957.

78. **Washio, H.,** Spontaneous miniature potentials in denervated coxal muscle fibres of the American cockroach, *J. Exp. Biol.*, 120, 143, 1986.

79. **Birks, R., Katz, B., and Miledi, R.,** Physiological and structural changes at the amphibian myoneural junctions in the course of nerve degeneration, *J. Physiol. (London)*, 150, 145, 1960.

80. **Miledi, R. and Slater, C. R.,** Electrophysiology and electron microscopy of rat neuromuscular junction after nerve degeneration, *Proc. R. Soc. London Ser. B*, 169, 289, 1968.

81. **Miledi, R. and Slater, C. R.,** On the degeneration of rat neuromuscular junction after nerve section, *J. Physiol. (London)*, 207, 507, 1970.

82. **Osborne, M. P.,** The ultrastructure of nerve-muscle synapses, in *Insect Muscle*, Usherwood, P. N. R., Ed., Academic Press, London, 1975, chap. 3.

83. **Washio, H. and Nihonmatsu, I.,** Structural changes in the denervated neuromuscular junction in coxal muscles of the cockroach *Periplaneta americana*, *Comp. Biochem. Physiol.*, 86A, 643, 1987.

84. **Rees, D. and Usherwood, P. N. R.,** Fine structure of normal and degenerating motor axons and nerve-muscle synapses in the locust, *Schistocerca gregaria*, *Comp. Biochem. Physiol.*, 43A, 83, 1972.

85. **Clark, R. B., Gration, K. A. F., and Usherwood, P. N. R.,** Relative 'trophic' influences of excitatory and inhibitory innervation of locust skeletal muscle fibres, *Nature (London)*, 280, 679, 1979.

86. **Pearson, K. G. and Iles, J. F.,** Innervation of coxal depressor muscles in the cockroach, *Periplaneta americana*, *J. Exp. Biol.*, 54, 215, 1971.

87. **Usherwood, P. N. R.,** Glutamate sensitivity of denervated insect muscle fibres, *Nature (London)*, 223, 411, 1969.

88. **Yamaguchi, T., Tsuru, A., and Washio, H.,** L-Glutamate and potassium induced contractures in denervated cockroach muscles, *Comp. Biochem. Physiol.*, 87C, 401, 1987.

89. **Curtis, D. R., Phillis, J. W., and Watkins, J. C.,** Chemical excitation of spinal neurones, *Nature (London)*, 183, 611, 1959.

90. **Hill, D. R. and Buwery, N. G.,** ^{3}H-Baclofen and ^{3}H-GABA bind to bicuculline-insensitive GABA sites in rat brain, *Nature (London)*, 290, 149, 1981.

91. **Simmonds, M. A.,** Multiple GABA receptors and associated regulatory sites, *Trends Neurosci.*, 6, 279, 1983.

92. **Scott, R. H. and Duce, I. R.,** Pharmacology of GABA receptors on skeletal muscle fibres of the locust (*Schistocerca gregaria*), *Comp. Biochem. Physiol.*, 86C, 305, 1987.

93. **David, J. A., Lummis, S. C. R., and Sattelle, D. B.,** Receptors for GABA and acetylcholine in the central nervous system of an insect, *Br. J. Pharmacol.*, 80, 432P, 1983.

94. **Lawrence, L. J. and Casida, J. E.,** Stereospecific action of pyrethroid insecticides on the γ-aminobutyric acid receptor-ionophore complex, *Science,* 221, 1399, 1983.
95. **Lummis, S. C. R. and Sattelle, D. B.,** Binding sites for 4-aminobutyric acid and benzodiazepines in the central nervous system of insects, *Pestic. Sci.,* 16, 695, 1985.

Chapter 10

ELECTRICAL AND MECHANICAL RECORDING METHODS

Moray Anderson, Karel Sláma, and Thomas Miller

TABLE OF CONTENTS

I. INTRODUCTION

In years past, the electrophysiological laboratory contained standard and predictable equipment; today, the versatility of instruments is much greater. The biggest changes have been the switch to digitalized designs and the advent of the microprocessor.

Previously, one needed a microelectrode puller and amplifier, an oscilloscope (most preferred the storage oscilloscope), and some form of recording device, which varied from frequency-modulated (FM) tape recorders to cameras. The Brush recorder became very popular because of its reliability. There were (and still are) any number of AC preamplifiers (preamps) available, and a number of designs for hand-crafted AC preamps are readily available to those who are inclined to "do it themselves".[1]

The choice of instrumentation for an electrophysiological laboratory will continue to change because of advances in microprocessors and associated software. Microelectrode pullers now come designed to make patch-clamp electrodes in addition to ordinary intracellular microelectrodes. Accordingly, microelectrode amplifiers can be purchased for ordinary recording with or without current injection provisions and can now be used for patch recording.

The choice for signal processing has become bewildering. The fainthearted might still purchase a storage oscilloscope and then decide how to make permanent records of traces. The more adventurous might buy a processor-based system (such as DATA-PAC from Run Technologies, 5768 W. Pico Blvd., Los Angeles, CA 90019) and add to it as advances are made in software. Tape recorders are still used, but the video cassette recorder (VCR) has now entered the electrophysiological laboratory. Also, a good pen recorder is still useful.

Electrophysiology, like electron microscopy, is not for the novice or occasional user. It is most efficiently used by those who can afford the time and attention necessary to perform tissue preparation and to manipulate signal processing information. This normally takes time to learn and must be accompanied by a continuing investment of time for retention as well as for keeping up with new developments.

II. EXTRACELLULAR RECORDING

A. WIRE ELECTRODES

When recording the nervous activity from any nerve in the cockroach, a great deal can often be learned from a recording using simple wire electrodes. The wire used varies according to the particular situation. For example, if the cuticle is to be penetrated, then it is best to use a tough stainless steel or tungsten wire, whereas if it is a delicate visceral nerve from which the recording is to be made, then it is better to use fine copper or silver wires.[1]

When using wire electrodes, the most common problem is the need to reduce the information arriving at the electrode. To keep out extraneous interference (information), it is necessary to shield the wire recording electrode by thorough insulation of all but the recording tip. This is best achieved by painting on a lacquer of some type, making the recording area as large or as small as desired.

The method of recording extracellular nerve activity using wire electrodes is well illustrated in the preparation where output from the tibial mechanoreceptive hairs on the cockroach leg can be recorded and their responses to mechanical distortion monitored. This simple yet extremely effective method uses stainless steel entomological pins as the recording electrodes. The pins are inserted into the tibia of the cockroach (Figure 1).

A large amount of nerve activity is often present initially, but after a short time this can be seen to decrease. By careful manipulation of the stimulating device, in this case a loudspeaker-driven probe, it is possible to single out the activity from one sensillum. Clear accommodation to the stimulating signal can often be illustrated (Figure 2).

FIGURE 1. A photograph of a device for stimulating and obtaining recordings via wire electrodes from an isolated cockroach leg. (1) Stainless steel recording electrodes; (2) retaining pins for attachment of the positive and indifferent recording wires; (3) magnets which allow for infinitely variable positioning of the leg when held by recording electrodes; (4) wooden stimulating rod driven by a small loudspeaker within the metal box; (5) control knob which alters degree of stimulation; (6) fine positioning control knob to line up stimulator with the leg spine.

When recording from nerves in the abdomen of the cockroach with wire electrodes, it is often essential to pick the loop of the nerve clear from the saline in order to obtain a clear recording. Desiccation problems can be overcome using various greases and light waxes, in a manner similar to the grease electrodes. (Grease electrodes are described below in Section II.C).

Recording from chemosensitive sensilla with wire electrodes is often possible. In this case, it is best to use sharpened tungsten wire. The fine tip is best produced by sharpening the tungsten wire electrolytically in hydrochloric acid, potassium hydroxide, or some other corrosive solution. The electrodes can then be placed into the sensillum close to the sensory cells, and clear recordings can subsequently be obtained (Figure 3).

B. SUCTION ELECTRODES

Suction electrodes, drawn-out glass or plastic tubes fitting onto nerves, have been used for a number of years by many research workers both to record nervous activity and to stimulate nerves. There are two ways in which the suction electrode can be used, either by drawing the cut end of the nerve into the electrode or by taking in a loop of the intact nerve; the latter is termed an "en passant" recording. In both cases, success of the method depends on the snugness of the fit of the nerve in the electrode tip, providing a high-resistance pathway from inside the electrode to the outside.

There are a number of detailed descriptions of producing good-fitting suction electrodes in the book by Miller,[1] including the best ways to achieve a good "suck" to hold the nerve in the electrode when recording or stimulation is taking place. Experience has shown that some of the very simple designs are, in fact, often the most suitable.

To produce the correct diameter of the glass suction electrode for the best fit, it is often

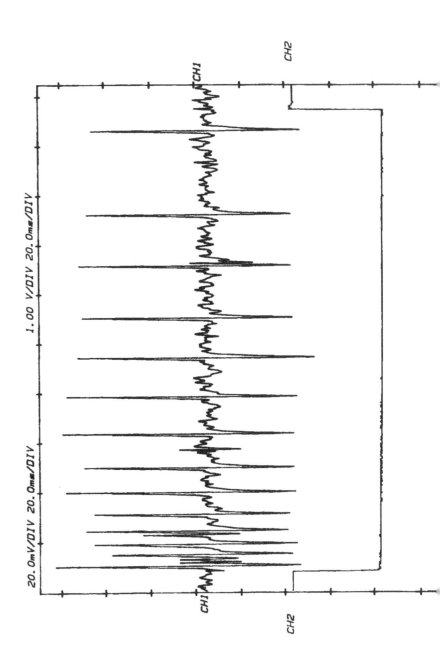

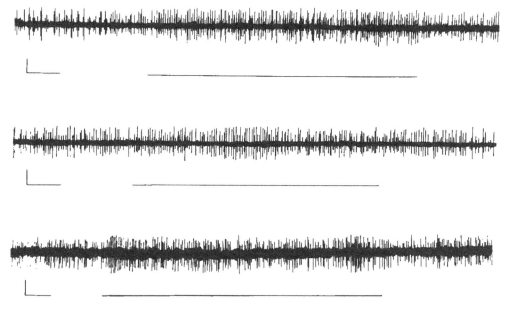

FIGURE 3.   Traces from a chemoreceptive sensillum on the antenna of a cockroach, recorded with sharpened tungsten wire electrodes. Lines below traces refer to where odor stimulus has been applied to antenna. Bars: vertical, 0.1 mV; horizontal, 0.4 s.

useful to melt the tip with some sort of furnace (Figure 4). The device in the illustration allows constant monitoring of the glass electrode as it is being closed by the heat from the nichrome wire loop. Satisfactory electrodes can be made in this way.

Those who have used suction electrodes know that in practice a great deal of time can be spent obtaining measurements of nerve diameter, making the electrode fit precisely that size only to find that a small piece of fat body gets into the electrode tip just as the nerve is being drawn in. Thus, no space remains for the nerve. Considered another way, the fat body can be a good "plug" to wedge a nerve into a slightly large electrode.

To produce a suction tip, many workers now prefer to use some sort of plastic/polyethylene, drawn out in a small flame and cut cleanly with a scalpel. These tips have the great advantage of being unbreakable and are easy to clean in an ultrasonic cleaner should the tip become blocked.

The suction can be applied most simply with a disposable syringe. The simple design shown in Figure 5 works effectively and has the great advantage of being quick to fabricate and service.

Within the suction electrode the signal is transmitted to the recording device by means of an Ag/AgCl wire. Sealing around the electrical connections for suction-proof junctions is very important. The modern silicone rubber adhesives and sealants are perfect for this function.

C. GREASE ELECTRODES

The grease electrode can be regarded as being a hybrid of the wire and suction electrodes. In this case, a metal hook is used for recording or stimulating, with the nerve being drawn into a nonconducting environment such as mineral oil. The advantage of this method is clear because there is a great reduction in the desiccation of the nerve, which makes extended recording or stimulating times possible.

D. ELECTROMYOGRAMS

Extracellular recording of muscle activity has been accomplished in a number of

FIGURE 4.   A photograph of a furnace and inspection microscope for producing glass suction electrodes. (1) Nichrome wire loop comprising the furnace; (2) drill chuck to hold glass of various diameters; (3) swivel holder with varying height adjuster to lower the glass into the furnace loop; (4) adjustable mirror which allows microscopic inspection of the glass as it is constricted by the furnace.

preparations using the cockroach. None of these are more elegant than the free-walking preparations of Gammon[2] and of Delcomyn and Usherwood.[3] These authors have used insulated wire (silver and copper) to record muscular activity while the animals are free-moving. Such techniques of relatively nonintrusive recording have distinct advantages when recordings are to be made of "normal" behavior before and altered behavior after the application of some test chemical.

The major difficulties with these techniques are those associated with the precise placement of the electrodes. It is relatively straightforward to insert the electrodes and to obtain a recording, but often it is not so easy to ascertain the exact muscle from which the recording has been obtained.

E. AMPLIFIERS FOR EXTRACELLULAR RECORDINGS

For electrophysiological experiments, the amplifiers can be divided into those required for intracellular recording and those needed for extracellular recording. Those for extracellular recording are much less demanding in their specifications. The important features are high-impedance differential inputs, low noise, and low drift. It is also desirable to have dual

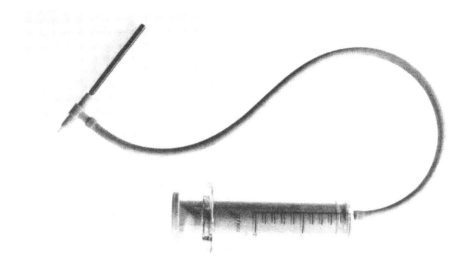

FIGURE 5. A simple suction electrode setup. The suction is produced by the syringe, and the plastic T-piece can accommodate the positive and negative electrodes. The joints are sealed with silicone sealer.

filters for high and low frequencies. There are a number of such amplifiers on the market, manufactured by Dagan (2855 Park Ave., Minneapolis, MN 55407), Grass Instruments (101 Old Colony Ave., Quincy, MA 02169), U.F.I. (545 Main, Suite C2, Morro Bay, CA 93442), and WPI (375 Quinnipiac Ave., New Haven, CT 06513), for example. The DAM 50 from WPI is shown in Figure 6.

III. INTRACELLULAR RECORDING

The first serious steps in the use of micropipettes for recording were taken by Gerard and co-workers[4,5] when they measured the membrane potentials of frog muscle. However, it was Hodgkin and Huxley[6] who first recorded from nerve in the giant squid axon using a large (0.1-mm) glass pipette as the recording electrode. Although these early attempts provided the impetus for intracellular recordings, they were limited to large cells. These large-diameter tips often gave unstable recordings. It was not until instrumentation was developed for the fabrication of fine electrodes that stable recordings for long time periods were obtainable, especially from small cells.

A. GLASS FOR MICROPIPETTES

Glass is an excellent electrical insulator, and when fabricated as a microelectrode it provides a cavity for confining various solutions as well as for isolating electrical signals. Once it penetrates a cell and seals the cell membrane against the glass, it provides a channel into the cell, allowing the stable measurement of internal electrical activity for considerable periods.

1. Tubing Design

In common use in neurophysiology is Corning® No. 7740 tubing, which contains a solid glass fiber 0.1 mm in diameter. Another type of tubing in common use is the Omega Dot tubing. Tubing with heavy Omega Dot or thick walls (high O.D./I.D. ratio) requires a greater heat output from a heating filament for drawing into micropipettes.

Capillary tubing is most often used for microinjection (microiontophoresis) or for cell penetration and intracellular recording. The tip opening can be enlarged by lowering the O.D./I.D. ratio. A standard O.D./I.D. ratio of 2 for Corning® glass is the norm for mi-

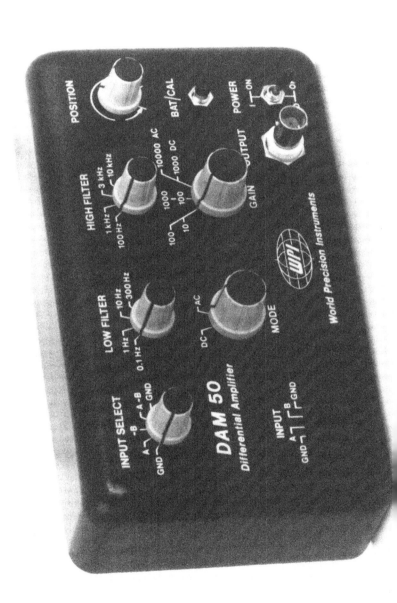

croinjection. Ratios too much higher give fragile tips. A short, stiff shank and very small-diameter tips aid in penetration. Thus, the O.D./I.D. ratio should be ≥ 2 to give a tip stiff enough to be able to penetrate cells and tissue.

B. MICROPIPETTE PULLERS

The preparation of the micropipette is the most fundamental step in preparing for intracellular work. An authoritative discussion of micropipette technology, written by Brown and Flaming,[7] is available and should be referred to by anyone starting this work. The following sections on pulling and beveling microelectrodes were written after a thorough reading of the Brown and Flaming work referred to above and also include the junior author's own experiences.

When drawing micropipettes, it is important to control the temperature of the glass. Glass temperatures that are too high give long, flimsy tips, while cool glass temperatures result in large, broken tips. Control of both timing and filament temperature is critical to achieve reproducibility in drawing micropipettes.[8] Consistent results are usually achieved only through the use of automated commercial instruments.

To form the fine tip, it is necessary that the pull begins slowly under weak tension and then accelerates with a stronger pull. By quickly removing the tip from the heating element the tip is allowed to cool, thus preventing the formation of tips too long to be usable. It is also important to hold the glass tubing straight during the pull in order to produce straight tips. The Du Bois micropipette puller[9] was one of the earliest automated microforges which incorporated the above characteristics in a horizontal puller, and it became a model for the design of other pullers. It is an excellent design for making microneedles for microdissection.

Another horizontal puller, designed by Alexander and Nastuk,[10] incorporates aspects of the older pullers as well as new features. It uses an electromagnetic solenoid to supply the pulling force and pulls reasonably straight and fine micropipettes. However, very fine tips for small cell work are difficult to make because insufficient velocity is obtained in the second pull. This instrument is still available with little modification through Industrial Science Associates, Inc. (25-44 163rd St., Flushing, NY 11358).

The Brown-Flaming micropipette puller incorporated many of the older characteristics which were desirable in a horizontal puller and minimized the variables by utilizing a microprocessor to control heat, pull velocity, cooling time, and pull strength.[7] These workers incorporated the use of an air stream similar to the Chowdhury puller in their design.[11] In the Chowdhury puller, air is blown over the softened glass to shorten the cooling time during the strong pull, this action resulting in two identical ultrafine (0.05-μm) tips with short shanks.

The Brown-Flaming puller as supplied by Sutter Instruments (P.O. Box 3592, San Rafael, CA 94912) is presently a very sophisticated puller capable of pulling a complete range of micropipettes including micropipettes for patch clamping. Two models are available, one incorporating the air-jet principle and a less expensive unit without this feature. These pullers now produce tips of a given size with high reliability. All the advantages of short tips (lower electrical resistance, decreased noise levels, greater stiffness for beveling) are obtainable in combination with ultrafine tips.

C. BEVELING

Beveling has been used by many workers to modify the tip of a micropipette to improve penetration into cells and tissues without causing significant damage. Early beveling methods consisted of breaking the micropipette tip to achieve an angle different from 90°, employing a rotating abrasive stone disk,[12] applying diamond dust as the grinding abrasive on a spinning quartz rod,[13] utilizing fine (0.05-μm) alumina particles suspended in a salt solution rotating on a wobble-free glass plate,[14] and using embedded particles of alumina or variously sized diamond dust in a polyurethane matrix rotating at a slow speed.[7,15]

In each case, progress was made toward the beveling of finer and more consistent micropipette tips. The necessary requirements in each case were sufficient contact of the tip with the abrasive surface and sufficient pressure to shape the tip. Long tips, such as those made in an Alexander-Nastuk puller, were sufficiently flexible so that tips did not break, with tip sizes of 0.2 μm obtained routinely. The short, stiff fine tips of the Brown-Flaming puller were initially troublesome to bevel, but using more uniform adhesive coating procedures now allows satisfactory beveling of these very fine micropipettes.

The K.T. Brown Micro-Pipette Beveler available through Sutter Instruments can conveniently bevel tips from 0.06 μm to very large sizes.[16] It is possible to bevel tips of 2 to 3 μm in a few minutes. This instrument uses a wet grinding process, so it is possible to continuously monitor the electrical resistance of the micropipette as the beveling proceeds. A gradual, continuous decrease in electrical resistance indicates a smooth abrasive action on the tip, while sharp drops in resistance indicate chipping of the tip.

The usual range of tip angles on beveled electrodes seldom exceeds 20 to 35°. It should be realized that beveling enlarges the tip opening and, thus, lowers the electrical resistance of a micropipette. The method of choice is to bevel the micropipette the minimum amount required for improved penetration of cells. Increases in resistance indicate plugging of the tip by abrasive particles. Some improvement in beveling of fine tips can be achieved by filling micropipettes overnight prior to actual beveling.

Methods for beveling of tips without instrumentation are also available. To obtain reasonably sharp tips, Baldwin[17] used sheets of tough plastic film (Imperial Lapping Film) with an adhesive backing containing 0.3-μm embedded alumina (Thomas Scientific, P.O. Box 779, Philadelphia, PA 19105-0779) on a rotating device. This method was used by Werblin[18] to selectively bevel one side of a dual-channel micropipette and decrease the electrical coupling between the two channels. Tauchi and Kikuchi[19] used ceric oxide, a polishing compound, in an agar matrix placed on a phonograph turntable to achieve tips of very nice quality. A jet stream microbeveler described by Ogden et al.[20] has also been used to obtain useful beveled tips. A very simple slurry beveling method[21] has also been shown to achieve bevel tips that allow penetration into small cells. The abrasives used in these methods can be obtained from Buehler Ltd. (2120 Greenwood St., Evanston, IL 60204).

The quality of a beveled tip depends on the type of abrasive and how it is applied to the micropipette tip. Abrasive suspended in saline results in slow beveling of tips because the abrasive particles are not anchored and, hence, can move about freely in the medium. Since there is freedom of movement, there is also difficulty in controlling the bevel angle. When the abrasive is concentrated in a slurry or other soft medium to restrict abrasive-particle movement, beveling time is reduced. However, the increased density of the medium causes greater forces on the tip, and bevel angle is again difficult to control. Embedded abrasive particles further decrease the beveling time, and accurate tip angles can be obtained by presetting the bevel angle. All three methods have their value, but the embedded abrasive method has been shown to produce a somewhat sharper cutting edge more reliably.

The advantage of beveling is that the micropipette can more easily penetrate cells without significant damage. The sharp cutting edge should allow a clean cut into the membrane and subsequent better sealing of the membrane to the tip, with higher membrane potentials and preparation stability the result.

Beveling can also reduce the signal/noise ratio by reducing the electrical resistance of the micropipette. This reduction in electrical resistance can also reduce the resistance experienced during iontophoretic injection of compounds onto or into the cell. Beveling has also been found to be advantageous for penetration through tough overlying membranes and for improving performance of dual-channel micropipettes.

D. MICROMANIPULATORS

When positioning a micropipette, a mechanism free of vibration and with the ability to control the movement of the micropipette in three dimensions is necessary. Inadvertent movements caused by hand motion are then not important. Although coarse controls are required for two dimensions, fine control is required when advancing toward a cell or tissue. Many types of micromanipulators have been described.[1,22,23]

IV. RECORDING MOVEMENT

Problems concerned with measuring the locomotory movements of cockroaches were described earlier,[24] and techniques used for measuring locomotory movements of insects in general were also described previously.[1]

In this section, we concentrate our attention on methods of recording movements of a different kind. It has been shown that insects periodically "pump" their abdomens, resulting in sharp increases and decreases in hemocoel pressure.[25,26] The change in pressure is thought to be caused by contractions of intersegmental muscles which connect adjacent flexible abdominal segments. The action of the intersegmental muscles, in turn, causes minute movements of the abdomen. These pressure pulses are thought to participate in the regulation of certain autonomic functions, such as hemolymph circulation through appendages, movements of visceral organs, and tracheal ventilation. Therefore, the recording of movements is another way of recording suspected autonomic functions.

Long ago, Harker[27] noticed that decapitated adult cockroaches could remain alive for many days or weeks. More recently, it was found that houseflies could be tested for 24-h mortality after treatment with organophosphate insecticides regardless of whether their heads were on or off.[28] This suggests that the head or brain of insects does not support a vital function aside from eating and orientation and that homeostasis may be maintained for some considerable time in a vegetative state.[29]

The body of the decapitated cockroach or of an immobilized cockroach is a very convenient model system to monitor changes in hemocoel pressure and other autonomic functions because of the extensive physiological information already available.[30]

A. ELEMENTS OF THE STRAIN GAUGE TECHNIQUE

Very small mechanical changes can be measured by strain gauge elements (tensiometers). These are usually made from a thin wire, metallic foil, or semiconductor silicon material. In principle, a mechanical deformation of the strain gauge shape would cause a corresponding change in its resistance. If the gauge were a part of a Wheatstone resistance bridge, the deformation would cause deviations in the balance current of the bridge. The imbalance could then be amplified and recorded.

B. GAUGE FACTOR

The efficiency of transformation of mechanical deformity into resistance of the strain gauge has been expressed conventionally as a gauge factor. The gauge factor usually ranges from 2 in the wire and foil gauges to 120 or more in the silicon gauges. The latter type thus provides a 60-fold more sensitive measurement than the former at the same amplification. This is especially convenient for the construction of strain or movement transducers, using the commercially available, low-cost, operational amplifiers.

C. WHEATSTONE BRIDGE AND AMPLIFIER CIRCUITRY

A simple circuit for an inexpensive strain gauge transducer and amplifier is shown in Figure 7. The circuit is suitable for recording changes in hemocoel pressure of insects. The matched pair of strain gauges, T_1 and T_2, form half of a Wheatstone resistance bridge. They are connected in neighboring branches of the bridge so that the effects of small temperature changes acting on both sensors should be eliminated. When the gauges are placed on opposite

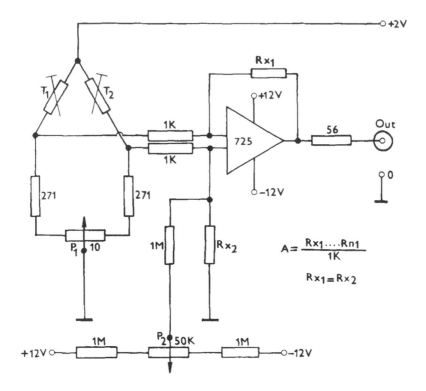

FIGURE 7.   Circuit diagram for an ordinary Wheatstone bridge amplifier of the type used in force transducer. T_1 and T_2 are strain gauges. P = potentiometer; A = the formula for amplification. Resistor values are shown, Rx resistors are selected, and 725 is the operational amplifier designation for analog devices or the equivalent.

sides of the foil, they are both "active" and additive; one causes an increase and the other a decrease in the current flowing through the bridge. A still more efficient system can be created from four gauges forming the full resistance bridge, where the sensors on one side alternate with the sensors on the other side of the foil.

The scheme described in Figure 7 uses a DC bridge-balancing current for simplicity. However, the amplifier will amplify the electrical "noise" along with the signal, and the sensitivity of this arrangement is limited. Greater stability and sensitivity are achieved using 5 kHz AC for the bridge-balancing current. When filtered, decoded (or demodulated), and converted into a DC output, the signal contains very little electrical "noise".

D. SOURCES

Manufacturers of electronic instruments offer many types of foil or silicon strain gauges, including prefabricated foil strain gauges. They can be purchased, for instance, from Hottinger Baldwin Measurements (139 Newbury St., Framingham, MA 01701, or P. O. Box 4235, D-6100 Darmstadt 1, Federal Republic of Germany), Philips GmbH (D-3500 Kassel, Federal Republic of Germany), Kyowa Electronic Instruments (Tokyo, Japan), and other suppliers. The silicon strain gauges used in the transducers described here were purchased from Podnik Služeb (76000 Gottwaldov, Czechoslovakia).

E. CONSTRUCTION TIP

The attachment of a strain gauge to the measuring membrane or foil is a very important step in fabricating the gauge. Any special instructions given by the manufacturer should be followed. In principle, since we mainly use dynamic (not static) measurements, any cyanoacrylic or hard epoxy resin adhesive with low hysteresis can be used. Foils made from beryllium bronze or stainless steel are suitable.

F. AN EXAMPLE OF AN ISOTONIC MOVEMENT TRANSDUCER

A very convenient strain gauge transducer that is capable of sensing small movements of the cockroach body has been described by Sláma.[26] This anisometric or partly isotonic transducer is shown on the right side of Figure 8. It is built from a flat box of Plexiglas®, with an internal width of 10 mm. The most important component is a thin beryllium-bronze foil (**r**), which contains at its base a matched pair of stain gauges (**s**) forming half of a resistance bridge.

The transducer records movements with minimum loading and can reliably detect displacements as small as 10 nm. Its advantage is that it can be used for prolonged recordings (for days and weeks) on unrestrained insects, without the necessity of penetrating the integument. Occasional twisting and large sideward movements of the abdomen can be minimized using a long wire to connect the transducer to the subject insect.

In most cases, the transducer foil is gently polarized initially by taking up the slack in the transducer connection to the insect and putting a few micrograms of tension on the foil. In certain cases, especially when the integument is very soft, the transducer can be used in a "pushing mode" with the membrane polarized in the opposite direction by pushing the transducer pin against the surface of the body.

G. IMPEDANCE CONVERTERS

Impedance converters remain among the most versatile of measuring instruments available to the physiologist. Originally designed for measuring pulmonary activity of crash victims during the ambulance ride to a hospital, the converter can be used for a number of other applications, especially in measuring movements or force in the cockroach.

The UFI® Model 2991 impedance converter is designed to measure impedance changes. A 1% change in source resistance, capacitance, or inductance causes the output of the converter to vary by about 0.5 V. Impedance changes at the input modulate the amplitude of a 50-kilocycle oscillator. The oscillator output is then demodulated and filtered. The output voltage is either directly proportional to the input changes and shows positional information, or it is filtered to show faster changes and is insensitive to very slow changes in impedance.

Compared to electrical measurements, impedance conversion has few electrical noise problems and is remarkably free from interference. The device is very sensitive in recording movements in insects, but the output recorded depends on the geometry of recording electrodes and on the physical location of the electrodes.

The strain gauges described above may be connected directly to the input of the impedance converter. As long as the output signal is calibrated, the need for amplifiers and Wheatstone bridge circuitry is eliminated. For rapid setup, versatility, and convenience, it is hard to improve upon the impedance converter. In addition to its other advantages, it is also rugged enough to be student proof.

H. THE VERO TRANSDUCER

The need to measure very small forces produced by visceral muscles in insects led to the design and construction of a transducer based on an inductance feedback principle. The basic idea for an inductance transducer was converted by Vero into a very sensitive device with a linear output.[31] This device is now being developed by U.F.I. as Microgage (Model 1630).

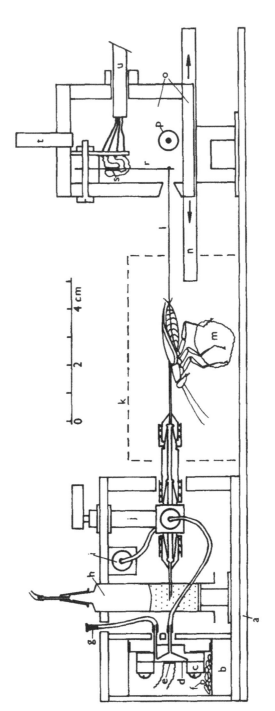

FIGURE 8.   Diagram of pressure transducer (left) and force transducer (right). (a) Supporting plate; (b) Plexiglas® body of the hydraulic transducer; (c) stainless steel drum of the transducer (25 mm O.D., 11 mm I.D.); (d) beryllium-bronze transducer arm; (e) silicon strain gauges, type AP 120-3-12, size 0.1 × 3 mm, gauge factor 120 (Podnik Služeb, Gottwaldov, Czechoslovakia); (f) silicic acid crystals; (g) outlet for washing out air bubbles; (h) reservoir of hydraulic fluid connected to water manometer by Teflon® tubing; (i) 701 N Hamilton® syringe; (j) 1XP Hamilton® inert valve; (k) light and humidity cover; (l) hooked wire connecting a small triangle cemented to the abdomen with transducer membrane; (m) polystyrene ball; (n) adjustable arm to position transducer and tighten with knurled knob; (o) 2-mm-thick side wall of a transducer box; (p) screw compressing the side wall for tightening; (r) beryllium-bronze membrane, 0.09 × 5 × 30 mm; (s) matched pair of silicon strain guages (AP 120-6-12), 6 mm length, 120 ohm resistance (Podnik Služeb, Gottwaldov, Czechoslovakia); (t) metallic rod for micromanipulator attachment; (u) shielded outlet cable. See text for explanation (Section IV.F).

V. RECORDING PRESSURE

A. HYDRAULIC TRANSDUCER

Prolonged monitoring of changes in hemocoel pressure in cockroaches may be achieved with the hydraulic transducer shown on the left side of Figure 8. The hemocoel cavity is connected with the hydraulic fluid of the transducer through a stainless steel syringe needle of a size appropriate for the measured object. The fluid may be distilled water, Ringer's solution, or light mineral oil, and it must be free from air bubbles (evacuated). A viscous material such as lanolin, adhesive gum, nonirritating cement, etc. should be applied around the tip of the needle to prevent bleeding during insertion of the needle into the body.

An essential component of the pressure transducer is a Hamilton® Model 1XP miniature inert valve, which connects the hemocoel with either the transducer or a syringe. The function of the syringe is to flush the needle used to penetrate the insect with hydraulic fluid.

In addition, solutions to be tested for their effects on hemocoel pressure can be taken into the syringe needle prior to its insertion into the hemocoel (usually separated on both sides with 1 µl of oil). Following a period of equilibration and control measurements, the Hamilton valve may be turned, switching the syringe and the attached insect from the pressure transducer to a test solution. After injecting the solution into the body, the valve may then be returned immediately to the transducer setting with very little interruption of measurements.

By moving the valve to its second position, the pressure sensors are connected to a reservoir of hydraulic fluid. The level of fluid in the reservoir (stippled portion of **h** in Figure 8) is adjusted to about 5 mm above the orifice of the needle. This hydrostatic pressure is arbitrarily taken as a zero (5 mm above local barometric) pressure. Using a manometric tube filled with water and connected to the reservoir, we can apply various hydrostatic pressures to the reservoir, which can be used for calibration or for direct manometric pressure control during the recordings. More details concerning the construction of the hydraulic transducers can be found in the paper by Sláma.[25]

VI. EXAMPLES OF RECORDING

Adult cockroaches are normally very active, especially when handled, and it may be necessary to narcotize them by submersion in water for 5 to 10 min before attachment of the transducers. For decapitation, the neck is ligated and the head separated from the body immediately after removal from the water. For direct recording with the hydraulic transducer, the needle is inserted into the body cavity, usually through the meso- or metanotal region.

For indirect recording of hemocoel pulsations with the isotonic transducer, a small wire triangle was cemented to the tip of the abdomen, using a quick-setting cyanoacrylate glue, before the cockroaches recovered from water narcosis. The triangle was hooked by a wire attached to the strain gauge foil. Decapitated cockroaches maintained their cleaning responses and by reflex wiping could use the hind-leg spines to remove any object not properly cemented to the abdomen.

After being immobilized by decapitation the cockroaches were placed in a horizontal position, either dorsal or ventral side uppermost. They could be affixed to a substrate between strips of cardboard and held in place by attachment to the mesonotum or at the tips of the wings. Females bearing oothecae could be studied by recording oothecal movements directly. Specimens to be measured continuously for several days should be isolated from light, sound, or air movement since these cause behavioral responses.

Recording from intact (nondecapitated) *Blattella germanica* adult cockroaches is difficult because of their struggling. However, during periods of quiescence, hemocoel pressure changes as a regular pattern of repeated bursts (Figure 9, top trace). The pressure pulses

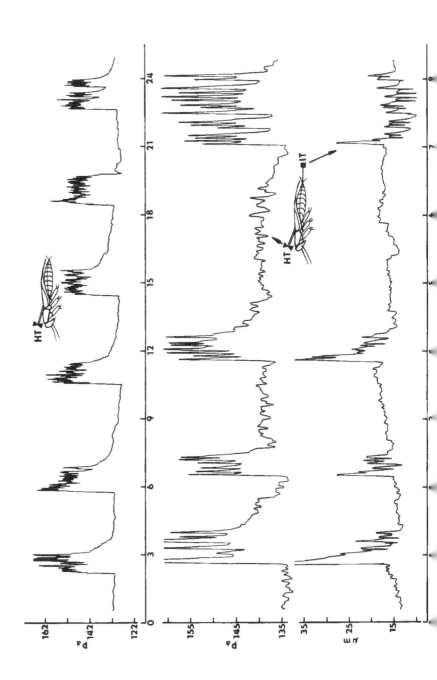

are shown to be associated with movements of the abdomen in simultaneous recordings (Figure 9 — middle trace, pressure; bottom trace, isotonic tension).

A difference in amplitude can be seen when comparing the pressure and movement records in Figure 9. This is thought to be due to spiracular activity during respiratory pumping. A 1-μm movement is roughly correlated with 4 Pa of pressure.

Nauphoeta cinerea is a convenient species with which to work. Unlike *Periplaneta americana*, it has no pungent odor when cultured. Pressure recordings from decapitated adults are similar to those from intact insects, i.e., a series of repeated pulse bursts (Figure 10, top trace). Movement recordings from *Nauphoeta* show that the bursts of abdominal pumping accompanying pressure bursts are sometimes interspersed with gross movements of the abdomen (Figure 10, middle and bottom traces, respectively).

Recordings from decapitated *P. americana* are very similar to those from the other insects. They show a series of repeated pulses that continue for as long as they are recorded (Figure 11, top trace). Simultaneous recording of pressure and movement also shows the coordination between these activities (Figure 11, bottom and middle traces, respectively). Although the amplitudes of the pressure pulses remain similar throughout the burst, the movement trace shows low-amplitude pulses first, then full-amplitude pulses, followed by low-amplitude pulses at the end of the burst.

The different amplitudes of movement in the middle trace (isotonic force record) of Figure 11 are thought to be due to the spiracles being either open or closed during the pressure pulses. If the spiracles are open during increasing pressure, air will be vented from the tracheal system much like from a bellows, with a corresponding reduction in the overt abdominal displacement.

In general, females produce pulsations more often or with larger amplitude than do males. This may be due to higher metabolic rates in the former. Any change in metabolic rate is immediately reflected in changes in the number and frequency of pressure pulses, such as would occur after a meal.

VII. SUMMARY

The cercal nerve/giant interneuron synapse in the sixth abdominal ganglion of the American cockroach has been used by zoologists as a model cholinergic synapse. The "giant" axons in this preparation produce enough electrical activity that they may be used *in situ* in the ventral nerve cord with an appropriately miniaturized nerve chamber to study the propagation of the nerve impulse.

The cockroach antenna has provided a model olfactory organ for the study of odor perception. Other unique tissue preparations from the cockroach, including salivary glands, heart, ventral diaphragm, and central nervous system, provide convenient model systems to study pharmacological responses to specific drugs.

Aside from necessary miniaturization in some cases, the techniques described here are identical to those used for studying any vertebrate; indeed, many are borrowed or modified directly from vertebrate studies and are therefore interchangeable. Force or movement transducers are possibly the most unique adaptation of all physiological techniques to study insect tissues, and these have been described here in great detail.

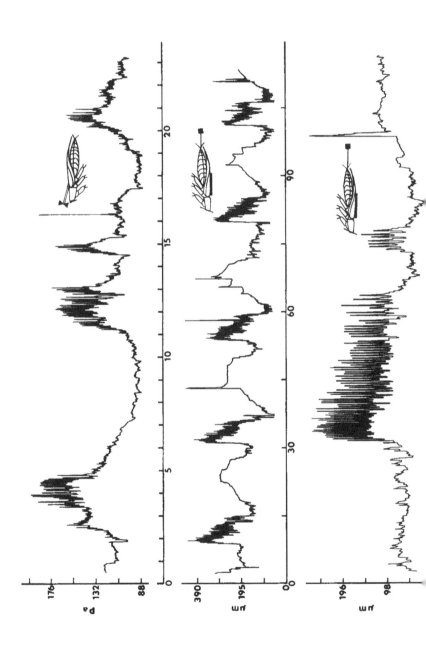

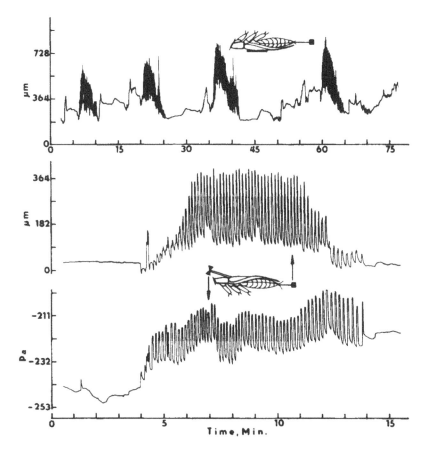

FIGURE 11. This is the same type of recording as Figure 10. See text for explanation (Section VI).

REFERENCES

1. **Miller, T. A.**, *Insect Neurophysiological Techniques*, Springer-Verlag, New York, 1979.
2. **Gammon, D.**, Nervous effects of toxins on an intact insect: a method, *Pestic. Biochem. Physiol.*, 7, 1, 1977.
3. **Delcomyn, F. and Usherwood, P. N. R.**, Motor activity during walking in the cockroach, *Periplaneta americana*, *J. Exp. Biol.*, 59, 629, 1973.
4. **Graham, J. and Gerard, R. W.**, Membrane potentials and excitation of impaled single muscle fibers, *J. Cell. Comp. Physiol.*, 28, 99, 1946.
5. **Ling, G. and Gerard, R. W.**, The normal membrane potential of frog sartorious fibers, *J. Cell. Comp. Physiol.*, 34, 383, 1949.
6. **Hodgkin, A. L. and Huxley, A. F.**, Resting and action potentials in single nerve fibers, *J. Physiol. (London)*, 104, 176, 1945.
7. **Brown, K. T. and Flaming, D. G.**, *Advanced Micropipette Technology for Cell Physiology*, John Wiley & Sons, New York, 1986.
8. **Flaming, D. G. and Brown, K. T.** Micropipette design: form of the heating filament and effects of filament width on tip length and diameter, *J. Neurosci. Methods*, 6, 91, 1982.
9. **Du Bois, D.**, A machine for pulling glass micropipettes and needles, *Science*, 73, 344, 1931.
10. **Alexander, J. T. and Nastuk, W. L.**, An instrument for the production of microelectrodes used in electrophysiological studies, *Rev. Sci. Instrum.*, 24, 528, 1953.
11. **Chowdhury, T. K.**, Fabrication of extremely fine glass micropipette electrodes, *J. Sci. Instrum.*, 2, 1087, 1969.

12. **Barrett, J. N. and Graubaud, K.**, Fluorescent staining of cat motoneurons *in vivo* with beveled micropipettes, *Brain Res.*, 18, 565, 1970.

13. **Barrett, J. and Whitlock, D. G.**, Technique for beveling glass microelectrodes, in *Intracellular Staining in Neurobiology*, Kater, S. B. and Nicholson, C., Eds., Springer-Verlag, New York, 1973.

14. **Kripke, B. R. and Ogden, T. E.**, A technique for beveling fine micropipettes, *Electroencephelogr. Clin. Neurophysiol.*, 36, 323, 1974.

15. **Brown, K. T. and Flaming, D. G.**, Technique for precision beveling of relatively large micropipettes, *J. Neurosci. Methods*, 1, 25, 1979.

16. **Brown, K. T. and Flaming, D. G.**, Instrumentation and technique for beveling fine micropipette electrodes, *Brain Res.*, 86, 172, 1975.

17. **Baldwin, D. J.**, Dry beveling of micropipette electrode, *J. Neurosci. Methods*, 2, 153, 1980.

18. **Werblin, F. S.**, Regenerative hyperpolarization in rods, *J. Physiol. (London)*, 244, 53, 1975.

19. **Tauchi, M. and Kikuchi, R.**, A simple method for beveling micropipettes for intracellular recording and current injection, *Pfluegers Arch.*, 368, 153, 1977.

20. **Ogden, T. E., Citron, M. C., and Pierantoni, R.**, The jet stream microbeveler: an inexpensive way to bevel ultrafine glass micropipettes, *Science*, 201, 469, 1978.

21. **Lederer, W. J., Spindler, A. J., and Eisner, D. A.**, Thick slurry bevelling, *Pfluegers Arch.*, 381, 287, 1979.

22. **Ellis, G. W.**, Piezoelectric micromanipulators, *Science*, 138, 84, 1962.

23. **Corey, D. P. and Hudspeth, A. J.**, Mechanical stimulation and micromanipulation with piezoelectric bimorph elements, *J. Neurosci. Methods*, 3, 183, 1980.

24. **Cornwell, P. B.**, *The Cockroach*, Vol. 1, Hutchinson, London, 1968, 391.

25. **Sláma, K.**, Insect haemolymph pressure and its determination, *Acta Entomol. Bohemoslov.*, 74, 362, 1976.

26. **Sláma, K.**, Recording of haemolymph pressure pulsations from the insect body surface, *J. Comp. Physiol.*, 154B, 635, 1984.

27. **Harker, J. E.**, Control of diurnal rhythms of activity in *Periplaneta americana* (L.), *Nature (London)*, 175, 733, 1955.

28. **Mengle, D. C. and Casida, J. E.**, Biochemical factors in the acquired resistance of house flies to organophosphate insecticides, *J. Agric. Food Chem.*, 8, 431, 1960.

29. **Sláma, K. and Miller, T. A.**, Insecticide poisoning: disruption of a possible autonomic function in pupae of *Tenebrio molitor*, *Pestic. Biochem. Physiol.*, 29, 25, 1987.

30. **Bell, W. J. and Adiyodi, K. G., Eds.**, *The American Cockroach*, Chapman and Hall, New York, 1981.

31. **Vero, M. and Miller, T.**, Sensitive tension and force transducer, *Med. Biol. Eng. Comput.*, 17, 662, 1979.

Chapter 11

CELLULAR RESPONSES TO NERVE INJURY IN THE COCKROACH NERVOUS SYSTEM

Robert M. Pitman

TABLE OF CONTENTS

I. INTRODUCTION

The limited ability of the human nervous system to achieve functional recovery following trauma presents a major clinical problem. Strictly speaking, for full recovery to occur, individual neurons should be able to re-form the connections that have been interrupted by a lesion. This, of course, is impossible if neurons have been damaged irreversibly, since neurons in the adult nervous system are no longer capable of cell division to replace any cell loss. However, even following a lesion that interrupts axonal pathways without causing any significant neuron destruction, recovery is normally far from complete; the reasons for this are apparently different in the human central and peripheral nervous systems. Although damaged mammalian peripheral nerves apparently possess considerable potential for axonal growth, functional recovery is often very poor after peripheral nerve trauma because the regenerating fibers show only a weak preference for their original target and, as a consequence, many inappropriate synapses may be formed.[1] Restricted regeneration in the central nervous system (CNS), on the other hand, appears to result largely from failure of central neurons to extend over long distances within the CNS. Although a number of hypotheses have been proposed to account for abortive regenerative growth within the CNS,[2] the causes have not been established. Failure of axons in the adult mammalian CNS to regenerate over long distances is not, however, a result of an inherent inability of central neurons to produce axonal elongation; axons of central neurons can extend over considerable distances when directed into peripheral nervous tissue grafts.[3-5]

In order to understand the mechanisms by which regenerative growth may occur, many research workers have directed their attention to animal preparations in which the nervous system has a well-developed capacity for regenerative recovery. Through studies on such preparations the mechanisms leading to restoration of appropriate connections may be elucidated. This knowledge may then be applied to determine the causes underlying abortive regeneration in man. Many invertebrates present a number of important advantages as experimental models in which to study regeneration. Their nervous systems are relatively simple by comparison with those of even the simpler vertebrates, even though they operate on most of the same mechanistic principles. Since the nervous systems of many invertebrates contain relatively few large neurons, it has become routine to adopt the "identified neuron" approach. In the context of regeneration, this approach allows a range of methods to be used to investigate responses to damage of a neuron in which the functional and morphological characteristics have been clearly defined. Thus, in such preparations, unlike most vertebrate systems, the need to work with populations of similar but nonidentical neurons is eliminated. Among the invertebrates, insects present a number of practical advantages for studying nerve regeneration. For example, they are extremely robust and can be easily manipulated experimentally. In addition, the insect nervous system is bilaterally symmetrical, allowing a neuron on one side to serve as a control for its contralateral homologue on the other in a variety of different types of experiments.

In this review, attention will be focused on the cockroach as an experimental model for studying cellular consequences of nerve damage. Although a neuron is a functional entity, axonal, dendritic, and somatic responses are discussed separately since each may be considered a different functional compartment within the neuron itself. Nerve lesions may operate through different mechanisms to generate appropriate responses in each compartment of the cell. It should be remembered, however, that different regions of a neuron interact with each other, so that the nature of a response in one compartment doubtless will have far-reaching consequences for other parts of that neuron. Since normal neural function can be maintained only if the chemical environment of the nervous system is controlled closely, changes in satellite cells during repair are also considered since these cells form the blood-brain barrier.

II. REGENERATIVE GROWTH OF AXONS

Although specific reinnervation of denervated targets does not occur in the mammalian nervous system, many lower vertebrates, such as amphibians, fish, and birds, are capable of extensive regeneration. Much work has been done on the re-formation of retinotectal connections after lesions to the optic pathways of lower vertebrates. Such studies have shown that following regeneration the topographical projection of retinal axons onto the optic tectum map is restored. It was originally proposed that this restoration could occur because retinal axons and their tectal target neurons carry complementary surface labels which confer upon them a specific chemoaffinity for each other.[6] However, several different lines of evidence, including experiments in which size disparities are introduced between the retina and tectum, now indicate that the topographical projection results from competition between retinal axonal terminals which tend to array themselves across the tectum such that they maintain their neighbor relationships.[7]

During regeneration of the neuromuscular system in lower vertebrates, as in the CNS, competition between neurons may be important in enabling muscles to become reinnervated by the "correct" motoneurons.[8-11] A "foreign" nerve can form neuromuscular junctions if it is directed into a urodele limb muscle after removal of the normal innervation. However, as the original nerve reinnervates the muscle, "foreign" synapses show a reduction in quantal content, indicating that release of neurotransmitter from presynaptic terminals of "foreign" motor axons has been suppressed.[12] However, the mechanisms responsible for determining the pattern of neuromuscular connections that are reestablished during regeneration in lower vertebrates are still unclear. While it is possible that there is an intrinsic tendency for muscles to be reinnervated by the "correct" motoneurons, the connections that form during regeneration may be controlled by other factors, such as whether or not a motoneuron is supporting an overexpanded peripheral field. According to the latter proposal, a motoneuron with an enlarged peripheral innervation field would be unable to compete successfully with a motoneuron having a reduced field.[13]

Evidence has been available for some time which indicates that damaged cockroach motor axons are able to regenerate successfully, producing a high degree of functional recovery in the majority of animals.[14-17] Return of function in the coxal depressor muscles is achieved by specific reinnervation with the appropriate "fast" motoneuron (D_f) and probably also by the original "slow" motoneuron (D_s).[18] The specificity with which neuromuscular junctions are formed during regeneration, however, is not absolute, since coxal depressor muscles of a metathoracic leg transplanted into the mesothoracic segment become innervated specifically by mesothoracic homologues of the motoneurons which would normally innervate them.[19] Coxal depressor motoneurons are also able to selectively innervate contralateral homologues of their normal target muscles if nerve 5 (which contains the axons of these neurons) is cut and directed across the midline.[20,21]

Although the studies described above indicate that the normal innervation pattern of coxal muscles may be reestablished eventually, they do not address the question of the manner in which this is achieved. Denburg et al.[22] studied reinnervation of coxal depressor muscles by backfilling regenerating nerves with cobalt at a range of time intervals after lesions were created at different locations on nerve 5 of the metathoracic leg (Figure 1). Using this method, they found little evidence of axonal growth within 12 d following nerve lesioning. Once axonal elongation had started, the growth rate appeared to be approximately 0.9 mm/d. These workers also presented evidence that regenerating axons of coxal depressor motoneurons frequently grew along nerve 5 some distance distal to the branch (nerve 5 ramus 1 [N5r1]) into which they normally would travel to their target muscles. In addition, some unidentified neurons whose axons normally would not enter N5r1 initially did so during regeneration. The number of motoneurons with "misdirected" axons fell off with increasing

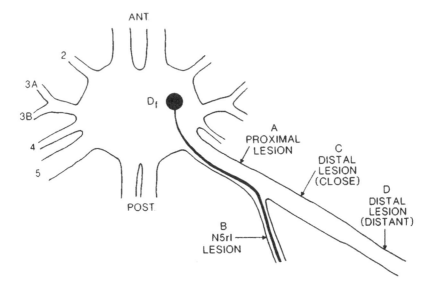

FIGURE 1.   Diagram showing the different positions at which nerve 5 of the cockroach metathoracic ganglion may be interrupted to gain information about the responses of coxal motoneurons to damage. The paths taken by regenerating axons of coxal motoneurons have been determined by cobalt backfilling through nerve 5 cut at each of the positions A to D. Cobalt staining was performed at different times after a nerve crush at position A.[22] Lesions also have been performed at positions A to D to study dendritic sprouting in the coxal depressor motoneuron D_f. (See text for details.) In normal, unoperated preparations, coxal depressor motoneurons would be axotomized by lesions at position A or B, while motoneurons to distal limb muscles would be axotomized by lesions at A, C, or D. A lesion at B would axotomize the coxal depressor motoneurons while sparing the axons of distal limb motoneurons. Lesions at C or D would spare the axons of coxal depressor motoneurons, but axotomize distal limb motoneurons. Abbreviations: D_f = "fast" coxal depressor motoneuron; N5rl = nerve 5 ramus 1 (the first nerve branch leaving the main trunk of nerve 5); ANT. = anterior connectives; POST. = posterior connectives. Numbers 2 to 5 indicate the numbering of peripheral nerves leaving the cockroach metathoracic ganglion.

time after the nerve lesion. Further evidence that regenerating cockroach leg motoneurons do not regenerate unerringly along the appropriate trajectory to reestablish connections with the correct target muscle has been provided by electrophysiological experiments.[23] In the initial stages of regeneration, both the D_f and D_s motoneurons innervated incorrect as well as correct coxal depressor muscles in the majority of preparations. In many cases, coxal depressor muscles also received synaptic contacts from unidentified "foreign" motoneurons that could elicit small excitatory junction potentials in these muscles. The results also indicated that the incidence of inappropriate connections made by coxal depressor and unidentified "foreign" motoneurons declined with increasing recovery time after nerve lesions. These observations, therefore, like those based on cobalt backfilling,[22] are consistent with the conclusion that regenerating motoneuron axons that take the wrong path or make incorrect synaptic connections may be eliminated selectively.

More recently, reinnervation of coxal depressor muscle 178[24] has been studied by determining which motoneuron cell bodies become stained after injecting horseradish peroxidase-conjugated wheat germ agglutinin (WGA-HRP) into the muscle.[25] Muscle 178 was selected since it normally receives innervation only from motoneuron D_f. Following nerve crushing, evidence for muscle reinnervation was obtained within 5 d, suggesting that the apparent 12-d lag before the onset of axonal outgrowth reported previously[22] was attributable to the relative insensitivity of the cobalt backfilling method. Another interesting observation made in this study was that after nerve 5 was crushed bilaterally, injection of WGA-HRP into muscle 178 on one side resulted in bilateral staining of motoneuron cell bodies in the metathoracic ganglion. Furthermore, bilateral cell body staining could also be seen if WGA-HRP was injected into a nondenervated muscle 178 contralateral to a crushed nerve 5. It was concluded that these observations could be accounted for most readily if motoneurons sent out processes bilaterally during regeneration. There was a progressive decrease in the number of inappropriate contralateral cell bodies that stained as the time interval between nerve crush and intramuscular injection of WGA-HRP was increased. The transience of the contralateral branches was attributed to the presence of the normal innervation on these muscles, with which the "foreign" branches presumably must compete. Perhaps relevant in this context is the observation that if the distal leg segments are removed, motoneurons which normally innervate muscles of distal limb segments via nerve 3B may send axonal branches into coxal depressor muscle 178. Such branches may develop even if nerve 5 is not crushed, so that the normal nerve supply to the muscle remains intact. Following distal limb amputation combined with nerve 5 crush, branches of "foreign" distal limb motoneurons in muscle 178 may persist indefinitely. Conversely, inappropriate contacts made by the D_s motoneuron upon this muscle are eliminated,[26] presumably because distal limb motoneurons have been deprived of their normal target muscles while motoneuron D_s has not.

The foregoing work indicates that reinnervation of cockroach muscles apparently is not achieved through absolute specificity between individual motoneurons and muscles. Instead, the "overextension hypothesis"[13] may account for many of the observations. Although a motoneuron may have some preference for a particular muscle, it may form and maintain neuromuscular synapses more avidly as the number of connections is reduced. Conversely, the ability to maintain synapses may fall as the total synaptic load increases. A motoneuron with a reduced target load would have a competitive advantage over a motoneuron with an enlarged target. Thus, if regenerating motoneurons grew out of their ganglion bilaterally, as proposed above,[25] they would have to form and maintain extra processes in contralateral limb muscles in addition to terminals in their correct (ipsilateral) target muscle. Muscles contralateral to a lesion are already innervated by the appropriate motoneurons, which would be at a competitive advantage because their undamaged axons would not have extended into "foreign" muscles. Damaged motoneurons which regenerate bilaterally will be overextended and, therefore, will tend to form stable connections only with denervated muscle, while their connections with muscles which already have an intact motor supply will be transient. On the other hand, a motoneuron deprived of its normal target would exhibit an increased tendency to branch into foreign muscles and form persistent connections. Thus, when distal segments of a limb are amputated, motoneurons that normally innervate distal limb muscles will have an increased tendency to form persistent contacts on "foreign" proximal muscles. Although the above results have been discussed in the context of overextension and interneuronal competition, it should be remembered that the observations depend upon neuronal uptake of intramuscularly injected WGA-HRP and that the mechanisms and conditions required for WGA-HRP uptake into nerves in this system are unknown. Therefore, although following injection of WGA-HRP into a particular muscle any motoneuron soma that is stained presumably has axonal branches in that muscle, those branches may not be functional. Thus, branches of different motoneurons that coexist anatomically in a muscle may not be competing functionally.

To directly observe axonal regeneration of individual identified motoneurons, the motoneuron D_f has been studied by intracellular injection of hexamminecobaltic chloride into the nerve cell body at different intervals after nerve lesioning.[27] It was found that at intervals of 1 to 2 d after crushing nerve 5 the crushed end of the axon was frequently indistinct, apparently because cobalt had leaked out into the interstitium. In those cases which allowed clear visualization, no sprouting was visible from the proximal axonal stump. The first sign of any outgrowth was seen on the third postoperative day and took the form of delicate varicose processes emanating from the crushed region of the axon. Thus, the delay between crushing nerve 5 and the onset of axonal outgrowth observed in this study was slightly longer than that reported by Denburg.[25] In the subsequent days a profusion of processes developed from the proximal axonal stump. Except in the immediate region of the lesion, fibers grew axially. In most preparations many processes could be seen growing distally, although a large number of fibers also extended toward the ganglion (Figures 2B and C). Therefore, in the initial stages of regenerative growth, there seems to be little proximodistal preference for growth along nerve 5 for the regenerating axonal processes of motoneuron D_f. By the sixth or seventh postoperative day, fine processes from the axon of motoneuron D_f could be seen entering N5r1, the branch into which this axon normally travels, although, as indicated by the observations of Denburg et al.,[22] many fibers normally grew beyond this point (Figure 2C). At postoperative periods of several weeks, processes frequently could be seen extending along nerve 5 more than 1 mm distal to N5r1. At progressively longer times after nerve crush, processes entering the appropriate nerve branch could be seen to increase in diameter, while "misdirected" processes remained thin and varicose (Figure 2D). The regenerated axon of motoneuron D_f never regained its normal appearance, even at intervals as long as 17 months after crushing nerve 5 (the longest postoperative period tested). Instead of sending a single axon into each subbranch of N5r1, many processes continued to travel in parallel in a single nerve branch. Observations on axonal outgrowth from unidentified motoneurons with axons in nerve 5 indicated that, although these neurons were able to send processes into N5r1, they were less likely to do so than motoneuron D_f (Figure 3). These observations directly confirm the conclusions of Denburg et al.[22] based on backfilling. However, in experiments where nerve 5 received a single crush lesion, contrary to the observations of Denburg,[26] axonal processes were never seen crossing the midline to exit through contralateral peripheral nerves. At this time the explanation for the difference between these two studies is unclear.

The above observations made by intracellular injection of motoneuron D_f suggest that during regeneration the axon produces a profusion of processes which do not grow in a highly directed manner. Processes that take the correct route and presumably form synaptic connections with the target muscle persist and become enlarged, while misdirected processes either remain thin or are lost. The apparent preference of motoneuron D_f for N5r1, therefore, does not appear to result from an ability of its regenerating neurites to grow unerringly along the appropriate path, but rather from being able to sprout profusely soon after damage and then consolidate appropriately directed processes. The fact that a relatively small proportion of regenerating processes from motoneurons to distal leg muscles enter N5r1, while many from motoneuron D_f do so, suggests that regenerating fibers may show some preference in the route they take. This effect, however, is most probably a result of the relative positions of the axons of different motoneurons in the main trunk of nerve 5. The axon of D_f runs on the margin of nerve 5 from which N5r1 originates. Its regenerating processes therefore will have easier access to this small nerve branch than will neurites growing from axons that are normally on the opposite side of nerve 5 from the exit point of N5r1. Thus, the relatively small proportion of "foreign" neurites that grow into N5r1 probably can be ascribed to geometric factors.

III. DENDRITIC CHANGES FOLLOWING LESIONS

In the vertebrate nervous system there is considerable evidence to indicate that dendrites may exhibit significant morphological plasticity both during development and in adulthood. Pyramidal cells of the mammalian cerebral cortex possess spines along their dendrites, each of which represents a synaptic specialization. Any change in the density of these dendritic

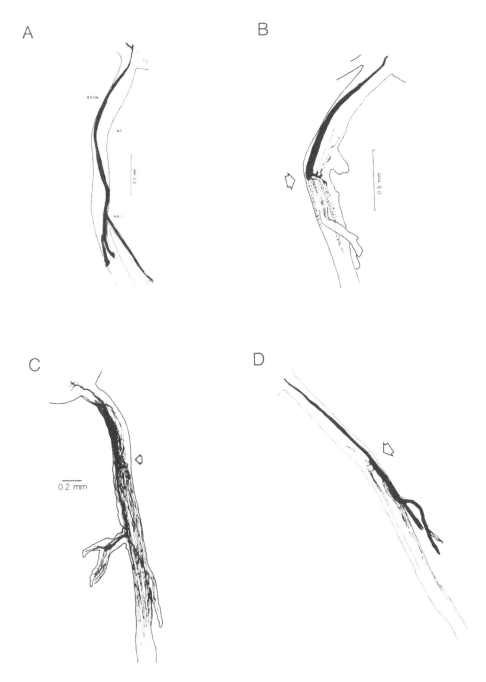

FIGURE 2.   Camera lucida drawings showing axonal outgrowth from motoneuron D_f after crushing metathoracic nerve 5 (N5), demonstrated by intracellular injection of hexamminecobaltic chloride. (A) Normal appearance of the axon in an uncrushed preparation. The axon travels along N5 and enters N5r1, in which it divides to enter the individual secondary branches of this nerve. (B) Axonal sprouting observed 7 d after crushing N5 (at arrow). Fine processes travel proximally and distally in the nerve; some fibers enter N5r1, while others extend beyond it. (C) Axonal outgrowth 14 d after nerve crush (at arrow); numerous fibers enter N5r1, while misdirected fibers can be seen growing beyond the branch point of N5r1. Processes also extend proximally as far as the ganglion (the margin of which is shown at the top of the figure). (D) Axon of D_f 103 d after crushing N5 (at arrow). Many fibers enter each branch of nerve N5r1. Those which have done so have a larger diameter than misdirected fibers. Calibrations for panels A and D are the same.

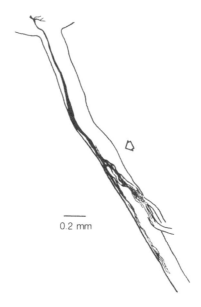

FIGURE 3.   Camera lucida drawing of axonal sprouting in an unidentified distal limb muscle motoneuron 42 d after nerve 5 was crushed (at arrow). The neuron was stained by intracellular injection of hexamminecobaltic chloride. Some processes can be seen near the junction of N5r1 with the main trunk of nerve 5; other nerve fibers extend beyond the branch point.

spines is therefore a measure of the number of presynaptic fibers that converge upon the pyramidal cell dendritic tree. At birth, dendritic spines are absent from pyramidal cells of the cerebral cortex in kittens and rats. The number of spines normally increases in postnatal life. This increase is, however, under environmental influence. There is a reduction in the number of dendritic spines on apical dendrites of pyramidal cells in the visual cortex of mice reared in darkness from birth until 22 to 25 d.[28,29] On the other hand, pyramidal cells in the visual and auditory cerebral cortices of young rats exposed to a diverse battery of sensory stimuli on a regular daily basis exhibited an increase in the number of spines relative to neurons of unstimulated controls. Sensory experience had a more marked effect upon basilar and oblique dendrites than upon apical dendrites.[30]

In addition to changes observed during postnatal development, the dendrites of vertebrate neurons may be affected morphologically by lesions suffered in adulthood. Axotomy is normally associated with dendritic retraction. Cerf and Chacko[31] showed that 8 to 11 d after bullfrog motoneurons were axotomized both the number and diameter of their dendrites were reduced. Similar observations have been made about mammalian motoneurons,[32,33] which show a similar dendritic response after their terminals have been treated with botulinum toxin (which produces a similar response in the neuron cell body to that of axotomy; see Section IV below). Reexpansion of the dendritic arbor was dependent upon reinnervation of muscle. If contact of the regenerating nerve with denervated muscle was delayed, dendritic spread did not occur until the time of reinnervation. This change in the extent of the dendritic arbor is accompanied by suppression of synaptic transmission to the injured neuron[34-37] and insertion of glial cells between the injured neuron and its presynaptic boutons. Synapses on the soma are affected to a greater extent than those on dendrites.[33,38]

Deafferentation in adult mammals, like axotomy, is normally associated with a reduction in the extent of dendritic arborizations. For example, lesions at different levels in the olfactory pathway produce significant alterations in the morphology of postsynaptic neurons. Observations on the rabbit accessory olfactory bulb, using Bodian's protargol staining method, have shown that dendrites of mitral and tufted cells undergo atrophy within 1 month of when they are deafferented by destruction of the olfactory mucosa. This atrophy consists of a reduction in number, length, and diameter of dendrites.[39] Interruption of fibers from the olfactory bulb to the prepiriform cortex of the rat causes a reduction in higher order dendritic branches of pyramidal cells in the denervated cortex.[40] Dendritic changes also have been reported following deafferentation of other regions of the mammalian CNS.[41]

Dendritic morphology of at least some neurons in immature insects, like those of mammals, is influenced by the connections they receive. For example, dendrites of cricket interneurons deprived of their presynaptic nerve fibers at nymphal stages remain smaller than normal.[42-45] The effects of partial deafferentation are very localized, since dendritic regions that are not deafferented are indistinguishable from those of control neurons.[42-44] At least one type of cricket interneuron, however, is capable of exhibiting dendritic growth in response to deafferentation.[46] If the distal segments of one prothoracic leg are amputated soon after hatching from the egg, the CNS will be unilaterally deprived of acoustic input, as the ears (which normally do not begin to develop until the third nymphal *instar*) are located on the tibial segment of the prothoracic legs. A pair of identified acoustic interneurons (interneurons 1) normally have a dendritic arbor restricted to one side of the prothoracic ganglion. This arbor receives input exclusively from the ipsilateral ear. After removal of this ear, the interneuron 1 deprived of its synaptic input projects dendritic branches contralaterally, where they selectively form functional connections with acoustic afferent neurons.[44]

Early experiments on neurons of adult insects, however, indicated that their dendrites apparently showed less plasticity than did dendrites of mammalian neurons. Although there was evidence that axotomy could be followed by changes in efficacy of synaptic input,[47] studies on the morphology of dendrites indicated that in adult animals they remained stable following either axotomy or deafferentation.[48] In fact, on the basis of such experiments, it had been suggested that neurons of adult insects may have become specialized for stability to enable the generation of stereotypical behaviors characteristic of many insects.[49] Cause to doubt these early observations first appeared when it was found that extensive dendritic sprouting could be induced in adult insect neurons by lesions involving a number of nerve trunks leaving a ganglion.[50] Under these circumstances it was observed that the motoneuron D_f could undergo extensive dendritic sprouting (Figure 4). Supernumerary dendrites frequently originated at or near the ends of branches of the normal dendritic arbor. These new branches could extend well outside the normal territory of this neuron, traveling into connectives or into segmental nerves which under normal circumstances would never contain either axonal or dendritic branches of this neuron. Near their origins, dendritic sprouts frequently had extremely convoluted trajectories, although at greater distances from their origin their course frequently straightened, suggesting that they were perhaps following preexisting pathways within the tissue. Although it was originally presumed that damage to a neuron would probably be the most effective stimulus of dendritic sprouting, it soon became apparent that some lesions which did not directly damage the axon of the motoneuron could serve as an equally effective stimulus.

Even though it is important to understand the factors which might trigger growth of supernumerary dendrites, the lesions used to induce dendritic sprouting in early experiments were sufficiently extensive to make a detailed examination virtually impossible. Fortunately, more recent work has indicated that extensive sprouting can take place after far more restricted lesions than those used originally. Although dendritic sprouting normally does not occur in motoneuron D_f after single cut or crush lesions to peripheral nerves through which the axon of this motoneuron runs, extensive sprouting can be observed after ligation or repetitive crushing of these peripheral nerves. It appears that development of supernumerary dendrites critically depends upon delaying the process of repair. If, for example, a single nerve trunk was simply crushed or cut, regeneration could occur relatively rapidly. In such preparations,

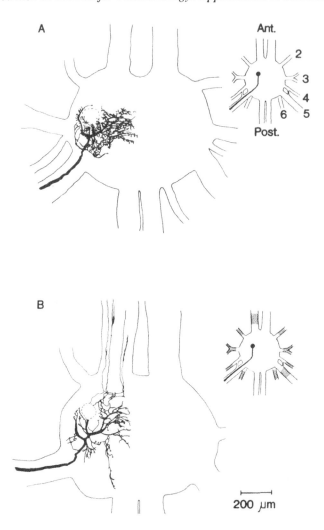

FIGURE 4.   Camera lucida drawings of motoneuron D_f stained by intracellular cobalt injection, showing dendritic sprouting after lesions were made on peripheral nerves. The motoneuron soma is indicated by a dotted line to avoid obscuring dendritic branches. (A) Neuron from a preparation in which the metathoracic ganglion had been exposed and the tracheal supply over the ganglion disrupted 35 d before staining. The dendritic arbor is indistinguishable from that of D_f in unlesioned preparations. (B) Neuron stained 21 d after cutting the nerve trunks, indicated by stippling in the inset. The axon of the stained neuron was not severed. Some dendrites of the motoneuron extend into the anterior connectives, while others extend within the ganglion beyond the limits of the normal arbor. Abbreviations in inset: Ant. = anterior connective; Post. = posterior connective. Peripheral nerves numbered according to standard nomenclature. (From Pitman, R. M. and Rand, K. A., *J. Exp. Biol.*, 96, 125, 1982. With permission.)

sprouting only occurred in a small minority of neurons studied, and even then its extent was very limited. On the other hand, if the same nerve trunk was ligated so that motor axons were physically prevented from regaining contact with their peripheral targets and sensory neurons (which in insects have peripherally located cell bodies) were kept from making connections with their central targets, extensive dendritic sprouting occurred in the majority

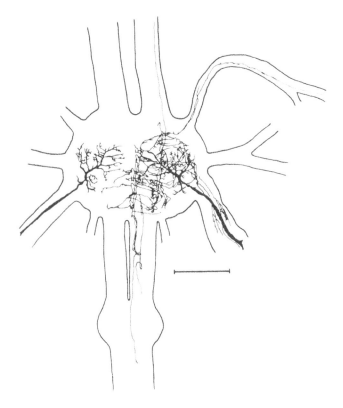

FIGURE 5. Camera lucida drawing of a pair of motoneurons D_f, stained by intracellular cobalt injection 44 d after a ligature was placed on the right nerve 5 (at position A in Figure 1). The dendrites of the motoneuron ipsilateral to the ligation show extensive intraganglionic sprouting and extend into the connectives and nerve 2 ipsilateral to the lesion. The dendrites of the motoneuron contralateral to the lesion show no sprouting. The somata of the two neurons have been omitted to provide an unobscured view of the dendritic branches. (Calibration: 0.5 mm.)

of neurons studied. Figure 5 shows a typical example of dendritic sprouting in motoneuron D_f following ligation of nerve 5 proximal to the exit point of ramus 1 containing the axons of the cell.

Using this model to routinely induce dendritic sprouting in adult neurons following relatively restricted lesions, efforts have been made to determine whether there was any one stimulus that could account for dendritic sprouting in this motoneuron. Therefore, the factors most likely to trigger dendritic sprouting have been listed to help establish which may be consistent with the experimental observations:

1.   Direct damage to the neuron itself
2.   Interruption of contact with target muscles
3.   Partial deafferentation of the motoneuron
4.   Release of "factors" from damaged muscle
5.   Chemical signals resulting from glial cell proliferation around the site of the nerve lesion
6.   General trauma resulting from damage associated with operations (e.g., disruption of the tracheal supply to the tissues)

To distinguish between these different alternatives, lesioning was performed as illustrated in Figure 1. Thus, confirmation was obtained for earlier observations that extensive dendritic

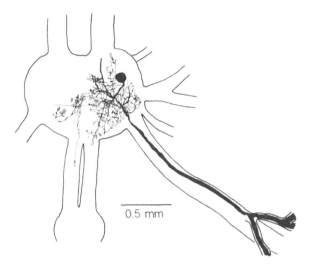

FIGURE 6.   Camera lucida drawing of motoneuron D_f stained by intracellular cobalt injection 45 d after a ligature was placed on the right-hand nerve 5 distal to N5r1 (at position C in Figure 1). Although this lesion did not axotomize the neuron, its dendrites have sprouted ipsilaterally and contralaterally within the metathoracic ganglion, and one process extends into the contralateral posterior connective.

sprouting could occur in the absence of direct surgical damage to the neuron itself.[50] A ligature placed on nerve 5 distal to the nerve branch (N5r1) through which the axon of motoneuron D_f exits (lesion C, Figure 1) was able to induce dendritic sprouting (Figure 6). Sprouting observed under these circumstances could not be attributed to trigger candidates 1, 2, or 4. Neither general trauma (option 6) nor any other systemic effect appeared to be primarily responsible for sprouting. No supernumerary dendrites were formed following sham operations in which the metathoracic ganglion and nerve 5 were exposed and the tracheal supply disturbed while all nerves were left intact (Figure 4A). If sprouting were triggered by a systemic mechanism, sprouting normally would occur bilaterally. However, dendritic sprouting only was found ipsilateral to unilateral lesions (see Figure 5). Furthermore, it is also hard to accept the proposal that partial deafferentation constitutes the only stimulus for sprouting (option 3). Ligation of the minor nerve branch, into which the axon of D_f travels (N5r1) and which contains few sensory axons, also can lead to dendritic sprouting. It therefore appears unlikely that partial deafferentation is the trigger for growth of supernumerary dendritic branches in these experiments. To gain further evidence in this respect, motoneuron 3 (a basalar motoneuron of the hindwing, identical to cell 3 in the map of Cohen and Jacklet[51]) also has been studied. The axon of this neuron travels in nerve 4, which has a purely motor function.[52] Therefore, if dendritic sprouting were observed in this neuron after ligation of nerve 4 it could not be attributed to partial deafferentation. In fact, ligation of nerve 4 constituted a very strong stimulus for dendritic sprouting in cell 3. Therefore, in the above series of experiments, the only single mechanism among the proposed candidates that could account for dendritic sprouting under all the experimental conditions would be a signal related to a glial injury reaction (option 5). An alternative proposal, however, is that sprouting may be initiated by more than one type of trigger mechanism. Lesions that affect neurons in different ways therefore may induce sprouting via differing mechanisms. Thus, axotomy may induce sprouting by one mechanism or a group of them, while partial deafferentation may operate through others.

Another example of dendritic sprouting in an adult insect neuron has been reported for an identified giant interneuron of the cricket *Acheta domesticus*. Dendritic sprouting has been observed several days after axotomy performed close to the terminal abdominal ganglion in which the cell body of this interneuron is located.[53] There is apparently a reciprocal relationship between the extent of dendritic and of axonal sprouting after this type of lesioning. Thus, neurons which exhibit extensive axonal sprouting generate little dendritic sprouting, while cells in which many supernumerary dendrites are present have few axonal sprouts. The authors concluded from their observations that axotomized neurons produce new membrane at a constant rate and that whether this is used to produce axonal or dendritic sprouts may be determined by the extent of retrograde degeneration of the axon after a lesion has been made. When axonal "die back" is marked, most new membrane will be used to produce new dendritic branches. At the present time it appears that dendritic sprouting may be regulated by different factors in giant interneurons of the cricket and in cockroach motoneurons. Dendritic sprouting only occurs in the majority of cricket giant interneurons when axotomy is performed very close (200 to 300 μm) to the cell body. Sprouting is never observed after an interganglionic connective has been cut at a distance >2.0 mm from the terminal ganglion, even if axonal growth is abortive. Conversely, in cockroach motoneuron D_f, dendritic sprouting is only seen after sustained lesioning (repeated nerve crushes or ligations). These, however, can be very effective even when made at distances >2.0 mm from the neuron soma.

IV. SOMATIC RESPONSES TO INJURY

When the axon of either a vertebrate or invertebrate neuron is damaged, there frequently follows a constellation of changes in the soma which may be termed the "axon reaction" or "cell body reaction". Changes may include a major redistribution of RNA within the cell body cytoplasm, an increase in nucleolar and nuclear volume, movement of the nucleus to the periphery of the cell body, and swelling of the cell body.[54-56] The relative timing of each of these somatic changes may vary between different types of neuron. Although the functional significance of most of the above changes is still not entirely clear, RNA changes have received a great deal of attention, as the cell body is the major site of protein synthesis within a neuron. Alterations in RNA handling may reflect changes in protein synthesis which may be significant for regeneration.

The exact nature of the change in distribution of RNA seen histologically varies considerably from one type of neuron to another. The cell body cytoplasm of many vertebrate neurons normally contains aggregations of basophilic "Nissl substance" which give it a granular appearance. This Nissl substance consists of ribosomes attached to cisternae of rough endoplasmic reticulum. Classically, the response of vertebrate neurons to axotomy is known as "chromatolysis" and consists of a dispersion of Nissl substance, causing the cell body cytoplasm to lose its granular appearance.[57] Ultrastructurally, these changes are seen as disorganization of rough endoplasmic reticulum and an increase in the number of free polyribosomes. In most neurons the response develops within 2 to 3 d of axonal injury, but is dependent upon the site and exact nature of the axonal lesion. At this time there is an increase in incorporation of nucleosides into RNA,[58-62] while RNA breakdown is also enhanced.[61] After axonal injury most neurons show an overall increase in protein synthesis. This, however, does not represent a global stimulation of synthetic activity, as synthesis of some individual proteins may undergo a decline. Among those proteins that show reduced synthesis are enzymes involved in neurotransmitter metabolism. In cholinergic neurons, acetylcholinesterase[63-65] and choline acetyltransferase[66] production may fall, while in aminergic neurons there may be a fall in dopamine β-hydroxylase, tyrosine hydroxylase, and monoamine oxidase synthesis.[67-69] On the other hand, synthesis of protein components of the cytoskeleton increases significantly.[70-72]

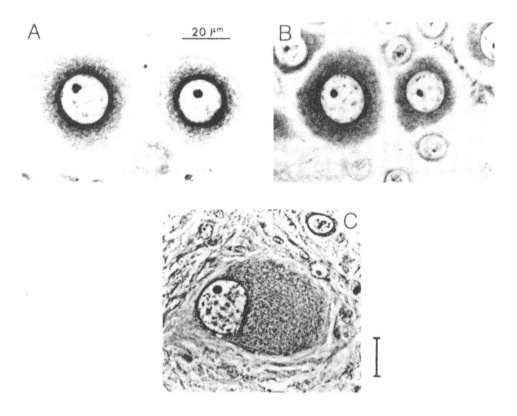

FIGURE 7.   Effect of axotomy on the histological appearance of cockroach motoneurons. Sections were stained with pyronine-malachite green to show RNA. (A) Cells which had been axotomized 2 d before fixation, showing a densely staining shell of material around the nucleus. (B) Bilaterally matched control cells, the axons of which remained intact up to the time of fixation. These cells lack any densely staining perinuclear shell. (C) Cell body of a neuron 24 d after axotomy. By this stage, there is no longer a perinuclear shell of dense staining and the nucleus has taken up an eccentric position. (Calibration: panels A and B are at the same magnification; scale bar in C = 20 μm.) (A and B from Cohen, M. J., in *Chemistry of Learning*, Corning, W. C. and Ratner, S. C., Eds., Plenum Press, New York, 1967, 407; C from Jacklet, J. W. and Cohen, M. J., *Science*, 156, 1640, Copyright 1967 by the AAAS. With permission.)

The response of insect nerve cell bodies to axotomy differs in some respects from the classical vertebrate pattern. Although it has been reported that insect neurons normally lack Nissl granules,[73-76] Byers[77] was able to observe Nissl granules in the cell bodies of a pair of identified neurons in the metathoracic ganglion of the cockroach *Diploptera punctata* using both light and electron microscopy. These granules are small by comparison with those observed in many types of vertebrate neuron. Byers[77] considered it likely that the use of relatively thick paraffin sections had caused Nissl granules to be overlooked. In such sections the cell body cytoplasm of intact neurons appears uniformly stained, with the exception of a very thin shell of higher density around the nucleus.[76] Within 12 h of axotomy the density and thickness of the perinuclear shell increase markedly, reaching a maximum after 2 to 3 d (Figure 7). After about 2 weeks the distribution of staining material in the cell body has returned to normal. The densely staining perinuclear shell is apparently RNA, since it is absent from sections exposed to ribonuclease before staining.[76] In addition to a change in distribution of basophilic material, at relatively short intervals (about 2 to 5 d) after axonal damage, cell bodies appear swollen and Golgi bodies decrease in size. At longer intervals (about 3 to 5 weeks), the cell body undergoes further changes in histological appearance. The perinuclear basophilia declines, the nuclear and nucleolar volumes increase, the nucleus moves toward the axon hillock region of the cell (Figure 7C), lysosomes become more numerous, and the Golgi bodies, which initially had decreased in size, now enlarge.[76,77]

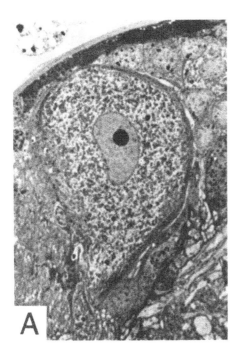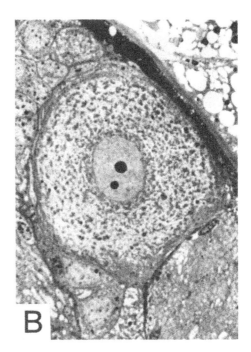

FIGURE 8.   Distribution of Nissl granules in cockroach motoneurons. (A) Normal cell, showing even distribution of Nissl granules in the cell body cytoplasm. (B) Soma of a neuron that was axotomized 3 d before fixation. Nissl granules are more dense near the nucleus and less dense near the cell membrane. (Calibration: 25 μm.) (From Byers, M. R., *Tissue Cell*, 2, 255, 1970. With permission.)

There are some differences in ultrastructural changes reported by different laboratories. Young et al.[79] found that although there was a generalized increase in the amount of endoplasmic reticulum in the cell body cytoplasm, accompanied by a dilatation of the individual endoplasmic reticulum cisternae, there was no preferential increase close to the nucleus as might have been expected from the changes in basophilia observed in light microscopy. Byers,[77] however, was able to observe an increase in the size and number of Nissl granules (rough endoplasmic reticulum) close to the nucleus and a corresponding decline at the periphery of the cell body cytoplasm (Figure 8). Autoradiographic data from the same laboratory indicate that the perinuclear basophilia and the change in distribution of Nissl granules are produced by a movement of available RNA rather than from accumulation of newly synthesized RNA around the nucleus. These conclusions, however, are not consistent with preliminary data reported by Cohen,[80] which indicated that the perinuclear shell of RNA seen after axotomy resulted from *de novo* synthesis.

Denburg and Hood[81] have used autoradiography to determine whether insect neurons undergo an increase in protein synthesis after axotomy. They found an increase in grain density over the cell body of axotomized coxal depressor motoneurons (D_f) compared to their contralateral unoperated partners. This difference in grain density first became statistically significant 12 d after axotomy and was still present at 16 d. It is surprising, however, that these authors were unable to detect any similar difference in grain density between the paired somata of a neuron (cell 27) adjacent to D_f. This should have been axotomized at the same time as D_f since, according to the ganglionic map of Cohen and Jacklet,[51] its axon travels in the same peripheral nerve as that of D_f. However, Denburg and Hood[81] state that the axon of this neuron runs in a different peripheral nerve trunk.

The significance of the above changes in the appearance of the cell body has attracted some attention and speculation. At the time when much of the work was done on the "axon reaction'' in insect neurons, it was believed that there was a delay of several weeks after an axonal lesioning before the onset of any axonal outgrowth.[14,15,17,22] On this basis, Jacklet and Cohen[17] suggested that the perinuclear shell of RNA had to disperse before any significant synthesis of new proteins could occur. At this stage axonal sprouting could begin. The observations of Denburg and Hood[81] confirm that the first detectable increase in protein synthesis occurs after the perinuclear accumulation of RNA has dispersed. On the other hand, the results on axonal sprouting of the motoneuron D_f, which have been discussed previously in this review, indicate that axonal sprouting begins within 3 d of axotomy, well before the dispersion of RNA or the first detectable signs of an increase in protein synthesis. Significant sprouting takes place within 1 week of injury, while at least some axonal processes may regain physical contact with their target muscle within 2 weeks. However, although extensive axonal outgrowth occurs relatively soon after axonal injury, most new processes are extremely fine at this stage. It is therefore likely that the increase in protein synthesis, which first becomes detectable 12 d after injury,[81] may be primarily associated with an increase in the diameter of regenerated fibers and the elaboration of new neuromuscular junctions rather than with supplying materials for the initial phase of axonal extension.

The nature of the signal or signals that trigger(s) the cell body reaction has been the subject of a great deal of interest and speculation (see References 82 and 83 for reviews). Although it was originally thought that axonal injury may produce a single signal that triggers the cell body reaction, more recently it has become clear that more than one signal may be involved. Bisby[83] has proposed that two main classes of signals are involved: (1) signals resulting from disconnection of the neuron from its target and (2) signals resulting from a fall in the total volume of axoplasm when distal parts of the axon are separated from the cell body. The first signal is assumed to be associated with changes in some trophic influences normally released from target structures. The best model for the operation of such a signal is provided by nerve growth factor (NGF) that in sympathetic neurons is able to prevent at least some, but not all, of the changes associated with axotomy.[84-87] It is possible that the amount of such trophic material available to a neuron is dependent upon the activity of the neuron and its target. The cell body reaction appears in motoneurons either following blockade of impulses with tetrodotoxin[88] or after treatment with botulinum toxin, which prevents acetylcholine release from motoneuron terminals.[69] Such results might be expected if the cell body reaction is normally prevented by trophic substances which are released from target structures, then internalized into neuronal terminals during recycling of synaptic vesicle membrane. If this mechanism were involved, suppression of transmitter release might reduce membrane recycling and sufficiently diminish uptake of trophic substances into nerve terminals to allow the cell body reaction to occur.

Bisby[83] considered the signal generated by a reduction in axoplasmic volume to most likely involve (1) reduction in the level of trophic materials derived from supporting cells, which normally would be transferred to the axon along its entire length; (2) influx of materials at the injury site; or (3) premature return of materials by retrograde axonal transport. Although there is little evidence for or against the first mechanism, it is known that injured axons may show an increase in the uptake of externally applied marker molecules.[90,91] However, uptake of actual signal molecules by this method has not been demonstrated. The observation that colchicine can stimulate the cell body reaction in neurons of the chick ciliary ganglion[92] or in cockroach motoneurons[93] without blocking axonal action potential conduction implies that under these conditions there is no generalized breakdown of membrane permeability. Entry of extracellular signal molecules into the axon would have to involve specific uptake processes. The effects of colchicine are more readily explained in terms of disruption of axonal transport.

Since cockroach motoneurons can exhibit dendritic sprouting in the absence of direct injury,[50] experiments have been performed to determine whether such sprouting is accom-

TABLE 1
Grain Densities in Control and Experimental Neurons Measured from Autoradiograms Obtained after Injection of ^{3}H-Uridine

Neuron	Grain density in perinuclear cytoplasm / Grain density in peripheral cytoplasm × 100
1. Control	8.7% ± 2.3 ($n = 16$)
2. Axotomized	30.8% ± 6.1 ($n = 11$)
3. Lesioned, but not axotomized	19.0% ± 4.4 ($n = 11$)

Note: Relative densities of grains in perinuclear and peripheral regions of soma cytoplasm were determined in motoneurons from (1) control animals; (2) animals which had received neural lesions that caused axotomy of the neurons; and (3) animals which had received lesions that did not axotomize the cells, but constituted a strong stimulus for dendritic sprouting.[50] ^{3}H-Uridine was injected 2 weeks before lesioning, and tissue was prepared for autoradiography 5 d after lesioning. Values expressed as means ± standard error; n = number of neurons contributing to each group.

panied by a cell body reaction. If this reaction is normally part of a response to meet an increase in the demand for axoplasmic and membrane components, dendritic sprouting might be expected to be accompanied by somatic changes similar to those observed after axotomy. The lesioning protocol employed was to section one anterior connective and all the segmental nerves exiting the metathoracic ganglion, with the exception of one nerve 5. This type of extensive lesion was used since it had previously been shown to constitute a powerful stimulus for dendritic sprouting.[50] To assess the cell body reaction in insect motoneurons, the distribution of RNA in the cell body cytoplasm of the metathoracic motoneuron D_f was measured autoradiographically. Animals were injected with ^{3}H-uridine 2 weeks before lesioning, since Byers[77] reported that if the labeled uridine was injected at this time the distribution of grains in autoradiograms reflected the perinuclear shell of basophilia seen after axotomy. Measurement of grain distribution constitutes a good index of the cell body reaction. Animals were sacrificed and tissues prepared for autoradiography 5 d after lesioning, at which time the axon reaction produced by axotomy was maximal. Autoradiogram grain densities were determined for the zone of cytoplasm immediately surrounding the nucleus (perinuclear) and for that just below the cell body membrane (peripheral). The grain density of the perinuclear zone was then expressed as a percentage above that in the peripheral cytoplasm (Table 1). Both axotomized neurons and those from ganglia that had received neural lesions which spared their axons showed a statistically significant increase in the relative grain density in the perinuclear cytoplasm relative to control neurons ($p < 0.001$ and 0.05, respectively). On the other hand, there was no statistically significant difference between the two experimental groups ($p > 0.1$). These experiments demonstrate that the cell body reaction can be stimulated in an identified insect motoneuron either by axotomy or by lesions which induce dendritic sprouting without axotomy. It therefore may be concluded that the cell body response studied here, like the cell body reaction observed in some other preparations, may be triggered by stimuli other than axonal damage. Although the exact nature of the stimulus is still uncertain, the observations are consistent with a trigger related to an increased biosynthetic demand.

V. ROLE OF SATELLITE CELLS IN REPAIR

There is evidence to indicate that, in at least some preparations, nonneuronal cells such as glia play a crucial role in determining the consequences of damage to the CNS. In mammals, for example, glia may have a significant influence on axonal growth. First, it has been proposed that glia may form a scar in the damaged tissue through which regenerating axons are unable to grow.[94] Second, normal glia in the CNS may provide a poor environment for axonal growth, since mammalian central neurons grow well when directed into peripheral nerve tissue, but exhibit only limited growth within the CNS.[3-5] The converse is also true. Growth of peripheral nerves may be inhibited when they are directed into CNS glial grafts.[95] In addition to influencing the extent of axonal growth, glial cells also may provide a substrate that assists in neuronal guidance. In fish, which have a good capacity for central regeneration, transplanted optic nerves and Mauthner neurons grow along stereotypical routes. It is proposed that they are guided along pathways provided by ependyma and glia.[96,97] After many types of neural lesions it is also important that the blood-brain barrier be restored so that the CNS is protected from materials in the blood circulation. To understand the overall process of CNS repair it is also essential to know the functions and origins of the different cell types involved.

The insect CNS serves as a good model for studying the function of supporting cells during recovery from neural damage, since these cells can undergo proliferation at lesion sites.[98] In addition, the insect CNS normally has a well-developed blood-brain barrier which can be disrupted by lesions, but which can subsequently be repaired. The insect CNS is surrounded by a nerve sheath which consists of two components, the outer acellular neural lamella and an inner cellular layer termed the perineurium. The blood-brain barrier is formed by perineurial glial cells which are coupled together by occluding junctional complexes.[99] The insect blood-brain barrier prevents access of water-soluble cations.[100,101] If undamaged nerve cords are bathed in high-potassium solutions, axonal action potentials can persist without any reduction in amplitude.

Ultrastructural observations have also demonstrated that the blood-brain barrier excludes lanthanum from undamaged nerve cords.[99,102] However, following sectioning of interganglionic connectives the blood-brain barrier is disrupted, enabling potassium and lanthanum to penetrate the CNS. The ability of the blood-brain barrier to exclude potassium and lanthanum was not restored until 4 to 6 months after the creation of such lesions.[103] During this time a number of ultrastructural changes were visible in the nerve cord. Among the most prominent of these changes were the appearance in the damaged region of large amounts of extracellular matrix material and of cells with prominent electron-dense granules in their cytoplasms. The granule cells, which resemble hemocytes, were particularly abundant in the superficial layers of the nerve cord, but also could be found more deeply. Some of these cells appear to be involved in phagocytosis. In order to further understand the role of different cell types in neural repair, glial cells have been disrupted selectively using ethidium bromide.[104] This drug selectively binds to RNA and DNA. When applied to interganglionic connectives it produces both nuclear and cytoplasmic damage in glia, but has relatively little effect upon the structure or function of axons. The cell bodies of these axons are located in ganglia and not connectives. Following selective glial disruption with ethidium bromide, as after surgical damage to connectives, granule cells appear. Initially, these cells seem to be involved in phagocytosis of cellular debris in the damaged region, but later they appear to become directly involved in repair, especially in the perineurial layers of the nerve sheath. However, responses to glial disruption and to surgical damage differ in several significant ways. Ethidium bromide treatment is followed by neither the increase in the amount of extracellular matrix material nor the extensive glial proliferation seen after surgical damage. Another important difference is that restoration of the ionic selectivity of the blood-brain barrier begins much more rapidly after glial disruption. Lanthanum is excluded as early as 4 d after ethidium bromide treatment, while potassium ion exclusion is reduced within 1 week, but takes about 4 weeks to return to normal. During the period in which the properties of the blood-brain barrier return to normal there are parallel changes in the ultrastructure of the perineurium. Initially, granule cells are present beneath the neural lamella, where they appear to contribute to the perineurium. However, the number of granule cells gradually falls, so that by 1 month after treatment with ethidium bromide these cells are absent and perineurial cells have gained cytological characteristics similar to those of cells in undamaged connectives. It is unclear, however, whether granule cells are gradually replaced by glia or are themselves transformed into glia during repair.

Whichever of the above possibilities turns out to be correct, experimental evidence indicates that granule cells do appear to be intimately involved in the process of repair. Perineurial repair is delayed if the antimitotic drug bleomycin is injected into the hemolymph before glial disruption with ethidium bromide. Under these conditions, cellular debris persists longer in damaged connectives, while the perineurium remains loosely packed and contains large extracellular spaces. Bleomycin also causes a concomitant marked reduction in the number of granule cells in damaged regions.[105] Restoration of the perineurium is also disrupted if inert microspheres are injected into the hemolymph before ethidium bromide treatment. Although the microspheres are rapidly removed from circulation by uptake into circulating hemocytes, they cause marked abnormalities in perineurial ultrastructure. A particularly notable feature is that granule cells are completely absent from the CNS during the repair following injection of microspheres into the hemolymph. On the other hand, other ultrastructural features of the perineurium are similar to those seen in preparations treated with bleomycin. Debris persists longer and cells are arranged loosely, with large extracellular spaces.[106] Since microspheres are unable to gain access to the CNS, they must exert their effects outside the nervous system. Because injecting microspheres into the hemolymph prevents the appearance of granule cells in the damaged perineurium, these cells presumably must originate from outside the CNS, where they would be accessible to the injected microspheres. When damage occurs they must then migrate into the CNS. It is likely that granule cells are derived from hemocytes, since the two cell types have similar cytological characteristics and since uptake of microspheres into hemocytes is associated with failure of granule cell recruitment into the CNS during repair. It is interesting that, despite the ultrastructural abnormalities of the repairing perineurium seen after injection of microspheres into the hemolymph, restoration of the blood-brain barrier follows a normal time course. This suggests that, although granule cells may normally play an important part in contributing to perineurial repair, other types of cells are capable of generating an adequate functional barrier. During repair of the insect blood-brain barrier, therefore, it appears that cells both endogenous and exogenous to the CNS may contribute to complete structural and functional restoration.

VI. CONCLUSION

From the material presented above, several conclusions can be made about the responses of cockroach neurons to nerve damage. The axons of insect motoneurons generate a profusion of new processes within 3 to 4 d of peripheral nerve injury. Initial growth apparently shows little directional preference so that, at relatively short postoperative periods, muscles may receive branches from appropriate and inappropriate motoneurons. As longer times elapse after lesioning, the number of motoneurons that branch into "foreign" muscles declines. Elimination of these contacts made by "foreign" motoneurons may be the result of competition with motoneurons that normally innervate the muscle.

The dendritic arbor of adult cockroach motoneurons is hardly altered by lesions such as nerve crushes that allow subsequent rapid recovery. Dramatic dendritic sprouting can occur, however, after sustained lesions such as ligation of peripheral nerves and under some circumstances that do not damage the cell directly. The effects of different types of experimental lesions suggest that there may be several alternative mechanisms by which dendritic sprouting can be triggered. However, the possibility that sprouting results from a stimulus released by glial cells in damaged peripheral nerves has not been eliminated.

Cockroach motoneuron cell bodies exhibit a histochemical response to axotomy which consists of a redistribution of RNA in the perinuclear cytoplasm. Although the stimulus for this reaction has not been identified, it is unlikely to result from loss of axoplasm following axotomy. The cell body reaction can be observed after treatment with colchicine, which does not break down the axonal membrane. The cell body reaction in cockroach neurons can be evoked not only by axotomy, but also by lesions which spare the axon yet act as powerful stimuli for dendritic sprouting. It is possible, therefore, that the axon reaction in this preparation reflects a response to a change in the biosynthetic needs of the neuron.

The insect nervous system normally has an efficient blood-brain barrier which is restored after neural lesioning. There is now evidence to indicate that cells both endogenous and exogenous to the CNS participate in repair. Glial cells provide the former category, while hemocytes appear to migrate into the CNS, where they may be involved both in the phagocytosis of cellular debris and in the actual repair of the blood-brain barrier.

ACKNOWLEDGMENTS

I thank Dr. K. A. Rand for her critical reading of a previous draft of this review.

REFERENCES

1. **Purves, D.**, Long term regulation in the vertebrate peripheral nervous system, *Int. Rev. Physiol.*, 10, 125, 1976.
2. **Kiernan, J. A.**, Hypotheses concerned with axonal regeneration in the mammalian nervous system, *Biol. Rev.*, 54, 155, 1979.
3. **Richardson, P. M., McGuinness, U. M., and Aguayo, A. J.**, Axons from CNS neurons regenerate into PNS grafts, *Nature (London)*, 284, 264, 1980.
4. **David, S. and Aguayo, A. J.**, Axonal elongation into peripheral nervous system "bridges" after central nervous system injury in adult rats, *Science*, 214, 931, 1981.
5. **Aguayo, A. J., David, S., and Bray, G. M.**, Influence of the glial environments on the elongation of axons after injury, *J. Exp. Biol.*, 95, 231, 1981.
6. **Sperry, R. W.**, Chemoaffinity in the orderly growth of nerve fiber patterns and connections, *Proc. Natl. Acad. Sci. U.S.A.*, 50, 703, 1963.
7. **Easter, S. S., Purves, D., Rakic, P., and Spitzer, N. C.**, The changing view of neural specificity, *Science*, 230, 507, 1985.
8. **Cass, D. T. and Mark, R. F.**, Re-innervation of axolotl limbs. I. Motor nerves, *Proc. R. Soc. London Ser. B*, 190, 45, 1975.
9. **Fangbonner, R. F. and Vanable, J. W.**, Formation and regression of inappropriate nerve sprouts during trochlear nerve regeneration in *Xenopus laevis*, *J. Comp. Neurol.*, 157, 391, 1974.
10. **Bennett, M. R. and Raftos, J.**, The formation and regression of synapses during the re-innervation of axolotl striated muscles, *J. Physiol. (London)*, 265, 261, 1977.
11. **Genat, B. R. and Mark, R. F.**, Electrophysiological experiments on the mechanism and accuracy of neuromuscular specificity in the axolotl, *Philos. Trans. R. Soc. London Ser. B*, 278, 335, 1977.
12. **Dennis, M. J. and Yip, J. W.**, Formation and elimination of foreign synapses on adult salamander muscle, *J. Physiol. (London)*, 274, 299, 1978.
13. **Landmesser, L. T.**, The generation of neuromuscular specificity, *Annu. Rev. Neurosci.*, 3, 279, 1980.
14. **Bodenstein, D.**, Studies on nerve regeneration in *Periplaneta americana*, *J. Exp. Zool.*, 136, 89, 1957.
15. **Guthrie, D. M.**, Regenerative growth in insect nerve axons, *J. Insect Physiol.*, 8, 79, 1962.
16. **Guthrie, D. M.**, The regeneration of motor axons in an insect, *J. Insect Physiol.*, 13, 1593, 1967.
17. **Jacklet, J. W. and Cohen, M. J.**, Nerve regeneration: correlation of electrical, histological and behavioral events, *Science*, 156, 1640, 1967.
18. **Pearson, K. G. and Bradley, A. B.**, Specific regeneration of excitatory motoneurons to leg muscles in the cockroach, *Brain Res.*, 47, 492, 1972.

19. **Young, D.,** Specific re-innervation of limbs transplanted between segments in the cockroach, *Periplaneta americana, J. Exp. Biol.,* 57, 305, 1972.
20. **Fourtner, C. R., Holtzmann, T. W., and Drewes, C. D.,** Specific contralateral afferent and efferent regeneration in the cockroach, *Am. Zool.,* 17, 962, 1977.
21. **Fourtner, C. R., Drewes, C. D., and Holtzmann, T. W.,** Specificity of afferent and efferent regeneration in the cockroach: establishment of a reflex pathway between contralaterally homologous target cells, *J. Neurophysiol.,* 41, 885, 1978.
22. **Denburg, J. L., Seecof, R. L., and Horridge, G. A.,** The path and rate of growth of regenerating motor neurons in the cockroach, *Brain Res.,* 125, 213, 1977.
23. **Whitington, P. M.,** The specificity of innervation of regenerating motor neurons in the cockroach, *J. Comp. Neurol.,* 186, 465, 1979.
24. **Carbonell, C. S.,** The thoracic muscles of the cockroach *Periplaneta americana* (L.), *Smithson. Misc. Collect.,* 107, 1, 1947.
25. **Denburg, J. L.,** Elimination of inappropriate axonal branches of regenerating cockroach motor neurons as detected by retrograde transport of horseradish peroxidase conjugated wheat germ agglutinin, *Brain Res.,* 248, 1, 1982.
26. **Denburg, J. L.,** Plasticity in the cockroach neuromuscular system, *Dev. Biol.,* 306, 1985.
27. **Brogan, R. T. and Pitman, R. M.,** Axonal regeneration in an identified insect motoneurone, *J. Physiol. (London),* 319, 34P, 1981.
28. **Valverde, F.,** Apical dendritic spines of the visual cortex and light deprivation in the mouse, *Exp. Brain Res.,* 3, 337, 1967.
29. **Ruiz-Marcos, A. and Valverde, F.,** The temporal evolution of the distribution of dendritic spines in the visual cortex of normal and dark raised mice, *Exp. Brain Res.,* 8, 284, 1969.
30. **Schapiro, S. and Vukovitch, K. R.,** Early experience effects upon cortical dendrites: a proposed model for development, *Science,* 167, 292, 1970.
31. **Cerf, J. A. and Chacko, L. W.,** Retrograde reaction in motoneuron dendrites following ventral root section in the frog, *J. Comp. Neurol.,* 109, 205, 1958.
32. **Sumner, B. E. H. and Watson, W. E.,** Retraction and expansion of the dendritic tree of motor neurones of adult rats induced *in vivo, Nature (London),* 233, 273, 1971.
33. **Sumner, B. E. H. and Sutherland, F. I.,** Quantitative electron microscopy on the injured hypoglossal nucleus in the rat, *J. Neurocytol.,* 2, 315, 1973.
34. **Eccles, J. C., Libet, B., and Young, R. R.,** The behaviour of chromatolysed motoneurones studied by intracellular recording, *J. Physiol. (London),* 143, 11, 1958.
35. **McIntyre, A. K., Bradley, K., and Brock, L. G.,** Responses of motoneurones undergoing chromatolysis, *J. Gen. Physiol.,* 42, 931, 1959.
36. **Shapovalov, A. I. and Grantyn, A. A.,** Nadsegmentarnye sinapticheskie vliianiia na khromatolizirovannye motoneurony, *Biofizika,* 13, 260, 1968.
37. **Kuno, M. and Llinas, R.,** Alterations of synaptic action in chromatolysed motoneurones of the cat, *J. Physiol. (London),* 210, 823, 1970.
38. **Blinzinger, K. and Kreutzberg, G.,** Displacement of synaptic terminals from regenerating motoneurons by microglial cells, *Z. Zellforsch. Mikrosk. Anat.,* 85, 145, 1968.
39. **Matthews, M. R. and Powell, T. P. S.,** Some observations on transneuronal cell degeneration in the olfactory bulb of the rabbit, *J. Anat.,* 96, 89, 1962.
40. **Jones, W. H. and Thomas, D. B.,** Changes in the dendritic organization of neurons in the cerebral cortex following deafferentation, *J. Anat.,* 96, 375, 1962.
41. **Cowan, W. M.,** Anterograde and retrograde transneuronal degeneration in the central and peripheral nervous system, in *Contemporary Research Methods in Neuroanatomy,* Nauta, W. J. H. and Ebbsson, S. O. E., Eds., Springer-Verlag, New York, 1970, 217.
42. **Murphey, R. K., Mendenhall, B., Palka, J., and Edwards, J. S.,** Deafferentation slows the growth of specific dendrites of identified giant interneurons, *J. Comp. Neurol.,* 159, 407, 1975.
43. **Murphey, R. K., Matsumoto, S. G., and Mendenhall, B.,** Recovery from deafferentation by cricket interneurons after reinnervation by their peripheral field, *J. Comp. Neurol.,* 169, 335, 1976.
44. **Murphey, R. K. and Levine, R. B.,** Mechanisms responsible for changes observed in response properties of partially deafferented insect interneurons, *J. Neurophysiol.,* 43, 367, 1980.
45. **Shankland, M., Bentley, D., and Goodman, C. S.,** Afferent innervation shapes the dendritic branching pattern of the medial giant interneuron in grasshopper embryos raised in culture, *Dev. Biol.,* 92, 507, 1982.
46. **Hoy, R. R., Nolen, T. G., and Casaday, G. C.,** Dendritic sprouting and compensatory synaptogenesis in an identified interneuron following auditory deprivation in a cricket, *Proc. Natl. Acad. Sci. U.S.A.,* 82, 7772, 1985.
47. **Horridge, G. A. and Burrows, M.,** Synapses upon motoneurons of locusts during retrograde degeneration, *Philos. Trans. R. Soc. London Ser. B,* 269, 95, 1974.

48. **Tweedle, C. D., Pitman, R. M., and Cohen, M. J.,** Dendritic stability of insect central neurons subjected to axotomy and deafferentation, *Brain Res.,* 60, 471, 1973.

49. **Cohen, M. J.,** Cellular events in the evolution of behavior, in *Simpler Networks and Behavior,* Fentress J. C., Ed., Sinauer Associates, Sunderland, MA, 1976, 39.

50. **Pitman, R. M. and Rand, K. A.,** Neural lesions can cause dendritic sprouting of an undamaged adult insect motoneurone, *J. Exp. Biol.,* 96, 125, 1982.

51. **Cohen, M. J. and Jacklet, J. W.,** The functional organization of motor neurons in an insect ganglion, *Philos. Trans. R. Soc. London Ser. B,* 252, 561, 1967.

52. **Dresden, D. and Nijenhuis, E. D.,** Fibre analysis of the nerve of the second thoracic leg in *Periplaneta americana, K. Ned. Akad. Versl. Gewone Vergud. Afd. Natuurk.,* 61C, 213, 1958.

53. **Roederer, E. and Cohen, M. J.,** Regeneration of an identified central neuron in the cricket. I. Control of sprouting from soma, dendrites and axon, *J. Neurosci.,* 3, 1835, 1983.

54. **Lieberman, A. R.,** The axon reaction: a review of the principal features of perikaryal responses to axonal injury, *Int. Rev. Neurobiol.,* 14, 49, 1971.

55. **Lieberman, A. R.,** Some factors affecting retrograde neuronal responses to axonal lesions, in *Essays on the Nervous System,* Bellairs, R. and Gray, E. G., Eds., Clarendon, Oxford, 1974, 71.

56. **Grafstein, B.,** The nerve cell body response to axotomy, *Exp. Neurol.,* 48, 32, 1975.

57. **Nissl, F.,** Uber die Veranderungen der Ganglienzellen am Facialiskern des Kaninchens nach Ausreissung der Nerven, *Allgem. Z. Psychiat. Ihre Grenzgeb.,* 48, 197, 1892.

58. **Brattgard, S.-O., Hyden, H., and Sjostrand, J.,** Incorporation of orotic acid-^{14}C and lysine-^{14}C in regenerating single nerve cells, *Nature (London),* 182, 801, 1958.

59. **Porter, K. R. and Bowers, M. B.,** A study of chromatolysis in motor neurons of the frog *Rana pipiens, J. Cell Biol.,* 19, 56A, 1963.

60. **Watson, W. E.,** An autoradiographic study of the incorporation of nucleic-acid precursors by neurones and glia during nerve regeneration, *J. Physiol. (London),* 180, 741, 1965.

61. **Watson, W. E.,** Observations on the nucleolar and total cell body nucleic acid of injured nerve cells, *J. Physiol. (London),* 196, 655, 1968.

62. **Haddad, A., Iucif, S., and Cruz, A. R.,** Synthesis of RNA in neurons of the hypoglossal nerve nucleus after section of the axon in mice, *J. Neurochem.,* 16, 865, 1969.

63. **Flumerfelt, B. A. and Lewis, P. R.,** Cholinesterase activity in the hypoglossal nucleus of the rat and the changes produced by axotomy: a light and electron microscopic study, *J. Anat.,* 119, 309, 1975.

64. **Watson, W. E.,** Quantitative observations upon acetylcholine hydrolase activity of nerve cells after axotomy, *J. Neurochem.,* 13, 1549, 1966.

65. **Sinicropi, D. V., Michels, K., and McIlwain, D. L.,** Acetylcholinesterase distribution in axotomized frog motoneurones, *J. Neurochem.,* 38, 1099, 1982.

66. **Frizell, M. and Sjostrand, J.,** Transport of proteins, glycoproteins, and cholinergic enzymes in regenerating hypoglossal neurons, *J. Neurochem.,* 22, 845, 1974.

67. **Cheah, T. B. and Geffen, L. B.,** Effects of axonal injury on norepinephrine, tyrosine hydroxylase and monoamine oxidase levels in sympathetic ganglia, *J. Neurobiol.,* 4, 443, 1973.

68. **Kopin, I. J. and Silberstein, S. D.,** Axons of sympathetic neurons: transport of enzymes *in vivo* and properties of axonal sprouts *in vitro, Pharmacol. Rev.,* 24, 245, 1972.

69. **Reis, D. J. and Ross, R. A.,** Dynamic changes in brain dopamine-beta-hydroxylase activity during anterograde and retrograde reactions to injury of central noradrenergic neurons, *Brain Res.,* 57, 307, 1973.

70. **Heacock, A. M. and Agranoff, B. W.,** Enhanced labeling of a retinal protein during regeneration of optic nerve in goldfish, *Proc. Natl. Acad. Sci. U.S.A.,* 73, 828, 1976.

71. **Lasek, R. J. and Hoffman, P. N.,** The neuronal cytoskeleton, axonal transport and axonal growth, in *Cell Motility, Book C, Microtubules and Related Proteins,* Pollard, T. and Rosenbaum, J., Eds., Cold Spring Harbor Laboratory, Cold Spring Harbor, NY, 1976, 1021.

72. **Hoffman, P. N. and Lasek, R. J.,** Axonal transport of the cytoskeleton in regenerating motor neurons: constancy and change, *Brain Res.,* 202, 317, 1980.

73. **Hess, A.,** Fine structure of nerve cells and fibers, neuroglia, and sheaths of the ganglion chain in the cockroach *(Periplaneta americana), J. Biophys. Biochem. Cytol.,* 4, 731, 1958.

74. **Wigglesworth, V. B.,** Axon structure and the dictysomes (Golgi bodies) in the neurones of the cockroach *(Periplaneta americana), Q. J. Microsc. Sci.,* 101, 381, 1960.

75. **Ashhurst, D. E.,** The cytology and histochemistry of the neurones of *(Periplaneta americana), Q. J. Microsc. Sci.,* 102, 399, 1961.

76. **Cohen, M. J. and Jacklet, J. W.,** Neurons of insects: RNA changes during injury and regeneration, *Science,* 148, 1227, 1965.

77. **Byers, M. R.,** Chromatolysis in a pair of identifiable metathoracic neurons in the cockroach *Diploptera punctata, Tissue Cell,* 2, 255, 1970.

78. **Cohen, M. J.,** Some cellular correlates of behavior controlled by an insect central ganglion, in *Chemistry of Learning,* Corning, W. C. and Ratner, S. C., Eds., Plenum Press, New York, 1967, 407.

79. **Young, D., Ashhurst, D. E., and Cohen, M. J.,** The injury response of the neurones of *Periplaneta americana, Tissue Cell,* 2, 387, 1970.

80. **Cohen, M. J.,** Correlations between structure, function and RNA metabolism in central neurons of insects, in *Invertebrate Nervous Systems. Their Significance for Mammalian Neurophysiology,* Wiersma, C. A. G., Ed., University of Chicago Press, Chicago, 1967, 65.

81. **Denburg, J. L. and Hood, N. A.,** Protein synthesis in regenerating motor neurons in the cockroach, *Brain Res.,* 125, 227, 1977.

82. **Cragg, B. G.,** What is the signal for chromatolysis?, *Brain Res.,* 23, 1, 1970.

83. **Bisby, M. A.,** Retrograde transport and regeneration studies, in *Axoplasmic Transport,* Iqbal, Z., Ed., CRC Press, Boca Raton, FL, 1986, 249.

84. **Banks, B. E. C. and Walter, S. J.,** The effects of postganglionic axotomy and nerve growth factor on the superior cervical ganglia of developing mice, *J. Neurocytol.,* 6, 287, 1977.

85. **Hamberger, V., Brunsobechtold, J. K., and Yip, J. W.,** Neuronal death in the spinal ganglia of the chick embryo and its reduction by nerve growth factor, *J. Neurosci.,* 1, 60, 1981.

86. **Hendry, I. A.,** The response of adrenergic neurons to axotomy and nerve growth factor, *Brain Res.,* 94, 87, 1975.

87. **Nja, A. and Purves, D.,** Effects of nerve growth factor and its antiserum on synapses in the superior cervical ganglion of the guinea pig, *J. Physiol. (London),* 227, 53, 1978.

88. **Czeh, G., Galego, R., Kudo, N., and Kuno, M.,** Evidence for the maintenance of motoneuron properties by muscle activity, *J. Physiol. (London),* 281, 239, 1978.

89. **Watson, W. E.,** The response of motor neurons to intramuscular injection of botulinum toxin, *J. Physiol. (London),* 202, 611, 1969.

90. **Kristensson, K. and Olsson, Y.,** Retrograde transport of horseradish peroxidase in transected axon. III. Entry into injured axons and subsequent localization in perikaryon, *Brain Res.,* 126, 154, 1977.

91. **Anderson, P. N., Mitchell, J., and Mayor, D.,** On the mechanism of the uptake of horseradish peroxidase into the retrograde transport system of ligated postganglionic sympathetic nerves *in vitro, J. Anat.,* 133, 371, 1981.

92. **Pilar, G. and Landmesser, L.,** Axotomy mimicked by localized colchicine application, *Science,* 177, 1116, 1972.

93. **Pitman, R. M., Tweedle, C. D., and Cohen, M. J.,** Electrical responses of insect central neruons: augmentation by nerve section and colchicine, *Science,* 178, 507, 1972.

94. **Windle, W. F.,** Regeneration of axons in the vertebrate central nervous system, *Physiol. Rev.,* 36, 426, 1956.

95. **Aguayo, A. J., Bray, G. M., Perkins, C. S., and Duncan, I. D.,** Axon-sheath cell interactions in peripheral and central nervous system transplants, *Soc. Neurosci. Symp.,* 4, 361, 1979.

96. **Katz, M. J. and Lasek, R. J.,** Substrate pathways which guide growing axons in *Xenopus* embryos, *J. Comp. Neurol.,* 183, 817, 1979.

97. **Katz, M. J. and Lasek, R. J.,** Substrate pathways demonstrated by transplanted Mauthner axons, *J. Comp. Neurol.,* 195, 627, 1981.

98. **Pitman, R. M.,** unpublished data, 1985.

99. **Lane, N. J. and Treherne, J. E.,** Studies on perineurial junctional complexes and sites of uptake of microperoxidase and lanthanum in the cockroach central nervous system, *Tissue Cell,* 4, 427, 1972.

100. **Treherne, J. E. and Schofield, P. K.,** Mechanisms of ionic homeostasis in the central nervous system of an insect, *J. Exp. Biol.,* 95, 61, 1981.

101. **Schofield, P. K. and Treherne, J. E.,** Localization of the blood-brain barrier of an insect: electrical model and analysis, *J. Exp. Biol.,* 109, 319, 1984.

102. **Lane, N. J.,** Invertebrate neuroglia: junctional structure and development, *J. Exp. Biol.,* 95, 7, 1981.

103. **Treherne, J. E., Harrison, J. B., Treherne, J. M., and Lane, N. J.,** Glial repair in an insect central nervous system: effects of surgical lesioning, *J. Neurosci.,* 4, 2689, 1984.

104. **Smith, P. J. S., Leech, C. A., and Treherne, J. E.,** Glial repair in an insect central nervous system: effects of selective glial disruption, *J. Neurosci.,* 4, 2698, 1984.

105. **Treherne, J. E., Howes, E. A., Leech, C. A., and Smith, P. J. S.,** The effects of an anti-mitotic drug, bleomycin, on glial repair in an insect central nervous system, *Cell Tissue Res.,* 243, 375, 1985.

106. **Smith, P. J. S., Howes, E. A., Leech, C. A., and Treherne, J. E.,** Haemocyte involvement in the repair of the insect central nervous system after selective glial disruption, *Cell Tissue Res.,* 243, 367, 1986.

Chapter 12

PERIPLANETA AMERICANA DISSOCIATED PRIMARY CULTURES

Isabel Bermudez and David J. Beadle

TABLE OF CONTENTS

I. INTRODUCTION

Cockroach embryos are ideal sources of both nerve and muscle tissue for the production of dispersed monolayer cultures. Techniques are now available for producing neuronal-, muscle-, and glial-enriched cultures from embryonic cockroach tissues, and these can be used for developmental, biochemical, and pharmacological studies. The cultures are ideal for biochemical investigations, as there is no possibility of biochemical processes from other cell types complicating the experimental results. The monolayer nature of the cultures and the lack of any barriers to the application of drugs also make them particularly suitable for electrophysiological studies. The cells are readily impaled with high-impedance glass microelectrodes, and drugs can be applied onto specific sites while the whole system is under continual observation. The clean plasma membrane and the architecture of the cells also make them suitable for the patch-clamp technique in both the cell-attached mode for single-channel analysis and in the whole-cell voltage-clamp mode for the analysis of macroscopic currents. It is also possible to grow mixed cultures such as nerve and muscle in which the individual cells interact in a normal physiological manner. These cultures provide ideal models for investigating the interactions of different cell types and the factors that govern them, since the investigator has almost total control over the environment in which the cells are growing.

II. METHODOLOGY

A. SOURCE OF TISSUE

Embryos of the cockroach *Periplaneta americana* are the source of all cells in our culture systems. The American cockroach is a particularly convenient insect to use for culturing because the eggs are laid in an ootheca that may contain up to 16 embryos. This facilitates sterilizing the eggs and permits the harvesting of relatively large quantities of developmentally homogeneous embryonic tissue. The female carries the ootheca around for about 24 h, so the oothecae can either be removed directly from her or collected every day to ensure that the eggs can be dated to within 24 h. The oothecae are stored at 29°C in plastic petri dishes in a moist atmosphere until required. Under these conditions, unused oothecae hatch after 30 to 31 d.

B. CULTURE TECHNIQUES

All operations are performed under aseptic conditions inside a laminar-flow cabinet. The walls and work surface of the cabinet are washed with absolute alcohol at the beginning and the end of culture work; instruments are sterilized by autoclaving or flaming. Culture media and all other solutions used are sterilized by filtration through 0.22-μm multipore filter units (Millipore, Bedford, MA), applying positive pressure from disposable syringes.

Cells are grown on Falcon® tight-seal petri dishes (50 × 12 mm) using a modification of the hanging column technique described by Shields et al.[1] Dissociated cultures will also grow on glass coated with polylysine (1 μg/ml) or protamine (0.1 mg/ml). The culture dishes are prepared by coating three pieces of broken glass microscope slide with Vaseline® and placing them in the dish as illustrated in Figure 1. The dissociated cells are plated between the three glass fragments, and then a sterile glass coverslip (40 × 20 mm) is lowered onto the glass fragments and gently pressed down until the medium is firmly positioned between the coverslip and the bottom of the dish (Figure 1). It is important to keep the culture drop in place, as excessive movement tends to dislodge the cells from the surface of the dish, thereby reducing cell density and growth. All cultures are grown in humidified air at 29°C until required for experimentation.

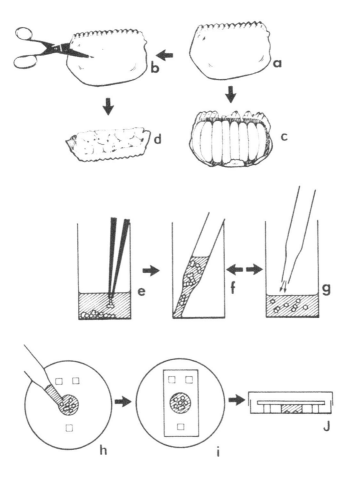

FIGURE 1.   Preparation of dissociated neuronal cultures. (a) *P. americana* ootheca, 23 to 26 d old. (b) Embryos consistently arrayed within oothecae (as shown in c) are decapitated using sterile scissors. (d) Each ootheca contains 12 to 16 heads from which cerebral ganglia are isolated using watchmakers' no. 5 forceps. (e to g) Intact cerebral ganglia are pooled and suspended in 5 + 4 medium prior to trituration by Pasteur pipette. (h) Four or three drops of the cell suspension are seeded onto a 50-mm Falcon® petri dish. (i and j) A glass coverslip is stuck down onto glass fragments to support the culture medium in the form of a hanging column. (From Beadle, D. J. and Lees, G., in *Neuropharmacology and Neurobiology*, Ellis Horwood, Chichester, England, 1986, 423. With permission.)

C. CULTURE MEDIA

Cockroach ganglia and muscle somites are dissociated in the 5 + 4 medium of Levi-Montalcini et al.[2] This consists of five parts Schneider's revised *Drosophila* medium and four parts Eagle's basal medium in Hepes buffer, containing 100 IU/ml penicillin, 100 μg/ml streptomycin (pH 6.8 to 7, 370 to 390 mosmol). Disaggregation of tissue in serum-containing medium should be avoided as it causes frothing of the cell suspension and loss of substrate adherence.

All cells are plated into the culture dishes in the 5 + 4 medium. After 1 h, during which time they attach to the floor of the culture dish, muscle cultures and glial cultures are transferred to their specific growth media. Neuronal cultures are usually grown for 4 to 6 d in the 5 + 4 medium before transfer to the neuronal culture medium.

The neuronal medium consists of equal parts of Leibovitz L-15 medium and Yunker's modified Grace's medium supplemented with 100 IU/ml penicillin and 100 μg/ml streptomycin. This medium appears to allow complete neuronal differentiation.

Muscle cultures and nerve-muscle cocultures are grown in a medium consisting of equal parts of Leibovitz L-15 medium and Yunker's modified Grace's medium, 10% horse serum, 10 μg/ml β-ecdysone, 100 IU/ml penicillin, and 100 μg/ml streptomycin. This medium not only permits full differentiation of muscle cells, but also inhibits the growth of fibroblast-like cells.

The growth medium of glial cultures consists of equal parts of Leibovitz L-15 medium and Grace's medium, supplemented with 10% fetal calf serum, 100 IU/ml penicillin, and 100 μg/ml streptomycin.

The medium for all cultures is changed every 7 or 14 d. More frequent replenishment of nutrients is deleterious to the differentiation of all cells, presumably due to the removal of trophic substances. The coverslip is removed gently under sterile conditions and the spent medium aspirated with a Pasteur pipette. Four drops of the fresh medium are added and a new coverslip placed in position.

D. NEURONAL CULTURES

For dissociated neuronal cultures the cerebral ganglia of 23- to 26-day-old embryos are used. The whole nerve cord can be dissected and used to produce cultures, but this is very time-consuming if large numbers of cultures need to be produced for biochemical studies. The technique illustrated in Figure 1 is recommended. The ootheca is sterilized in alcohol and then cut with sharp sterile scissors approximately one third of the distance from the dorsal edge, producing a fragment that contains 12 to 16 sacs, each containing one brain. The sacs are removed with sterile forceps and stored in Schneider's revised *Drosophila* medium at room temperature until the required number has been obtained. These can be dissected rapidly to provide large numbers of brains that can be dissociated to produce neuronal cultures.

To dissociate the brains a sterile Pasteur pipette is wetted thoroughly by sucking the 5 + 4 culture medium up and down to prevent the nerve tissue from sticking to the glass, and this is used to transfer the brains to a bijou bottle containing one drop of 5 + 4 medium (0.025 ml) for every two brains. A second Pasteur pipette is then wetted, and the medium containing the tissue is drawn into it and expelled vigorously against the bottom of the bottle. This is repeated four or five times. A third Pasteur pipette is then flame polished by rotating its tip in a Bunsen flame to reduce the aperture by about 50%, cooled, wetted, and the dissociation procedure repeated five or six times. Finally, a fourth pipette is used to place three drops of the suspension in the center of a Falcon® petri dish prepared as described in Section II.B. After the coverslip has been lowered onto the culture drop, the culture is left for at least 1 h to allow the neurons to attach to the bottom of the dish. After this time they are inverted and stored in large containers as discussed above. The inversion of the dishes prevents fine debris from settling on the growing surface as this inhibits neurite growth. Cultures prepared in this way can be maintained for at least 6 weeks. The culture medium is changed once a week, the first change being into the neuronal culture medium. The dissociation process and the initial growth of the cells in the 5 + 4 medium destroy the vast majority of the glial cells in the ganglion, while the neurons survive (Figure 2a). Cultures prepared in this manner are comprised of at least 90% neurons.[3] Such preparations have many advantages compared to vertebrate preparations, which are often contaminated by other types of cell.

E. GLIAL CELL CULTURES

For dissociated glial cell cultures the cerebral ganglia of 18- to 20-day-old embryos are used. The ganglia are dissected and stored in 5 + 4 medium as described for neuronal

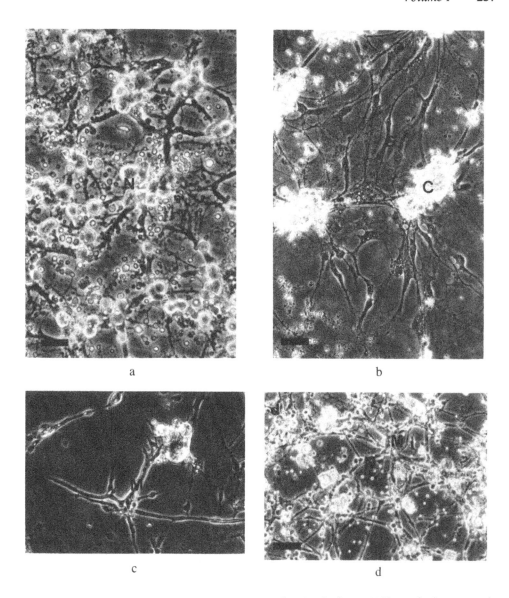

a

b

c

d

FIGURE 2.   Phase-contrast micrographs of *P. americana* dissociated cultures. (a) Neuronal culture prepared from 23-day-old embryonic brains after 26 d *in vitro*. Small clumps or single spherical somata (N) are connected by bundles of neuronal processes. No other type of cell is present. (Scale bar: 100 μm.) (b) Glial cell cultures after 7 d of growth in culture, showing the glial cells (G) flattened out on the floor of the culture dish. Clumps of cells (C) are apparent and these contain neuronal cell bodies. (Scale bar: 20 μm.) (c) Muscle culture after 17 d *in vitro* showing the morphology of the cultured muscle cells (M). (Scale bar: 30 μm.) (d) A nerve-muscle coculture after 14 d of growth in culture showing neurons (N) and muscle cells (M) connected via neuronal processes (P). (Scale bar: 20 μm.)

cultures. The ganglia are then treated in 5 + 4 medium containing collagenase/dispase at 2 mg/ml (Boehringer-Mannheim Corp.) for 1 h at room temperature prior to dissociation by gentle trituration with a Pasteur pipette. The enzyme treatment allows the ganglia to be dissociated with much less force than that used during the preparation of neuronal cultures, and this results in the survival of the glial cells (Figure 2b). The dissociated cells are plated onto Falcon® petri dishes and allowed to attach to the substratum for 1 h. After this time the medium is replaced with the glial culture medium.[4] The glial cells that develop in this preparation do not cross-react with any of the monoclonal antibodies that have been used as markers for vertebrate glia.[4a] Thus, the relationship between insect and vertebrate glial cells needs to be elucidated in more detail.

F. MUSCLE CULTURES

For dissociated muscle cultures, thoracic and abdominal muscle somites of 10- to 11-day-old embryos are used. It is also possible to prepare cultures from the legs of 11-day-old embryos, but the dissection is quite time-consuming and is only suitable for the production of a small number of cultures containing ≤10 μg of protein for electrophysiological studies.

When 10- to 11-day-old oothecae are available, they are removed from the incubator and surface sterilized in absolute ethanol for 5 to 10 min. They are then allowed to dry, and the outer case is removed with sterile forceps under a dissecting microscope in a laminar-flow cabinet. The embryos, in their sacs, are transferred to a petri dish containing sterile calcium- and magnesium-free phosphate-buffered saline (dissecting saline). This saline is used to facilitate dissociation of the tissue.[5] The egg sacs are removed carefully with sterile fine forceps to reveal the embryonic cockroaches, and these must then be oriented with their dorsal midline uppermost.

The embryos are opened along their midline, exposing the yolk sac, which is cleaned away carefully. Below the yolk sac are the thoracic and abdominal muscle somites running along both sides of the nerve cord. The muscle somites are excised and then incubated for 15 min at 37°C in 2 to 3 ml of dissecting saline containing 0.05% trypsin (Figure 3). The tissue is then washed with 3 to 5 ml of Leibovitz L-15 medium supplemented with 20% fetal calf serum to arrest trypsinization and then with 3 ml of 5 + 4 medium to remove the serum.

The trypsinized tissue is then resuspended in 5 + 4 medium at a rate of one drop of medium per dissected embryo, and the cells are dissociated by repeated trituration with a fire-polished Pasteur pipette until no large tissue fragments are visible (usually ten times). The cell suspension is then plated onto culture dishes at a rate of three drops per dish and kept at room temperature for 50 to 60 min, when the medium is removed gently and replaced with three to four drops of muscle medium. This procedure removes unattached cells and debris. Cultures prepared in this manner are comprised of approximately 80% myogenic cells (Figure 2c).

The plating density is approximately 6 to 7×10^4 cells per drop of suspension, and each embryo yields ca. 6×10^4 cells. At lower densities, cell viability and differentiation are impaired. Higher densities can be used, although the visualization of individual cells becomes more difficult.

G. NERVE-MUSCLE CULTURES

These cultures are obtained by plating myoblastic cells onto culture dishes which have previously been seeded with cerebral neurons of 23-day-old *P. americana* embryos as described in Section II.D. The neuronal cultures are grown in 5 + 4 medium for 4 to 5 d and then seeded with three drops of freshly prepared muscle cell suspension. The 5 + 4 medium is removed gently 1 h after seeding and replaced with muscle medium. After 14 d of growth *in vitro*, the neurons and muscle cells can be seen connected to each other, forming a complex cellular network (Figure 2d). It is also possible to obtain nerve-muscle cultures by dissociating whole 11-day-old cockroach embryos, although this type of culture will also contain a number of other cell types.

III. APPLICATIONS

A. DEVELOPMENTAL STUDIES

Dissociated cultures prepared from embryonic *P. americana* myoblasts and/or neurons are particularly useful for developmental studies, where problems such as complex cellular

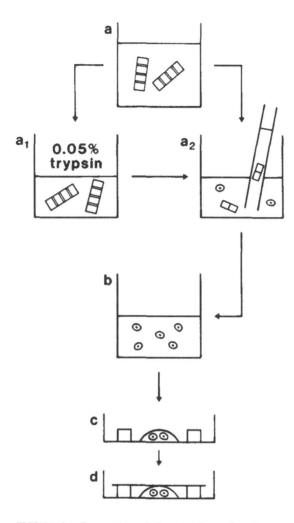

FIGURE 3. Preparation of dissociated muscle cultures.
(a) Thoracic and abdominal muscle somites from 11-day-old *P. americana* embryos are suspended in a calcium/magnesium-free phosphate-buffered saline. The tissue can be dissociated by trypsin treatment (a_1) followed by trituration (a_2) or by trituration only (a_2). (b) The resulting cell suspension is dispensed onto culture dishes (three drops per culture dish) and kept at room temperature for 1 h. (c) After 1 h the medium is removed and replaced by muscle medium. (d) A coverslip is lowered onto the glass support to keep the culture drop firmly in place.

networks and accessibility can make *in vivo* experiments difficult to perform and can complicate the interpretation of experimental data. The cultured cells can be followed closely throughout development using biochemical, electrophysiological, or morphological techniques. In addition, the conditions of growth (culture media, cell density, etc.) can be easily altered and their effects on differentiation readily assessed.

Muscle cultures have been used successfully to investigate the differentiation of rounded, mononucleated myoblasts into contractile, multinucleated muscle fibers.[6] The sequence of events accompanying this process is similar to that characterizing the differentiation of vertebrate muscle and consists of one or two cycles of cell division followed by elongation, fusion, and synthesis and assembly of the contractile apparatus of the muscle cells.[6] In nerve-muscle cocultures, muscle cells are connected to neurons via neuromuscular synapses.

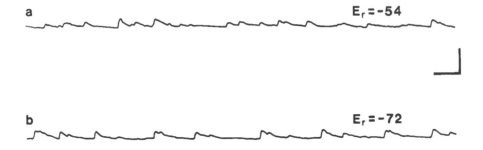

FIGURE 4.   Intracellular current-clamp recordings showing spontaneous excitatory synaptic potentials. (a) These synaptic potentials were recorded from a muscle cell present in a 16-day-old nerve-muscle coculture. (Calibration: 20 mV, 100 ms.) (b) Synaptic potentials recorded from a neuron after 26 d of growth in a neuronal culture. (Calibration as in a.)

The *in vitro* synapses are morphologically similar to those seen in insect muscles *in vivo*[6] and are physiologically functional (Figure 4a). We have not studied in detail the effect of coculturing on the properties of both muscle cells and neurons. Preliminary studies have shown that coculturing improves the survival of muscle cells and increases the number of both L-glutamate-sensitive neurons and γ-aminobutyric acid (GABA) transport sites on neurons.

The development of neurons has been investigated morphologically[7] and electrophysiologically.[8] The cultured neurons possess a typically neuronal fine structure.[7] The cells have a rounded soma with a prominent nucleus and nucleolus, as well as small clumps of heterochromatin that are not associated with the nuclear envelope. The cytoplasm contains large numbers of ribosomes and polysomes, mitochondria, lysosomes, and Golgi bodies. The cells produce long, narrow processes that contain microtubules, mitochondria, and elements of smooth endoplasmic reticulum, and they terminate in regions that contain synaptic vesicles. Synaptic profiles are usually encountered in neuronal cultures. The presynaptic terminals contain large numbers of spherical, clear vesicles, 40 to 50 nm in diameter, and presynaptic dense bars associated with clusters of vesicles at points of contact with the postsynaptic membrane thickenings.

Conventional intracellular recording studies have shown that the development of the active membrane properties of cultured *P. americana* neurons is similar to that of insect neurons during normal embryogenesis.[9] During the first 12 d in culture the cells cannot be stimulated to produce action potentials. The characteristic membrane response of these young cells to the injection of depolarizing current pulses is the activation of outward, voltage-dependent potassium channels. However, after this time, increasing numbers of cells can be stimulated (by appropriate current pulses) to produce sodium action potentials. Spontaneous activity can only be seen after about 3 weeks in culture. At this time excitatory, spontaneous synaptic potentials are frequently encountered (Figure 4b).

B. UPTAKE STUDIES

High-affinity uptake systems may play a role in transmitter inactivation (e.g., amino acid neurotransmitters) and/or the synthesis of neurotransmitter (e.g., acetylcholine). We therefore have taken advantage of the cellular homogeneity of our culture systems to characterize the uptake of choline[10] and amino acid neurotransmitters[4,11] in cockroach neurons, myotubes, and glial cells. Cultured neurons show both high- and low-affinity uptake of ^{3}H-choline.[10] The high-affinity transport mechanism is sodium dependent (Figure 5a,b), temperature sensitive, and inhibited by hemicholinium-3, but relatively insensitive to metabolic inhibitors. This uptake system is associated with a considerable degree of acetylcholine synthesis, indicating that the cultures contain a high proportion of cholinergic neurons.

Our studies on amino acid neurotransmitter uptake systems have advanced our knowledge

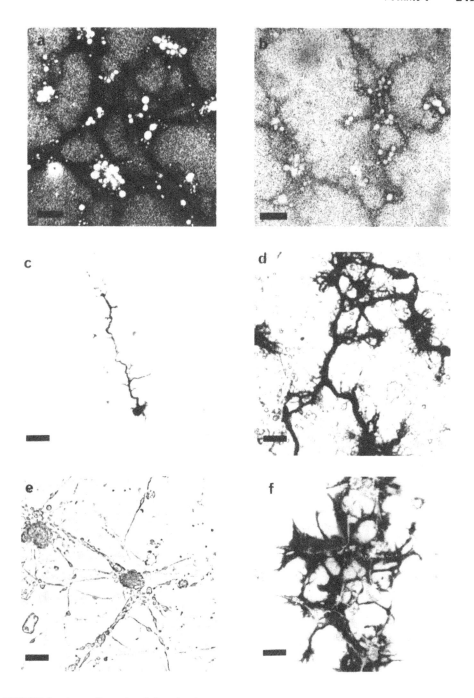

FIGURE 5. Autoradiographs of dissociated *P. americana* cultures. (a) Neuronal culture after incubation in a sodium phosphate-buffered saline containing 0.5 μ*M* ³H-choline. Label is confined to the neuronal processes of a large population of neurons comprising approximately 90% of the cultured neurons. As shown in (b), ³H-choline labeling decreases markedly in the absence of sodium ions. (Scale bars for a and b: 50 μm.) (c) Neuronal culture (14 d old) after incubation in a sodium phosphate-buffered saline containing 1 μ*M* ³H-GABA. (Scale: 20 μm.) (d) Neuronal culture (18 d old) after incubation in a sodium-containing saline supplemented with 1 μ*M* ³H-glutamate. (Scale bar: 30 μm.) (e) Muscle culture (18 d old) after incubation in saline containing sodium ions and 1 μ*M* ³H-GABA. These cells have not accumulated the amino acid. (Scale bar: 15 μm.) (f) Glial cells after incubation in ³H-GABA; all glial cells have accumulated the amino acid. (Scale bar: 20 μm.)

in this area significantly, since it has been possible for the first time to analyze the glial, neuronal, and muscle uptake systems of insects separately. The most significant findings indicate that a high-affinity, sodium-dependent GABA uptake system resides in a small subpopulation of the cultured neurons and that muscle cells do not seem to be endowed with such a system[11] (Figure 5c,e). Both neurons and muscle cells show sodium-dependent, saturable uptake of L-glutamate, but their properties differ markedly in terms of kinetic parameters. Thus, for example, the neuronal uptake is of high affinity ($K_m = 1.39 \ \mu M$; Figure 5d), whereas the muscular transport system is of low affinity ($K_m = 29.2 \ \mu M$). GABA and L-glutamate are also transported into glial cells via sodium-dependent transport mechanisms[4] (Figure 5f). The structural requirements of the glial and neuronal transport systems are similar, suggesting that there may not be marked differences in the mechanisms subserving amino acid transport in these two cell types.

C. PHARMACOLOGICAL STUDIES

The majority of studies of neurotransmitters using these culture systems have concentrated on acetylcholine, the major excitatory transmitter in the insect central nervous system;[12] GABA, the transmitter at insect inhibitory central nervous synapses[3] and neuromuscular junctions;[13] and L-glutamate, the transmitter at the insect excitatory neuromuscular junction.[13]

The majority of the neurons in these cultures are depolarized by acetylcholine and nicotine when these agonists are applied to the cells by pressure ejection at concentrations as low as 0.1 to 5 μM (Figure 6a). The pharmacological properties of the receptors mediating these responses are nicotinic, with α-bungarotoxin and tubocurare being the most potent antagonists and atropine and quinuclidinyl benzilate being less so.[3] The neurons in these cultures are suitable for patch clamping, and studies have revealed unitary currents, evoked by acetylcholine and carbamylcholine, that have two conductance states. The neurons are inhibited by low concentrations of GABA, and the responses are antagonized by picrotoxin, but not by bicuculline (Figure 6b). The GABA responses are potentiated by the benzodiazepine flunitrazepam, suggesting that they resemble the vertebrate GABA$_A$ receptor.[14] A small number of neurons are also sensitive to L-glutamate[15] (Figure 6c).

L-Glutamate (100 μM) depolarizes all cultured myotubes when applied by pressure ejection, the depolarization being accompanied by a marked decrease in membrane input resistance (Figure 6d). L-Glutamate potentials are readily desensitized in the absence of concanavalin A and reverse at about -3 mV. Qualitatively similar responses are obtained to pressure-applied 1 mM quisqualate. The muscle cells are also hyperpolarized by GABA with a pharmacology similar to that observed with the neurons (Figure 6e).

IV. CONCLUDING REMARKS

P. americana glial cells, myotubes, and neurons can now be successfully grown and maintained in culture for several weeks. These cultured cells differentiate and display morphological and physiological characteristics that are essentially similar to those of their *in vivo* counterparts. Although at present it is not possible to determine the completeness of the *in vitro* differentiation of these cells, the presence of physiologically active synapses and neuromuscular junctions in neuronal and nerve-muscle cultures, respectively, indicates that an advanced level of development is reached under our culture conditions. Therefore, these cultures should provide a powerful tool with which to examine the properties of insect glial cells, myotubes, and neurons. Since they also possess neurotransmitter receptor and uptake mechanisms that are essentially similar to those of their vertebrate counterparts, they provide ideal models for comparing the pharmacology of the nervous systems of these two important animal groups.

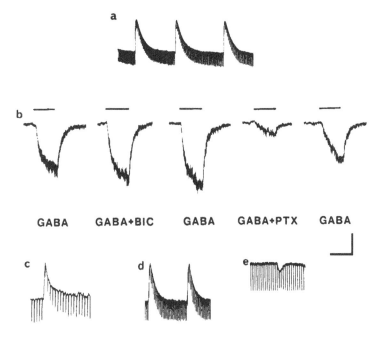

FIGURE 6.   Response types evoked by application of putative neurotransmitters onto 14- to 20-day-old cultured muscle cells and neurons. (a) Typical neuronal response to 0.1 m*M* ACh pressure ejected onto the neuronal somata for 1 s. The resting potential (E_R) of the cell was -54 mV. (Calibration: 20 mV, 10 s.) (b) GABA (1 m*M*) applied onto the somata of whole-cell voltage-clamped neurons evokes an inward current which reverses at the equilibrium potential predicted by the Nernst equation for chloride ions. Bicuculline (10 μM) has no effect on this current, whereas 10 μM picrotoxin decreases it markedly. (Clamping potential: -50 mV; calibration: 50 pA, 10 s.) Bars indicate pulse duration. (c) Response of a whole-cell voltage-clamped neuron to pressure-applied 1 m*M* L-glutamate. $E_R = -60$ mV, calibration $= 20$ mV, 5 s. (d) Current-clamp record of a typical L-glutamate response in a muscle cell. The amino acid (0.1 m*M*) was applied onto the cell by pressure ejection. Clamping potential $= -50$ mV; calibration as in a. (e) Typical response to 1 m*M* GABA applied by pressure ejection onto a muscle cell. $E_R = -56$ mV; calibration $= 10$ mV, 10 s.

REFERENCES

1. **Shields, G., Dubendorfer, A., and Sang, J. H.,** Differentiation *in vitro* of larval cell types from early embryonic cells of *Drosophila melanogaster, J. Embryol. Exp. Morphol.,* 33, 159, 1975.
2. **Levi-Montalcini, R., Chen, J. S., Seshan, K. R., and Aloe, L.,** An *in vitro* approach to the insect nervous system, in *Developmental Neurobiology of Arthropods,* Young, D., Ed., Cambridge University Press, London, 1973, 5.
3. **Beadle, D. J. and Lees, G.,** Insect neuronal cultures — a new tool in insect neuropharmacology, in *Neuropharmacology and Pesticide Action,* Ford, M. G., Lunt, G. G., Reay, R. C., and Usherwood, P. N. R., Eds., Ellis Horwood, Chichester, England, 1986, 423.
4. **Beadle, C. A., Bermudez, I., and Beadle, D. J.,** Amino acid uptake by neurones and glial cells from embryonic cockroach brain growing *in vitro, J. Insect Physiol.,* 33, 761, 1987.
4a. **Bermudez, I. and Beadle, D. J.,** unpublished observation, 1987.
5. **Moscona, A., Trowell, O. A., and Willmer, E. N.,** Chemical dissociation, in *Cells and Tissues in Culture,* Vol. 1, Willmer, E. N., Ed., Academic Press, New York, 1965, 52.
6. **Bermudez, I., Lees, G., Botham, R. P., and Beadle, D. J.,** Myogenesis and neuromuscular junction formation in cultures of *Periplaneta americana* myoblasts and neurones, *Dev. Biol.,* 116, 467, 1986.

7. **Beadle, D. J., Hicks, D., and Middleton, C.,** Fine structure of neurones from embryonic *Periplaneta americana* growing in long term culture, *J. Neurocytol.,* 11, 611, 1982.
8. **Lees, G., Beadle, D. J., Botham, R. P., and Kelly, J. S.,** Excitable properties of insect neurones from the central nervous system of embryonic cockroaches, *J. Insect Physiol.,* 31, 135, 1985.
9. **Goodman, C. S. and Spitzer, N. C.,** The development of electrical properties of identified neurones in grasshopper embryos, *J. Physiol.,* 313, 369, 1981.
10. **Bermudez, I., Lees, G., Middleton, C., Botham, R. P., and Beadle, D. J.,** Choline uptake by cultured neurones from the central nervous system of embryonic cockroaches, *Insect Biochem.,* 15, 427, 1985.
11. **Bermudez, I., Botham, R. P., and Beadle, D. J.,** High- and low-affinity uptake of amino acid transmitters in cultured neurones and muscle cells of the cockroach *Periplaneta americana, Insect Biochem.,* 18, 249, 1988.
12. **Sattelle, D. B.,** Acetylcholine receptors in insects, *Adv. Insect Physiol.,* 15, 215, 1980.
13. **Piek, T.,** Neurotransmission and neuromodulation of skeletal muscle, in *Comprehensive Insect Physiology, Biochemistry and Pharmacology* , Vol. 11, Kerkut, G. A. and Gilbert, L. I., Eds., Pergamon Press, Oxford, 1985, 55.
14. **Shimahara, T., Pichon, Y., Lees, G., Beadle, C. A., and Beadle, D. J.,** Gamma-aminobutyric acid receptors on cultured cockroach brain neurones, *J. Exp. Biol.,* 131, 231, 1987.
15. **Horseman, B. G., Seymour, C., Bermudez, I., and Beadle, D. J.,** The effects of L-glutamate on cultured insect neurones, *Neurosci. Lett.,* 85, 65, 1988.

Index

INDEX

VOLUMES I AND II

peripheral, II:23
Peptidergic signaling, II:13—24, see also specific
    types
    interneurons and, II:23
    motoneurons and, II:21—23
    neurohemal organs and, II:14—21
    neurosecretory cells and, see Neurosecretory cells
    peripheral peptidergic cells and, II:23
Peptides, I:104; II:4, see also Hormones; Neuropep-
    tides; specific types
    adipokinetic hormone-like, II:6, 11, 14, 15
    brain-gut, II:54, 67
    as class of neurotransmitters, II:36
    development, II:11—13
    hyperglycemic, see Hyperglycemic hormones
    insulin-like, II:12
    metabolic, II:11
    myotropic, see Myotropins
    poly-, see Polypeptides
    purification of, II:67
    RF, II:10
    sequence analysis of, II:76
    verification of, II:73, 74
    vertebrate, II:13
Perchloric acid, II:68
Pericardial sinus, I:47
Perineural glial cells, I:81
Perineurium, I:81, 226, 227
Perinuclear RNA, I:74
Peripheral inhibitory innervation, I:170
Peripheral nervous system, I:69, 74, 81
Peripheral peptidergic cells, II:23
Peripheral regeneration, I:82
Periplaneta
    americana, I:7, 11
        abdominal ganglia in, I:150
        abdominal nerve cord in, I:155
        aging and, II:298, 300
        avoidance experiments on, II:218—220
        brain in, I:108
        bursicon and, II:12
        calcium-potassium interactions in, I:175
        central nervous system in, I:74, 108, 136
        chemosensilla in, II:272—274
        circadian rhythms of, II:217, 218, 228, 230
        circulatory system in, I:47
        compound eye anatomy in, II:204—208
        compound eye physiology in, II:210—212
        conditioning experiments on, II:218
        corpus cardiacum in, II:55
        culturing of, I:14
        degeneration and, I:115
        depressor muscles in, I:181, 182
        developmental studies of, I:238—240
        digestive system in, I:43
        dissociated primary cultures of, see Dissociated
            primary cultures
        excitatory neuromuscular junctions in, I:170
        feeding and mating activity in, II:191
        GABA and, I:178
        giant axons in, I:134

giant interneurons of, I:150
head of, I:34
hyperglycemic factor in, II:56
hyperglycemic hormones in, II:68, 72
interneurons in, I:132
juvenile hormones in, I:91
leg muscles in, I:172
Malpighian tubules in, II:115
muscles in, I:46, 172, 180, 184
myotropins in, II:56
nerve cord in, II:116
neuroendocrine organs in, II:55
neuromuscular system of, I:181
neuropeptides in, II:68
ocellus of, II:220—222
olfactory system in, II:270, 277—280
optic lobe physiology in, II:212—216
ovaries in, II:190
peptidergic fibers in, II:97
pharmacological studies of, I:242
postembryonic development in, II:208—210
proctolin and, II:6
pyrethroids and, II:127
reproductive system in, I:54
retractor unguis muscle of, I:184
sex pheromones in, II:182—186, 188, 193
simple eye of, II:220—222
sources of, for culturing, I:14, 15
staining techniques and, I:109
synaptic transmission in, I:159
toxicology and, II:126
trehalose and, II:150
uptake studies of, I:240—242
visually guided behaviors in, II:217—220
vitellogenesis in, II:162, 163
spp., I:7, 10, 25, 27; II:184
    abdominal ganglia of, II:21
    action potential in, I:134
    circulatory system in, I:48
    giant axons in, I:134
    as pests in culturing, I:31
    reproductive system in, I:56
    resting potential in nerve cell bodies of, I:132
Periplanetin, II:11, 13, 20, 27, 28
Periplanetin-like immunoreactivity, II:22
Periplanone-A, II:182—184
Periplanone-B, II:182—184
Peritrophic membrane, I:42
Permethrin, II:128
Peroxidase-antiperoxidase (PAP) method, I:111
Pesticides, see also Acaricides; Insecticides; specific
    types
Pests, I:30, 31, see also specific types
    cockroaches as, in culturing, I:31
    control of, I:30, 31
Phagocytosis, I:226
Phallic lobes, I:50
Phallobase, I:50
Phallomeres, I:50
Phallotreme, I:50
Pharmacology, I:242; II:65, see also specific aspects

Printed and bound by CPI Group (UK) Ltd, Croydon, CR0 4YY

22/10/2024

01777633-0015